Birkhäuser

Trends in Mathematics

Trends in Mathematics is a series devoted to the publication of volumes arising from conferences and lecture series focusing on a particular topic from any area of mathematics. Its aim is to make current developments available to the community as rapidly as possible without compromise to quality and to archive these for reference.

Proposals for volumes can be submitted using the Online Book Project Submission Form at our website www.birkhauser-science.com.

Material submitted for publication must be screened and prepared as follows:

All contributions should undergo a reviewing process similar to that carried out by journals and be checked for correct use of language which, as a rule, is English. Articles without proofs, or which do not contain any significantly new results, should be rejected. High quality survey papers, however, are welcome.

We expect the organizers to deliver manuscripts in a form that is essentially ready for direct reproduction. Any version of TeX is acceptable, but the entire collection of files must be in one particular dialect of TeX and unified according to simple instructions available from Birkhäuser.

Furthermore, in order to guarantee the timely appearance of the proceedings it is essential that the final version of the entire material be submitted no later than one year after the conference.

More information about this series at http://www.springer.com/series/4961

Hypercomplex Analysis: New Perspectives and Applications

Swanhild Bernstein
Uwe Kähler
Irene Sabadini
Frank Sommen
Editors

Birkhäuser

Editors
Swanhild Bernstein
Institute of Applied Analysis
TU Bergakademie Freiberg
Freiberg, Germany

Uwe Kähler
Departamento de Matemática
Universidade de Aveiro
Aveiro, Portugal

Irene Sabadini
Dipartimento di Matematica
Politecnico di Milano
Milano, Italy

Frank Sommen
Dept. Mathematical Analysis
University of Gent
Gent, Belgium

ISSN 2297-0215 ISSN 2297-024X (electronic)
ISBN 978-3-319-08770-2 ISBN 978-3-319-08771-9 (eBook)
DOI 10.1007/978-3-319-08771-9
Springer Cham Heidelberg New York Dordrecht London

Library of Congress Control Number: 2014952606

Mathematics Subject Classification (2010): 30G35, 30G25, 22E46, 32A50, 68U10

Printed on acid-free paper

Springer Basel is part of Springer Science+Business Media (www.birkhauser-science.com)

Contents

Preface

At the 9th International ISAAC Congress (International Society for Analysis, its Applications, and Computations), held at the Pedagogical University of Krakow, Krakow, Poland from August 5 to August 9, 2013, one of the largest sessions was on "Clifford and Quaternionic Analysis" with around 40 speakers coming from all parts of the world: Belgium, Cech Republic, China, Finland, Germany, Israel, Italy, Mexico, Portugal, Russia, Turkey, Venezuela, Ukraine, United Kingdom and the United States.

While there are official congress proceedings, the success of the session led the organizers to ask the participants to present their most recent and promising achievements in a special volume to promote the exciting field of hypercomplex analysis. This volume contains a careful selection of 15 of these papers which cover several different aspects of hypercomplex analysis going from function theory over quaternions, Clifford numbers and multicomplex numbers, operator theory, monogenic signals, to the recent field of fractional Clifford analysis. Additionally, applications to image processing, crack analysis, and the theory of elasticity are covered. All contributed papers represent the most recent achievements in the area. We hope that anybody interested in the field can find many new ideas and promising new directions in these papers.

The Editors are grateful to the contributors to this volume and to the referees, for their painstaking and careful work. They also would like to thank the Pedagogical University in Krakow for hosting the Conference and Vladimir Mityushev, in particular, as Chairman of the local organising committee.

May 2014, Swanhild Bernstein
Uwe Kähler
Irene Sabadini
Frank Sommen

Hypercomplex Analysis: New Perspectives and Applications
Trends in Mathematics, 1–18

Symmetries and Associated Pairs in Quaternionic Analysis

Ricardo Abreu Blaya, Juan Bory Reyes, Alí Guzmán Adán and Uwe Kähler

Abstract. The present paper is aimed at proving necessary and sufficient conditions on the quaternionic-valued coefficients of a first-order linear operator to be associated to the generalized Cauchy–Riemann operator in quarternionic analysis and explicitly we give the description of all its nontrivial first-order symmetries.

Mathematics Subject Classification (2010). 30G35.

Keywords. Quaternionic analysis, generalized Cauchy–Riemann operator, symmetries and associated pairs.

1. Motivation and basic facts of quaternionic analysis

Approaches by symmetry operators and methods based on associated pairs not only play an important role for finding explicit solutions to systems of partial differential equations, see for instance [5, 9, 10, 11, 15, 16, 17], but are also closely linked to invariance groups of operators. One of the important points is that first-order symmetries form a Lie algebra where the action of the transformation group is induced by the Lie derivatives [12]. This was used quite extensively in the past in Clifford Analysis [23, 21, 22, 3]. Furthermore, the study of first-order symmetries of the Cauchy–Riemann–Fueter operator, as well as the description of all its associated pairs, has been done recently by Y. Krasnov [9] and by T.V. Nguyen [17].

Quaternionic analysis offers a function theory related to Cauchy–Riemann–Fueter operator, which represent a generalization of classical complex analysis to

This work was supported by Portuguese funds through the CIDMA – Center for Research and Development in Mathematics and Applications, and the Portuguese Foundation for Science and Technology ("FCT – Fundação para a Ciência e a Tecnologia"), within project PEst-OE/MAT/UI4106/2014.

higher-dimensional Euclidean space. For a more detailed account on this matter along more traditional lines we refer the reader to [2, 7, 8].

Starting from a definition of a generalized Cauchy–Riemann operator (related to some fixed orthonormal basis in $\mathbb{R}^4$) proposed by Vasilevski and Shapiro ([24]) quaternionic analysis shows important advantages in the effort of solving partial differential equations in the hypercomplex framework. Their approach is based on the notion of a structural set ψ^k, representing a general orthonormal frame. While the obtained function theory is in most aspects the same as for the generalized Cauchy–Riemann operator with respect to the standard basis this notion is the starting point for addressing problems which arise when more than one structural set is involved. Such cases arise, for example, when one studies derivatives of monogenic functions with respect to a different orthonormal frame. While in the quaternionic setting Mitelman/Shapiro [13] showed that the conjugate generalized Cauchy–Riemann operator can play the role of a ψ-hyperholomorphic derivative, it cannot separate all directions (which later led to the notion of a hyperholomorphic constant), which means that to recover all partial derivatives like in the complex case one would need three differential operators where each of them is obtained by conjugating only one of the elements of the structural set.

Another example is the study of Beltrami equations, in particular the study of the monogenic part of its solution. Hereby, one major problem arises. While one structural set can be transformed into another by an orthogonal transformation, in general this transformation will not be a rotation. But a generalized Cauchy–Riemann operator itself is only invariant under rotations, not under general orthogonal transformations. This leads to the notion of left- (right-) equivalent structural sets as the subclass of structural sets which can be transformed into each other by an appropriate rotation. This leads to such interesting properties that while for left- (right-) equivalent structural sets the corresponding sets of ψ-hyperholomorphic functions are coinciding that is not anymore true for general structural sets.

Additionally, while in the framework of the quaternionic algebra it is natural to represent rotations as multiplications of quaternions, this is not possible, for example, in case of a reflection. Therefore, the treatment of general orthogonal transformations requires the embedding of the quaternions in a higher-dimensional algebra. This can be done either by considering quaternions as 4×4-dimensional real matrices and applying a general orthogonal transformation to this matrix or by embedding the quaternions into a higher-dimensional algebra with signature $(4, 4)$. In both cases the resulting algebra of endomorphisms has to be isomorphic to the full matrix algebra of 4×4-matrices, not just the sub-algebra of quaternions.

The present paper is aimed at proving necessary and sufficient conditions on the quaternionic-valued coefficients of a first-order linear operator to be associated to the generalized Cauchy–Riemann operator in quarternionic analysis and we explicitly describe all its nontrivial first-order symmetries. This is done at first by linking with matrix description of our operator. In the last section we will show that the traditional approach by Sommen/van Acker [21, 22] via identifying the algebra of endomorphisms with a complexified Clifford algebra does also work in this case. This work can be regarded as a continuation of that in [1], where the authors deal with simultaneous null solutions of two different generalized Cauchy–Riemann operators.

In the following, we review briefly the basic facts of quaternionic analysis needed throughout the paper.

Let $\mathbb{H}$ be the set of real quaternions with a unit (denoted by 1), generated by $\{i, j, k\}$. This means that any element x from $\mathbb{H}$ is of the form $x = x_0 + x_1 i + x_2 j + x_3 k$, where $x_m \in \mathbb{R}, m \in \mathbb{N}_3 \cup \{0\}$; $\mathbb{N}_3 := \{1, 2, 3\}$.

In this paper we denote the generators by $1 =: e_0$, $i =: e_1$, $j =: e_2$, and $k =: e_3$ subject to the multiplication rules

$$e_m^2 = -1, \ m \in \mathbb{N}_3,$$

$$e_1 \, e_2 = -e_2 \, e_1 = e_3; \ e_2 \, e_3 = -e_3 \, e_2 = e_1; \ e_3 \, e_1 = -e_3 \, e_1 = e_2,$$

more suited for a future extension.

If $x = \sum_{m=0}^{3} x_m \, e_m$ is a quaternion then $\vec{x} := \begin{pmatrix} x_0 \\ x_1 \\ x_2 \\ x_3 \end{pmatrix} = (x_0, x_1, x_2, x_3)^T \in \mathbb{R}^4$,

where the index T denotes transposition.

With natural operations of addition and multiplication $\mathbb{H}$ is a non-commutative, associative skew-field. There is the quaternionic conjugation, which plays an important role and is defined as follows:

$$\overline{e}_m := -e_m, \ \forall m \in \mathbb{N}_3.$$

This involution extends onto the whole $\mathbb{H}$ as an $\mathbb{R}$-linear mapping: If $x \in \mathbb{H}$ then

$$\overline{x} := x_0 - x_1 e_1 - x_2 e_2 - x_3 e_3.$$

We have

$$\overline{x \cdot y} := \overline{y} \cdot \overline{x} \ \text{ and } \ x \cdot \overline{x} = \overline{x} \cdot x = |x|^2 \in \mathbb{R}.$$

This norm of a quaternion coincides with the usual Euclidean norm in $\mathbb{R}^4$. Therefore, for $x \in \mathbb{H} \setminus \{0\}$, the quaternion $x^{-1} := \overline{x}/|x|^2$ is a multiplicative inverse of x.

The quaternion $x \cdot y$, coincides with the result of multiplying $\vec{y}$ by the left regular matrix representation of x given by:

$$\begin{pmatrix} x_0 & -x_1 & -x_2 & -x_3 \\ x_1 & x_0 & -x_3 & x_2 \\ x_2 & x_3 & x_0 & -x_1 \\ x_3 & -x_2 & x_1 & x_0 \end{pmatrix} \cdot \begin{pmatrix} y_0 \\ y_1 \\ y_2 \\ y_3 \end{pmatrix} =: B_l(x) \cdot \vec{y}.$$

We will denote by $B_l := \{B_l(x) : x \in \mathbb{H}\}$ the set of all matrices of left regular representations of real quaternions.

One can compute directly that:

- $B_l(x+y) = B_l(x) + B_l(y)$,
- $B_l(xy) = B_l(x)B_l(y)$,
- $B_l(e_0) = I_4$, the identity matrix 4×4,
- $B_l(\lambda x) = \lambda B_l(x)$, $\lambda \in \mathbb{R}$,
- $B_l(\bar{x}) = B_l(x)^T$,
- $\det B_l(x) = |x|^4$.

In this way,

$$x \in \mathbb{H} \mapsto B_l(x) \in B_l \tag{1.1}$$

is an isomorphism of real algebras.

We consider functions f defined in a domain $\Omega \subset \mathbb{R}^4$ and taking values in $\mathbb{H}$. Such a function may be written as $f = f_0 + f_1e_1 + f_2e_2 + f_3e_3$ and each time we assign a property such as continuity, differentiability, integrability, and so on, to f it is meant that all $\mathbb{R}$-components f_m share this property. Thus notations $f \in C^p(\Omega, \mathbb{H})$, $p \in \mathbb{N} \cup \{0\}$, will have the usual component-wise meaning.

Let $M_{n\times m}(\mathbb{R})$ be the set of real $n \times m$ matrix, $(n, m \in \mathbb{N})$. We can identify any $f : \Omega \to \mathbb{H}$ with two specific matrix functions:

$$\vec{f} := \begin{pmatrix} f_0 \\ f_1 \\ f_2 \\ f_3 \end{pmatrix} : \Omega \to M_{4\times 1}(\mathbb{R})$$

and

$$B_l(f) := \begin{pmatrix} f_0 & -f_1 & -f_2 & -f_3 \\ f_1 & f_0 & -f_3 & f_2 \\ f_2 & f_3 & f_0 & -f_1 \\ f_3 & -f_2 & f_1 & f_0 \end{pmatrix} : \Omega \to M_{4\times 4}(\mathbb{R}).$$

Thus notations $\vec{f} \in C^p(\Omega, M_{4\times 1}(\mathbb{R}))$ and $B_l(f) \in C^p(\Omega, M_{4\times 4}(\mathbb{R}))$, $p \in \mathbb{N} \cup \{0\}$ might be understood directly.

Denote for a quaternionic constant c,

$$c_\Omega : x \in \Omega \to c \in \mathbb{H}.$$

By abuse of notation, we continue to write c_Ω for the case of a constant matrix c.

Let $\psi := \{\psi^0, \psi^1, \psi^2, \psi^3\} \in \mathbb{H}^4$ be a system of quaternions such that the conditions

$$\psi^m \cdot \bar{\psi}^n + \psi^n \cdot \bar{\psi}^m = 2\langle \vec{\psi^m}, \vec{\psi^n} \rangle_{\mathbb{R}^4} = 2\delta_{n,m} \qquad \forall n, m \in \mathbb{N}_3 \cup \{0\}, \tag{1.2}$$

be fulfilled, where $\delta_{n,m}$ is the Kronecker symbol and $\langle \cdot, \cdot \rangle_{\mathbb{R}^4}$ denotes the scalar product.

For abbreviation, $\vec{\psi} := \{\vec{\psi^0}, \vec{\psi^1}, \vec{\psi^2}, \vec{\psi^3}\}$ can be thought of as an orthonormal (in the usual Euclidean sense) basis in $\mathbb{R}^4$. In this way we obtain what is known as structural set, see [13, 18, 19, 20, 24].

Let us now introduce the usual definition of equivalent structural sets.

Definition 1.1. *Two structural sets φ, ψ are said to be left equivalent (resp. right) if there exists $h \in \mathbb{H}$, $|h| = 1$ such that*

$$\psi = h\varphi \quad (resp. \quad \psi = \varphi h).$$

This name is shortened if misunderstanding is excluded.

Remark. Observe that the left equivalence (also the right) represents an equivalence relation on the collection of all structural sets. Moreover, each established equivalence class has a unique representative structural set of type $\psi = \{1, \psi^1, \psi^2, \psi^3\}$.

In geometric terms Definition 1.1 means that there exists a rotation which maps the orthonormal basis φ into the orthonormal basis ψ.

The following properties of the structural sets of type $\psi = \{1, \psi^1, \psi^2, \psi^3\}$ are established by direct computation.

Proposition 1.2. *Let $\psi = \{1, \psi^1, \psi^2, \psi^3\}$ be a structural set. Then:*

i) $\overline{\psi^n} = -\psi^n$, $\forall n \in \mathbb{N}_3$,
ii) $(\psi^n)^2 = -1$, $\forall n \in \mathbb{N}_3$,
iii) $\psi^n \cdot \psi^m = -\psi^m \cdot \psi^n$, $\forall n, m \in \mathbb{N}_3$, $n \neq m$,
iv) $\psi^1 \cdot \psi^2 \cdot \psi^3 = \pm 1$.

Taking into account the non-commutativity of $\mathbb{H}$, every structural set ψ generates Cauchy–Riemann operators (left or right), which are defined in $C^1(\Omega, \mathbb{H})$ by the following equalities:

$$^{\psi}D[f] := \sum_{n=0}^{3} \psi^n \cdot \partial_{x_n}[f]; \qquad D^{\psi}[f] := \sum_{n=0}^{3} \partial_{x_n}[f] \cdot \psi^n, \tag{1.3}$$

where $\partial_{x_n} := \partial / \partial x_n$.

Let $\Delta_{\mathbb{H}}[f] = \sum_{n=0}^{3} \Delta_{\mathbb{R}^4}[f_n] e_n$, where $\Delta_{\mathbb{R}^4} = \sum_{n=0}^{3} \partial^2_{x_n^2}$. Then in $C^2(\Omega, \mathbb{H})$ the equalities

$$^{\psi}D \cdot {}^{\bar{\psi}}D = {}^{\bar{\psi}}D \cdot {}^{\psi}D = D^{\psi} \cdot D^{\bar{\psi}} = D^{\bar{\psi}} \cdot D^{\psi} = \Delta_{\mathbb{H}} \tag{1.4}$$

hold.

For fixed ψ and Ω we introduce the set of the so-called ψ-hyperholomorphic functions (left or right respectively), which are given by

$$^{\psi}\mathfrak{M}(\Omega; \mathbb{H}) := \ker {}^{\psi}D = \{f \in C^1(\Omega; \mathbb{H}) : \ {}^{\psi}D[f] = 0_\Omega\},$$
$$\mathfrak{M}^{\psi}(\Omega; \mathbb{H}) := \ker D^{\psi} = \{f \in C^1(\Omega; \mathbb{H}) : \ D^{\psi}[f] = 0_\Omega\}.$$

In [1] are described all different classes of hyperholomorphic functions, see also [13, 18, 19, 20, 24].

Theorem 1.3. $^{\psi}\mathfrak{M}(\Omega, \mathbb{H}) = {}^{\varphi}\mathfrak{M}(\Omega, \mathbb{H})$ *if and only if φ, ψ are left equivalent.*

We emphasize that the set of all classes of hyperholomorphy and the quotient of the collection of structural sets by the left equivalence relation are isomorphic.

2. Necessary and sufficient conditions for associated pairs

Let $\Omega \subset \mathbb{R}^4$ a domain and let $f = \sum_{n=0}^{3} f_n e_n$ be ψ-hyperholomorphic in Ω. We check at once that ${}^{\psi}D[f] = 0$ is equivalent to ${}^{\Psi}\mathfrak{D}\vec{f} = 0$, where ${}^{\Psi}\mathfrak{D}$ is defined on $C^1(\Omega, M_{4\times 1}(\mathbb{R}))$ and is given by

$$ {}^{\Psi}\mathfrak{D}\vec{f} := \sum_{n=0}^{3} \Psi^n \frac{\partial \vec{f}}{\partial x_n}, \quad \text{where} \quad \Psi^n = B_l(\psi^n) \quad \text{and} \quad \frac{\partial \vec{f}}{\partial x_n} = \begin{pmatrix} \frac{\partial f_0}{\partial x_n} \\ \frac{\partial f_1}{\partial x_n} \\ \frac{\partial f_2}{\partial x_n} \\ \frac{\partial f_3}{\partial x_n} \end{pmatrix}. $$

In the same manner we can see the action of ${}^{\Psi}\mathfrak{D}$ on $C^1(\Omega, M_{4\times 4}(\mathbb{R}))$ and consider the relation between both realization of ${}^{\Psi}\mathfrak{D}$.

Let $F \in C^1(\Omega, M_{4\times 4}(\mathbb{R}))$, then $F = (f_0|f_1|f_2|f_3)$ where $f_n \in C^1(\Omega, M_{4\times 1}(\mathbb{R}))$, $n \in \mathbb{N}_3 \cup \{0\}$ denotes the nth column of F. Using the properties of the matrix product the following correlation is valid

$$ {}^{\Psi}\mathfrak{D}F = \left({}^{\Psi}\mathfrak{D}f_0 | {}^{\Psi}\mathfrak{D}f_1 | {}^{\Psi}\mathfrak{D}f_2 | {}^{\Psi}\mathfrak{D}f_3\right). \tag{2.5} $$

Hence, ${}^{\Psi}\mathfrak{D}F = 0$ yields the ψ-hyperholomorphicity of $f_n, n \in \mathbb{N}_3 \cup \{0\}$.

Introduce the first-order differential operator $\mathfrak{L}$ on $C^1(\Omega, M_{4\times 1}(\mathbb{R}))$ as follows,

$$ \mathfrak{L}\vec{f} = \sum_{n=0}^{3} A_n \frac{\partial \vec{f}}{\partial x_n} + B\vec{f} + C, \tag{2.6} $$

where A_n and B are $M_{4\times 4}(\mathbb{R})$-valued functions on Ω meanwhile C is one $M_{4\times 1}(\mathbb{R})$-valued.

Definition 2.1. *A pair of operators* ${}^{\Psi}\mathfrak{D}, \mathfrak{L}$ *is said to be associated if* ${}^{\Psi}\mathfrak{D}\vec{f} = 0$ *implies* ${}^{\Psi}\mathfrak{D}\left(\mathfrak{L}\vec{f}\right) = 0$.

We will denote by $S^1_{{}^{\Psi}\mathfrak{D}}$ the set of all operators (2.6) which are associated to ${}^{\Psi}\mathfrak{D}$.

Proposition 2.2. *Let ψ and φ be left equivalent structural sets. Then,*

$$ \mathcal{S}^1_{{}^{\Psi}\mathfrak{D}} = \mathcal{S}^1_{{}^{\Phi}\mathfrak{D}}. \tag{2.7} $$

Proof. It suffices to make the following observation

$$ \mathfrak{L} \in \mathcal{S}^1_{{}^{\Psi}\mathfrak{D}} \quad \Leftrightarrow \quad \mathfrak{L}\left({}^{\psi}\mathfrak{M}(\Omega, \mathbb{H})\right) \subset {}^{\psi}\mathfrak{M}(\Omega, \mathbb{H}) = {}^{\varphi}\mathfrak{M}(\Omega, \mathbb{H}) \quad \Leftrightarrow \quad \mathfrak{L} \in \mathcal{S}^1_{{}^{\Phi}\mathfrak{D}}. \quad \square $$

Theorem 2.3. *Let $\psi = \{1, \psi^1, \psi^2, \psi^3\}$ be a structural set. Then, the operator* (2.6) *is associated to ${}^{\Psi}\mathfrak{D}$ if and only if there exists a first-order linear differential operator $\mathfrak{L}'$ (not necessarily associated to ${}^{\Psi}\mathfrak{D}$) such that on $C^2(\Omega, M_{4\times 1}(\mathbb{R}))$ we have*

$$ {}^{\Psi}\mathfrak{D}\,\mathfrak{L} = \mathfrak{L}'\,{}^{\Psi}\mathfrak{D}. \tag{2.8} $$

Proof. The sufficiency in this case is clear. Indeed, if there exists $\mathfrak{L}'$ satisfying (2.8) we obtain, for every f such that ${}^{\Psi}\mathfrak{D}\vec{f} = 0$, that ${}^{\Psi}\mathfrak{D}\left(\mathfrak{L}\vec{f}\right) = \mathfrak{L}'\left({}^{\Psi}\mathfrak{D}\vec{f}\right) = 0$, by the linearity of $\mathfrak{L}'$.

Suppose now that $\mathfrak{L}$ is associated to ${}^{\Psi}\mathfrak{D}$. By (2.6) we have:

$$\mathfrak{L}\vec{f} = A_0{}^{\Psi}\mathfrak{D}\vec{f} + \sum_{n=1}^{3}\left(A_n - A_0\Psi^n\right)\frac{\partial \vec{f}}{\partial x_n} + B\vec{f} + C.$$

To shorten notation we write $X_0 := A_0$, $X_n := A_n - A_0\Psi^n$ $(n \in \mathbb{N}_3)$, $X_4 := B$ y $X_5 := C$, then $\mathfrak{L}$ takes the form

$$\mathfrak{L}\vec{f} = X_0{}^{\Psi}\mathfrak{D}\vec{f} + \sum_{n=1}^{3} X_n\frac{\partial \vec{f}}{\partial x_n} + X_4\vec{f} + X_5, \tag{2.9}$$

where X_n (n=0,1,...,4) are $M_{4\times 4}(\mathbb{R})$-valued functions in Ω meanwhile X_5 is an $M_{4\times 1}(\mathbb{R})$-valued one.

Defining the linear operator $\mathfrak{L}^*$ given by

$$\mathfrak{L}^*\vec{f} = {}^{\Psi}\mathfrak{D}\left(X_0\vec{f}\right) + \sum_{n=1}^{3} X_n\frac{\partial \vec{f}}{\partial x_n} + X_4\vec{f},$$

we will calculate the precise expression of $R_{\mathfrak{L}} = {}^{\Psi}\mathfrak{D}\ \mathfrak{L} - \mathfrak{L}^*\ {}^{\Psi}\mathfrak{D}$ to make it act on $C^2(\Omega, \mathbb{H})$.

Let $f \in C^2(\Omega, \mathbb{H})$, then

$$\begin{aligned} R_{\mathfrak{L}}\vec{f} &= {}^{\Psi}\mathfrak{D}\left(\mathfrak{L}\vec{f}\right) - \mathfrak{L}^*\left({}^{\Psi}\mathfrak{D}\vec{f}\right) \\ &= \sum_{n=1}^{3}\left[{}^{\Psi}\mathfrak{D}\left(X_n\frac{\partial \vec{f}}{\partial x_n}\right) - X_n\frac{\partial\left({}^{\Psi}\mathfrak{D}\vec{f}\right)}{\partial x_n}\right] \\ &\quad + \left[{}^{\Psi}\mathfrak{D}\left(X_4\vec{f}\right) - X_4\left({}^{\Psi}\mathfrak{D}\vec{f}\right)\right] + {}^{\Psi}\mathfrak{D}X_5. \end{aligned} \tag{2.10}$$

Note that if $F = [f_{ij}]_{i,j=0,\ldots,3}$ and $\vec{f}$ represent $M_{4\times 4}(\mathbb{R})$- and $M_{4\times 1}(\mathbb{R})$-valued functions respectively, then

$$\frac{\partial\left(F\vec{f}\right)}{\partial x_n} = \frac{\partial F}{\partial x_n}\vec{f} + F\frac{\partial \vec{f}}{\partial x_n}, \qquad \text{where} \qquad \frac{\partial F}{\partial x_n} = \left[\frac{\partial f_{ij}}{\partial x_n}\right]_{i,j=0,\ldots,3}.$$

Thus

$${}^{\Psi}\mathfrak{D}\left(F\vec{f}\right) = \left({}^{\Psi}\mathfrak{D}F\right)\vec{f} + \sum_{n=0}^{3}\Psi^n F\frac{\partial \vec{f}}{\partial x_n}, \qquad \text{and} \qquad \frac{\partial\left({}^{\Psi}\mathfrak{D}\vec{f}\right)}{\partial x_n} = \sum_{m=0}^{3}\Psi^m\frac{\partial^2 \vec{f}}{\partial x_n \partial x_m}.$$

Combining the last two equalities in (2.10) yields

$$
\begin{aligned}
R_{\mathfrak{L}}\vec{f} &= \sum_{n=1}^{3}\left[\left({}^{\Psi}\mathfrak{D}X_n\right)\frac{\partial \vec{f}}{\partial x_n} + \sum_{m=0}^{3}\Psi^m X_n\frac{\partial^2 \vec{f}}{\partial x_m \partial x_n} - \sum_{m=0}^{3} X_n \Psi^m \frac{\partial^2 \vec{f}}{\partial x_n \partial x_m}\right] \\
&\quad + \left({}^{\Psi}\mathfrak{D}X_4\right)\vec{f} + \sum_{n=0}^{3}\left[\Psi^n X_4 \frac{\partial \vec{f}}{\partial x_n} - X_4 \Psi^n \frac{\partial \vec{f}}{\partial x_n}\right] + {}^{\Psi}\mathfrak{D}X_5 \\
&= \sum_{n,m=1}^{3}\left(\Psi^m X_n - X_n \Psi^m\right)\frac{\partial^2 \vec{f}}{\partial x_m \partial x_n} \qquad (2.11)\\
&\quad + \sum_{n=1}^{3}\left({}^{\Psi}\mathfrak{D}X_n + \Psi^n X_4 - X_4\Psi^n\right)\frac{\partial \vec{f}}{\partial x_n} \\
&\quad + \left({}^{\Psi}\mathfrak{D}X_4\right)\vec{f} + {}^{\Psi}\mathfrak{D}X_5.
\end{aligned}
$$

Then,

$$
R_{\mathfrak{L}}\vec{f} = \sum_{1\le n\le m\le 3}\mathcal{A}_{nm}\frac{\partial^2 \vec{f}}{\partial x_m \partial x_n} + \sum_{n=1}^{3}\mathcal{B}_n\frac{\partial \vec{f}}{\partial x_n} + \mathcal{C}\vec{f} + \mathcal{D}, \qquad (2.12)
$$

where,

$$
\mathcal{A}_{nm} = \begin{cases} \Psi^n X_n - X_n \Psi^n & n = m, \\ \left(\Psi^n X_m - X_m \Psi^n\right) + \left(\Psi^m X_n - X_n \Psi^m\right) & n \neq m, \end{cases} \qquad (2.13)
$$

$$
\mathcal{B}_n = {}^{\Psi}\mathfrak{D}X_n + \Psi^n X_4 - X_4 \Psi^n \qquad (2.14)
$$

$$
\mathcal{C} = {}^{\Psi}\mathfrak{D}X_4, \qquad (2.15)
$$

$$
\mathcal{D} = {}^{\Psi}\mathfrak{D}X_5. \qquad (2.16)
$$

with $n, m \in \mathbb{N}_3$.

Observe that $R_{\mathfrak{L}}$ does not depend on ∂_{x_0}. As $\mathfrak{L}$ is associated to ${}^{\Psi}\mathfrak{D}$, the linearity of $\mathfrak{L}^*$ gives $R_{\mathfrak{L}}[f] = 0_\Omega$ for all $f \in {}^{\psi}\mathfrak{M}(\Omega, \mathbb{H})$. Having disposed of this preliminary step, we proceed to show the nullity of $R_{\mathfrak{L}}$ on $C^2(\Omega, \mathbb{H})$. To do that, we need to consider the following sequence of assertions:

- When $\vec{f} = \vec{0}_\Omega$ is substituted in (2.12) we have

$$
\vec{0}_\Omega = R_{\mathfrak{L}}\vec{f} = \mathcal{D}. \qquad (2.17)
$$

- Taking in (2.12) ψ-hyperholomorphic functions $f^k \equiv e_k$ for $k \in \mathbb{N}_3 \cup \{0\}$ yields

$$
\vec{0}_\Omega = R_{\mathfrak{L}}\vec{f^k} = \mathcal{C}\vec{e_k} \qquad k \in \mathbb{N}_3 \cup \{0\}.
$$

  Then,

$$
\mathcal{C} \equiv 0. \qquad (2.18)
$$

- For $n \in \mathbb{N}_3$ let us consider the ψ-hyperholomorphic functions

$$f_n^k(x) = (x_0 + \psi^n x_n)\, e_k, \quad \text{i.e.,} \quad \vec{f_n^k}(x) = x_0 \vec{e_k} + B_l(\psi^n)\, x_n \vec{e_k},$$

for $k \in \mathbb{N}_3 \cup \{0\}$. Then $\partial \vec{f_n^k}/\partial x_n = B_l(\psi^n)\, \vec{e_k}$ and,

$$\vec{0}_\Omega = R_{\mathfrak{L}} \vec{f_n^k} = \mathcal{B}_n B_l(\psi^n)\, \vec{e_k} \qquad k \in \mathbb{N}_3 \cup \{0\}.$$

The last means that the kth column of $\mathcal{B}_n B_l(\psi^n)$ is $\vec{0}_\Omega$ for $k \in \mathbb{N}_3 \cup \{0\}$ and

$$\mathcal{B}_n B_l(\psi^n) \equiv 0 \Rightarrow \mathcal{B}_n \equiv 0 \qquad n \in \mathbb{N}_3. \tag{2.19}$$

- Now, for $n \in \mathbb{N}_3$ consider the ψ-hyperholomorphic functions $f_{nn}^k(x) = (x_0^2 - x_n^2 + \psi^n 2x_0 x_n)e_k$ for $k \in \mathbb{N}_3 \cup \{0\}$, for which $\partial^2 \vec{f_{nn}^k}/\partial x_n^2 = -2\vec{e_k}$. So we find

$$\vec{0}_\Omega = R_{\mathfrak{L}} \vec{f_{nn}^k} = -2\mathcal{A}_{nn}\vec{e_k}, \qquad k \in \mathbb{N}_3 \cup \{0\}.$$

One concludes that the kth column of $\mathcal{A}_{nn}$ is $\vec{0}_\Omega$ for $k \in \mathbb{N}_3 \cup \{0\}$. This implies

$$\mathcal{A}_{nn} \equiv 0 \qquad n \in \mathbb{N}_3. \tag{2.20}$$

- Finally, for each pair m, n, $1 \le n < m \le 3$ by choosing the ψ-hyperholomorphic functions $f_{nm}^k(x) = \left(\psi^n (x_m^2 - x_n^2) + \psi^m 2x_n x_m\right) e_k$ for $k \in \mathbb{N}_3 \cup \{0\}$, we get $\partial^2 \vec{f_{nm}^k} \big/ \partial x_n \partial x_m = 2B_l(\psi^m)\, \vec{e_k}$ and thus

$$\vec{0}_\Omega = R_{\mathfrak{L}} \vec{f_{nm}^k} = 2\mathcal{A}_{nm} B_l(\psi^m)\, \vec{e_k} \qquad k \in \mathbb{N}_3 \cup \{0\}.$$

The result is that the kth column of $\mathcal{A}_{nm} B_l(\psi^m)$ is $\vec{0}_\Omega$ for $k \in \mathbb{N}_3 \cup \{0\}$ obtaining

$$\mathcal{A}_{nm} B_l(\psi^m) \equiv 0 \Rightarrow \mathcal{A}_{nm} \equiv 0, \qquad 1 \le n < m \le 3. \tag{2.21}$$

Adding up (2.12), (2.17), (2.18), (2.19), (2.20) and (2.21) we can assert that $R_{\mathfrak{L}} \equiv 0$ on $C^2(\Omega, \mathbb{H})$. Then, exists $\mathfrak{L}' = \mathfrak{L}^*$ for which (2.8) is satisfied. □

Next we formulate a criterion under which the operator (2.6) and ${}^{\Psi}\mathfrak{D}$ form an associated pair.

Theorem 2.4. *Let $\psi = \{1, \psi^1, \psi^2, \psi^3\}$ be a structural set. Then, the operator (2.6) is associated to ${}^{\Psi}\mathfrak{D}$ if and only if the following conditions are satisfied:*

$$\Psi^n X_m + \Psi^m X_n = X_m \Psi^n + X_n \Psi^m,, \tag{2.22}$$

$$ {}^{\Psi}\mathfrak{D} X_n + \Psi^n B - B\Psi^n = 0, \tag{2.23}$$

$$ {}^{\Psi}\mathfrak{D} B = 0, \tag{2.24}$$

$$ {}^{\Psi}\mathfrak{D} C = 0, \tag{2.25}$$

where $X_n = A_n - A_0 \Psi^n$ and $n, m \in \mathbb{N}_3$.

3. First-order symmetries of the generalized Cauchy–Riemann operator

This section will be devoted to the study of first-order symmetries of the quaternionic Cauchy–Riemann ψ-operator. In particular, to those symmetries which represent also φ-operators of Cauchy–Riemann for some structural set φ.

Definition 3.1. *A quaternionic first-order partial differential operator L is said to be a symmetry of ${}^{\psi}D$ if ${}^{\psi}D[f] = 0_\Omega$ implies that ${}^{\psi}D\left[L[f]\right] = 0_\Omega$.*

Taking in (2.6) $A_n = B_l(\alpha_n)$, $n \in \mathbb{N}_3 \cup \{0\}$, $B = B_l(\beta)$ and $C = \vec{\gamma}$, where α_n ($n \in \mathbb{N}_3 \cup \{0\}$), β and γ are $\mathbb{H}$-valued functions, we can write $\mathfrak{L}$ in a quaternionic form L defined by

$$L[f] = \sum_{n=0}^{3} \alpha_n \cdot \partial_{x_n}[f] + \beta \cdot f + \gamma. \tag{3.26}$$

From this transformation and (1.1), Theorem 2.3 shows the following result.

Theorem 3.2. *Let $\psi = \{1, \psi^1, \psi^2, \psi^3\}$ be a structural set. Then, the operator (3.26) is a symmetry of ${}^{\psi}D$ if and only if there exists a quaternionic first-order linear partial differential operator L' (not necessarily a symmetry of ${}^{\psi}D$) such that on $C^2(\Omega, \mathbb{H})$ it holds*

$${}^{\psi}D\, L = L'\, {}^{\psi}D. \tag{3.27}$$

We can now proceed analogously to the proof of the following criterion:

Theorem 3.3. *Let $\psi = \{1, \psi^1, \psi^2, \psi^3\}$ be a structural set. Then, the operator (3.26) is a symmetry of ${}^{\psi}D$ if and only if the following identities hold:*

$$\psi^n \cdot \mathcal{X}_m + \psi^m \cdot \mathcal{X}_n = \mathcal{X}_m \cdot \psi^n + \mathcal{X}_n \cdot \psi^m, \tag{3.28}$$

$${}^{\psi}D[\mathcal{X}_n] + \psi^n \cdot \beta - \beta \cdot \psi^n = 0_\Omega, \tag{3.29}$$

$${}^{\psi}D[\beta] = 0_\Omega, \tag{3.30}$$

$${}^{\psi}D[\gamma] = 0_\Omega, \tag{3.31}$$

where $\mathcal{X}_n = \alpha_n - \alpha_0 \cdot \psi^n$ and $m, n \in \mathbb{N}_3$.

Remark. Likewise, we can see that for $x \in \mathbb{H}$ we have $B_l(x) = \left(\vec{x}\middle|\vec{xe_1}\middle|\vec{xe_2}\middle|\vec{xe_3}\right)$. Then, (2.5) enables us to write for $f \in C^1(\Omega, \mathbb{H})$

$$\begin{aligned} {}^{\Psi}\mathfrak{D}(B_l(f)) &= \left({}^{\Psi}\mathfrak{D}\left(\vec{f}\right) \middle| {}^{\Psi}\mathfrak{D}\left(\vec{fe_1}\right) \middle| {}^{\Psi}\mathfrak{D}\left(\vec{fe_2}\right) \middle| {}^{\Psi}\mathfrak{D}\left(\vec{fe_3}\right)\right) \\ &= \left(\overrightarrow{{}^{\psi}D[f]} \,\middle|\, \overrightarrow{{}^{\psi}D[f]e_1} \,\middle|\, \overrightarrow{{}^{\psi}D[f]e_2} \,\middle|\, \overrightarrow{{}^{\psi}D[f]e_3}\right) = B_l\left({}^{\psi}D[f]\right). \end{aligned}$$

This gives (3.29) and (3.30) as direct consequences of (2.23) and (2.24) respectively.

3.1. ψ-symmetries given by φ-operators

Let $\psi = \{1, \psi^1, \psi^2, \psi^3\}$ be a structural set. Our next goal is to determine all the structural sets $\varphi := \{\varphi^0, \varphi^1, \varphi^2, \varphi^3\}$ such that ${}^{\varphi}D$ be a symmetry of ${}^{\psi}D$.

According to Theorem 3.3 for $L = {}^{\varphi}D$, we have that ${}^{\varphi}D$ is a symmetry of ${}^{\psi}D$ if and only if

$$\psi^n \cdot \mathcal{X}_m + \psi^m \cdot \mathcal{X}_n = \mathcal{X}_m \cdot \psi^n + \mathcal{X}_n \cdot \psi^m, \qquad \forall n, m \in \mathbb{N}_3, \tag{3.32}$$

where $\mathcal{X}_n = (\varphi^n - \varphi^0 \cdot \psi^n)_\Omega$, $n \in \mathbb{N}_3 \cup \{0\}$. Observe that in this case $\gamma = \beta = 0_\Omega$ and $\mathcal{X}_n$ are constant functions in Ω. Then, for all ${}^{\psi}D$, the conditions (3.29), (3.30) and (3.31) hold. From this, our next concern will be only (3.32).

Writing $\varphi^n = h^n \cdot \psi^n$, $n \in \mathbb{N}_3 \cup \{0\}$, ($|h^n| = 1$) yields

$$\mathcal{X}_n = \varphi^n - \varphi^0 \cdot \psi^n = (h^n - h^0) \cdot \psi^n, \qquad \forall n \in \mathbb{N}_3. \tag{3.33}$$

Substituting (3.33) into (3.32) and assuming first $n = m$ and later on $n \neq m$, $n, m \in \mathbb{N}_3$, we deduce that

$$\psi^n \cdot (h^n - h^0) \cdot \psi^n = (h^n - h^0) \cdot \psi^n \cdot \psi^n \tag{3.34}$$

$$\begin{aligned}\psi^n \cdot (h^m - h^0) \cdot \psi^m + \psi^m \cdot (h^n - h^0) \cdot \psi^n = {} & (h^m - h^0) \cdot \psi^m \cdot \psi^n \\ & + (h^n - h^0) \cdot \psi^n \cdot \psi^m.\end{aligned} \tag{3.35}$$

Applying Proposition 1.2 to (3.34)–(3.35) we obtain that ${}^{\varphi}D$ is a symmetry of ${}^{\psi}D$ if and only if

$$\psi^n \cdot (h^n - h^0) = (h^n - h^0) \cdot \psi^n, \tag{3.36}$$

$$\psi^m \cdot \psi^n \cdot (h^m - h^n) = -(h^m - h^n) \cdot \psi^m \cdot \psi^n, \tag{3.37}$$

for all $m, n \in \mathbb{N}_3$ with $n \neq m$.

According to **iv)** in Proposition 1.2 we need to consider two cases, but we will give the proof for $\psi^1 \cdot \psi^2 \cdot \psi^3 = -1$, the other case is left to the reader because it is completely analogous and shares the majority of the results. The structural set ψ under study has the multiplication properties.

$$\psi^1 \cdot \psi^2 = -\psi^2 \cdot \psi^1 = \psi^3; \quad \psi^2 \cdot \psi^3 = -\psi^3 \cdot \psi^2 = \psi^1; \quad \psi^3 \cdot \psi^1 = -\psi^1 \cdot \psi^3 = \psi^2.$$

These allow us to write (3.37) in the form:

$$\psi^k \cdot (h^m - h^n) = -(h^m - h^n) \cdot \psi^k \qquad m, n, k \in \mathbb{N}_3,\ m \neq n,\ m \neq k,\ n \neq k. \tag{3.38}$$

We next turn to finding the elements of $\mathbb{H}$ which simultaneously commute and anti-commute with ψ^n, $n \in \mathbb{N}_3$. For this to happen, it is enough to examine only the case ψ^1, the other cases are left to the reader.

It is easy to check if $a = a_0 + a_1\psi^1 + a_2\psi^2 + a_3\psi^3$ that,

$$a \cdot \psi^1 = \psi^1 \cdot a \Leftrightarrow a = a_0 + a_1\psi^1, \quad \text{and} \quad a \cdot \psi^1 = -\psi^1 \cdot a \Leftrightarrow a = a_2\psi^2 + a_3\psi^3.$$

This established the following characterization

Proposition 3.4. *Let $\psi = \{1, \psi^1, \psi^2, \psi^3\}$ be a structural set and let $a = a_0 + a_1\psi^1 + a_2\psi^2 + a_3\psi^3$, with $a_n \in \mathbb{R}$, $n \in \mathbb{N}_3 \cup \{0\}$. Then, for all each $m \in \mathbb{N}_3$ we have*

$$\cdot\psi^m = \psi^m \cdot a \Leftrightarrow a = a_0 + a_m\psi^m \quad \textit{and} \quad a \cdot \psi^m = -\psi^m \cdot a \Leftrightarrow a_0 + a_m\psi^m = 0.$$

By the above results and (3.36) we have $h^n - h^0 = l_n + k_n\psi^n$, where $n \in \mathbb{N}_3$ and $l_n, k_n \in \mathbb{R}$. On the other hand, (3.38) implies for each pair $m, n \in \mathbb{N}_3$ $(m \neq m)$ that

$$h^m - h^n = (l_m - l_n) + k_m\psi^m - k_n\psi^n = k_m\psi^m - k_n\psi^n \Rightarrow l_m = l_n.$$

Taking $k = l_1 = l_2 = l_3$ we can write:

$$h^n = (k + k_n\psi^n) + h^0, \qquad \forall n \in \mathbb{N}_3. \tag{3.39}$$

We have thus proved that relation (3.39) between $h^n, n \in \mathbb{N}_3 \cup \{0\}$ is necessary and sufficient to give (3.36) y (3.37). Let us now indicate the real numbers k, k_1, k_2 and k_3 can be taken for $\varphi := \{h^0, h^1 \cdot \psi^1, h^2 \cdot \psi^2, h^3 \cdot \psi^3\}$ to be a structural set.

Set a unitary quaternion $h^0 = q_0 + q_1\psi^1 + q_2\psi^2 + q_3\psi^3$, i.e., $1 = |h^0|^2 = q_0^2 + q_1^2 + q_2^2 + q_3^2$. Then, $\varphi^n \cdot \overline{\varphi^0} + \varphi^0 \cdot \overline{\varphi^n} = 0$ for $n \in \mathbb{N}_3$, or equivalently,

$$0 = h^n \cdot \psi^n \cdot \overline{h^0} + h^0 \cdot \overline{h^n \cdot \psi^n} = h^n \cdot \psi^n \cdot \overline{h^0} - h^0 \cdot \psi^n \cdot \overline{h^n} \Leftrightarrow h^n \cdot \psi^n \cdot \overline{h^0} = h^0 \cdot \psi^n \cdot \overline{h^n}.$$

Replacing (3.39) above

$$\begin{aligned} (k + k_n\psi^n) \cdot \psi^n \cdot \overline{h^0} &= h^0 \cdot \psi^n \cdot (k - k_n\psi^n) \\ \Leftrightarrow \quad (-k_n + k\psi^n) \cdot \overline{h^0} &= h^0 \cdot (k_n + k\psi^n), \end{aligned}$$

or equivalently $k\left(\psi^n \cdot \overline{h^0} - h^0 \cdot \psi^n\right) = k_n\left(h^0 + \overline{h^0}\right) = 2k_n q_0$.

It follows easily that $\psi^n \cdot \overline{h^0} - h^0 \cdot \psi^n = 2q_n$, $n \in \mathbb{N}_3$. Then

$$\varphi^n \cdot \overline{\varphi^0} + \varphi^0 \cdot \overline{\varphi^n} = 0 \quad \Leftrightarrow \quad kq_n = k_n q_0, \qquad \forall n \in \mathbb{N}_3. \tag{3.40}$$

Assuming $q_0 \neq 0$, we can assert that $k = \lambda q_0$, hence that $k_n = \lambda q_n$ for $n \in \mathbb{N}_3$, and finally that

$$h^n = \lambda(q_0 + q_n \cdot \psi^n) + h^0, \qquad \forall n \in \mathbb{N}_3. \tag{3.41}$$

But on account of $|h^n| = 1$ we have $(\lambda^2 + 2\lambda)\, q_0^2 + (\lambda^2 + 2\lambda)\, q_n^2 = 0$. Therefore

$$(\lambda^2 + 2\lambda)\,(q_0^2 + q_n^2) = 0 \Rightarrow \lambda = 0 \text{ or } \lambda = -2,$$

since $q_0 \neq 0$ so $q_0^2 + q_n^2 > 0$.

If $\lambda = 0$, then $h^n = h^0$ for $n \in \mathbb{N}_3$, which implies that φ is a structural set equivalent to ψ.

We next show that for $\lambda = -2$ φ become a structural set. It is sufficient to prove that

$$0 = \varphi^n \cdot \overline{\varphi^m} + \varphi^m \cdot \overline{\varphi^n} = 2\left\langle \overrightarrow{h^n \cdot \psi^n}, \overrightarrow{h^m \cdot \psi^m} \right\rangle_{\mathbb{R}^4} \qquad \forall n, m \in \mathbb{N}_3, n \neq m, \tag{3.42}$$

because the cases $n = m \in \mathbb{N}_3 \cup \{0\}$ and $m = 0$, $n \in \mathbb{N}_3$ have been already considered.

If this is so, we have

$$\begin{cases} h^1 \cdot \psi^1 = (-q_0 - q_1\psi^1 + q_2\psi^2 + q_3\psi^3) \cdot \psi^1 = q_1 - q_0\psi^1 + q_3\psi^2 - q_2\psi^3, \\ h^2 \cdot \psi^2 = (-q_0 + q_1\psi^1 - q_2\psi^2 + q_3\psi^3) \cdot \psi^2 = q_2 - q_3\psi^1 - q_0\psi^2 + q_1\psi^3, \\ h^3 \cdot \psi^3 = (-q_0 + q_1\psi^1 + q_2\psi^2 - q_3\psi^3) \cdot \psi^3 = q_3 + q_2\psi^1 - q_1\psi^2 - q_0\psi^3, \end{cases} \tag{3.43}$$

which gives for each pair $m, n \in \mathbb{N}_3$ $(m \neq n)$

$$\left\langle \overrightarrow{h^m \cdot \psi^m}, \overrightarrow{h^n \cdot \psi^n} \right\rangle_{\mathbb{R}^4} = 0$$

and (3.42) is proved. Conversely, suppose that $q_0 = 0$, then (3.40) yields $k = 0$. From this $h^n = k_n\psi^n + h^0$. But $|h^n| = 1 \Rightarrow k_n^2 + 2k_nq_n = 0 \Rightarrow k_n = 0$ or $k_n = -2q_n$, $n \in \mathbb{N}_3$.

If for every $n \in \mathbb{N}_3$ $k_n = 0$ or $k_n = -2q_n$, the analysis is reduced to the cases appeared when $q_0 \neq 0$ and φ is a structural set as claimed.

Assume that k_1, k_2, k_3 are not all zero. As before, conditions (3.42) are the only needed to be proved in order that φ became a structural set. The proof falls naturally into two parts.

Only one k_n assume a non-zero value, e.g., $k_1 \neq 0$. Then

$$k_1 = 0, \quad k_2 = -2q_2 \neq 0, \quad k_3 = -2q_3 \neq 0,$$

and

$$\begin{cases} h^1 \cdot \psi^1 = (q_1\psi^1 + q_2\psi^2 + q_3\psi^3) \cdot \psi^1 = -q_1 + q_3\psi^2 - q_2\psi^3, \\ h^2 \cdot \psi^2 = (q_1\psi^1 - q_2\psi^2 + q_3\psi^3) \cdot \psi^2 = q_2 - q_3\psi^1 + q_1\psi^3, \\ h^3 \cdot \psi^3 = (q_1\psi^1 + q_2\psi^2 - q_3\psi^3) \cdot \psi^3 = q_3 + q_2\psi^1 - q_1\psi^2. \end{cases} \tag{3.44}$$

Combining these equalities with conditions (3.42) yields

$$\left\langle \overrightarrow{h^1 \cdot \psi^1}, \overrightarrow{h^2 \cdot \psi^2} \right\rangle_{\mathbb{R}^4} = -2q_1q_2, \quad \left\langle \overrightarrow{h^2 \cdot \psi^2}, \overrightarrow{h^3 \cdot \psi^3} \right\rangle_{\mathbb{R}^4} = 0,$$
$$\left\langle \overrightarrow{h^3 \cdot \psi^3}, \overrightarrow{h^1 \cdot \psi^1} \right\rangle_{\mathbb{R}^4} = -2q_1q_3.$$

Then, φ should be an structural set only if $0 = q_1q_2 = q_1q_3 \Leftrightarrow q_1 = 0$, since the assumption $q_2 \neq 0$ and $q_3 \neq 0$. Hence, in this case $k_1 = -2q_1 = 0$, to conclude $k_n = -2q_n$ for $n \in \mathbb{N}_3$, which is one of the already analyzed cases.

On the other hand, if in the set $\{k_1, k_2, k_3\}$ two zero elements appeared, e.g., k_1 and k_2, then

$$k_1 = 0, \quad k_2 = 0, \quad k_3 = -2q_3 \neq 0,$$

and so

$$\begin{cases} h^1 \cdot \psi^1 = (q_1\psi^1 + q_2\psi^2 + q_3\psi^3) \cdot \psi^1 = -q_1 + q_3\psi^2 - q_2\psi^3, \\ h^2 \cdot \psi^2 = (q_1\psi^1 + q_2\psi^2 + q_3\psi^3) \cdot \psi^2 = -q_2 - q_3\psi^1 + q_1\psi^3, \\ h^3 \cdot \psi^3 = (q_1\psi^1 + q_2\psi^2 - q_3\psi^3) \cdot \psi^3 = q_3 + q_2\psi^1 - q_1\psi^2. \end{cases} \tag{3.45}$$

As before, conditions (3.42) and the last equalities give

$$\left\langle \overrightarrow{h^1 \cdot \psi^1}, \overrightarrow{h^2 \cdot \psi^2} \right\rangle_{\mathbb{R}^4} = 0, \quad \left\langle \overrightarrow{h^2 \cdot \psi^2}, \overrightarrow{h^3 \cdot \psi^3} \right\rangle_{\mathbb{R}^4} = -2q_2q_3,$$
$$\left\langle \overrightarrow{h^3 \cdot \psi^3}, \overrightarrow{h^1 \cdot \psi^1} \right\rangle_{\mathbb{R}^4} = -2q_1q_3.$$

Then, φ should be a structural set when $0 = q_2q_3 = q_1q_3 \Leftrightarrow 0 = q_1 = q_2$, because the assumption $q_3 \neq 0$. Therefore, in this case we have $k_1 = -2q_1 = 0$ and $k_2 = -2q_2 = 0$, and $k_n = -2q_n$ for $n \in \mathbb{N}_3$, hence we are in exactly the same situation as before.

We have actually proved that given a structural set ψ, with $\psi^0 = 1$, the only structural set φ being ${}^{\varphi}D$ a symmetry of ${}^{\psi}D$ are those left equivalent to ψ or are of the form

$$\varphi = \{h, h^1 \cdot \psi^1, h^2 \cdot \psi^2, h^3 \cdot \psi^3\}, \tag{3.46}$$

where

$$h^1 = -q_0 - q_1\psi^1 + q_2\psi^2 + q_3\psi^3, \quad h^2 = -q_0 + q_1\psi^1 - q_2\psi^2 + q_3\psi^3,$$
$$h^3 = -q_0 + q_1\psi^1 + q_2\psi^2 - q_3\psi^3,$$

for some unitary quaternion $h = q_0 + q_1\psi^1 + q_2\psi^2 + q_3\psi^3$.

Finally, note that

$$\begin{cases} h^1 \cdot \psi^1 = \left(-q_0 - q_1\psi^1 + q_2\psi^2 + q_3\psi^3\right) \cdot \psi^1 \\ h^2 \cdot \psi^2 = \left(-q_0 + q_1\psi^1 - q_2\psi^2 + q_3\psi^3\right) \cdot \psi^2 \\ h^3 \cdot \psi^3 = \left(-q_0 + q_1\psi^1 + q_2\psi^2 - q_3\psi^3\right) \cdot \psi^3 \end{cases}$$

which leads to

$$\begin{cases} \psi^1 \left(-q_0 - q_1\psi^1 - q_2\psi^2 - q_3\psi^3\right) = -\psi^1 \cdot h, \\ \psi^2 \left(-q_0 - q_1\psi^1 - q_2\psi^2 - q_3\psi^3\right) = -\psi^2 \cdot h, \\ \psi^3 \left(-q_0 - q_1\psi^1 - q_2\psi^2 - q_3\psi^3\right) = -\psi^3 \cdot h. \end{cases}$$

The structural set φ satisfies the relation (3.46) if and only if it is right equivalent to $\overline{\psi} = \{1, -\psi^1, -\psi^2, -\psi^3\}$. Hence the following theorem has been proved.

Theorem 3.5. *Let $\psi = \{1, \psi^1, \psi^2, \psi^3\}$ be a structural set. Then, the Cauchy–Riemann operator ${}^{\varphi}D$ should be a symmetry of ${}^{\psi}D$ if and only if φ is either left equivalent to ψ or right equivalent to $\overline{\psi}$.*

Remark. Due to Proposition 3.4 the classes of structural sets involved in the conclusion of the above theorem are disjoint.

4. Endomorphisms over the quaternions

Let $\{1, \psi^1, \psi^2, \psi^3\}$ be a structural set. We want to consider the algebra of endomorphisms $\mathrm{End}(\mathbb{H})$, i.e., the algebra of linear maps $T : \mathbb{H} \mapsto \mathbb{H}$. Hereby we follow [21]. Obviously, $\mathrm{End}(\mathbb{H})$ has to be isomorphic to the full matrix algebra of 4×4-matrices. Now, let $a \in \mathbb{H}$ and we consider the left multiplication operators

$$\psi^k \quad : \quad a \mapsto \psi^k a$$

as well as the conjugation operators

$$\tilde{\psi}^k : a \mapsto \overline{a}\psi^k.$$

Clearly, these operators satisfy the relations

$$\psi^k\psi^j + \psi^j\psi^k = -2\delta_{jk}, \qquad \tilde{\psi}^k\tilde{\psi}^j + \tilde{\psi}^j\tilde{\psi}^k = 2\delta_{jk}, \qquad \psi^j\tilde{\psi}^k = -\tilde{\psi}^k\psi^j$$

and generate an (ultra-hyperbolic) space with bilinear form

$$B(\underline{x} + \tilde{\underline{x}}, \underline{y} + \tilde{\underline{y}}) = \sum x_j y_j - \tilde{x}_j \tilde{y}_j.$$

This bilinear form is invariant under $O(3,3)$ as well as $SO(3,3)$ respectively. The vector space can also be decomposed into

$$E^4 \times \tilde{E}^4$$

where $E^4 = \mathrm{span}\{\psi^k\}$ and $\tilde{E}^4 = \mathrm{span}\{\tilde{\psi}^k\}$. Another representation is given by the so-called Witt basis

$$f_j = \frac{1}{2}(\psi^j - \tilde{\psi}^j) \quad f_j' = \frac{1}{2}(\psi^j + \tilde{\psi}^j)$$

with the decomposition $V^4 \times V^{4'}$, whereby $V^4 = \mathrm{span}\{f_k\}$ and $V^{4'} = \mathrm{span}\{f_k'\}$.

In the last case we can also introduce the primitive idempotent

$$I = I_1, \ldots, I_m, I_k = -f_k f_k',$$

for which we have $\tilde{\psi}^k I = \psi^k I = f_k' I$.

For more details we refer to [21].

We consider now the algebra $\mathrm{End}(\Pi_4) \otimes \mathrm{End}(\mathbb{H})$. Hereby, the algebra of scalar polynomial operators $\mathrm{End}(\Pi_4)$ contains as subalgebra the algebra of scalar differential operators with polynomial coefficients $D(4)$. The resulting algebra $D(4) \otimes \mathrm{End}(\mathbb{H})$ of quaternionic differential operators with polynomial coefficients is generated by

- the multiplication operators $x_j : g(x) \mapsto x_j g(x)$,
- the differential operators $\partial_j : g(x) \mapsto \partial_j g(x)$,
- the left multiplication operators $\psi^k : g(x) \mapsto \psi^k g(x)$,
- as well as the conjugation operators $\tilde{\psi}^k : g(x) \mapsto \overline{g(x)}\psi^k$.

We remark that this algebra contains differential operators acting from both sides.

Let us take a short look which operators are preserving the action

$$L(s)g = sg(\overline{s}xs),$$

like our operator ${}^{\psi}D$. That means that

$$[P(x, {}^{\psi}D), L(s)] = 0$$

or in other words, that the operator $P(x, {}^{\psi}D)$ is $\tilde{GL}(m)$-invariant. Since one can associate to each differential operator $P(x, {}^{\psi}D) \in D(4) \otimes \mathrm{End}(\mathbb{H})$ a symbol $P(x,t)$ determined by

$$P(x, {}^{\psi}D)e^{\langle x,t\rangle} = P(x,t)e^{\langle x,t\rangle}$$

the question is which symbols $P(x,t)$ are $\tilde{SO}(m)$ invariant, i.e.,

$$SP(\overline{S}xS, \overline{S}tS)\overline{S} = P(x,t) \qquad \forall S \in \tilde{SO}(m).$$

Since the algebra of $\tilde{SO}(m)$-invariant scalar-valued polynomial operators $P(x,t)$ are generated by $|x|^2 = x\overline{x}$, $|t|^2 = t\overline{t}$ and $\langle x,t\rangle = \frac{1}{2}(\overline{x}t + t\overline{x})$ and the subgroup which leaves x and t invariant is isomorphic to $\tilde{SO}(m-2)$, i.e., it is the group which leaves

$$\mathrm{span}\{x,t\} = \mathrm{span}\{\psi^1, \psi^2\}$$

invariant, we get (cf. [21]) the following theorem.

Theorem 4.1. *The algebra of $SO(m)$-invariant quaternionic differential operator with polynomial coefficients is generated by the left and right vector variable operator ${}^{\psi}x\cdot$ and $\cdot x^{\overline{\psi}}$, the left and right Cauchy–Riemann operator ${}^{\psi}D$ and $D^{\overline{\psi}}$.*

For more details we refer to [21] and, of course, the book [4].

Now, let us take a look at the commutator of ${}^{\psi}D$ with the generators of our algebra. Here, simple calculations give us

$$[{}^{\psi}D, \partial_k] = 0, \qquad [{}^{\psi}D, X_k] = \psi^k, \qquad [{}^{\psi}D, \psi^k] = -2\sum_{l\neq 0,k} \psi^l \partial^l,$$

$$[{}^{\psi}D, \tilde{\psi}^k] = -\frac{1}{2}\tilde{\psi}^k\left(\sum_{l=1}^{3}(\psi^l f_0' + \psi^0 f_l')\partial_l\right) = -\frac{1}{2}\tilde{\psi}^k\left(\sum_{l=1}^{3}(2\psi^l + \psi^l\tilde{\psi}^0 + \psi^0\tilde{\psi}^l)\partial_l\right).$$

In the same way as in the previous sections we can get as conditions for $[{}^{\psi}D, L] = M^{\psi}D$ with $L = \sum a_l\partial_l + b$ and $M = \sum c_l\partial_l + d$:

$$\sum_{k<l}(\psi^k a_l - a_l\psi^k + \psi^l a_k - a_k\psi^l)\partial_k\partial_l = \sum_{k<l}(c_l\psi^k + c_k\psi^l)\partial_k\partial_l$$

$${}^{\psi}Da_l + \psi^k b - b\psi^k = d\psi^k, \qquad {}^{\psi}Db = 0$$

The principal identities we have is $\{{}^{\psi}D, {}^{\overline{\psi}}x\} = {}^{\psi}D^{\overline{\psi}}x + {}^{\overline{\psi}}x^{\psi}D = 2E + 4$ as well as $[{}^{\psi}D, E] = {}^{\psi}D$, where $E = \sum_{k=0}^{3} x_k\partial_k$ is the Euler operator. From this and the above relations we obtain as generators of the Lie algebra of symmetries

$$\partial_k \tag{4.47}$$

$$E + 2 \tag{4.48}$$

$$x_k\partial_l - \partial_l x_k - \frac{1}{2}\psi^k\psi^l \tag{4.49}$$

$$2x_kE - x\overline{x}\partial_k + 5x_k - \overline{x}\psi^k. \tag{4.50}$$

Corollary 4.2. *It is easy to see that the above generators are in fact the infinitesimal generators of the conformal group.*

Acknowledgment

This work was supported by Portuguese funds through the CIDMA – Center for Research and Development in Mathematics and Applications, and the Portuguese Foundation for Science and Technology (FCT-Fundação para a Ciência e a Tecnologia), within project PEst-OE/MAT/UI4106/2014.

References

[1] R. Abreu-Blaya; J. Bory-Reyes; A. Guzmán-Adán; U. Kähler (2013), On some structural sets and a quaternionic (φ,ψ)-hyperholomorphic function theory. Submitted for publication.

[2] F. Brackx, R. Delanghe and F. Sommen, Clifford Analysis, Pitman Publishers, Boston-London-Melbourne, 1982.

[3] F. Brackx, N. de Schepper, F. Sommen, Clifford–Hermite-monogenic operators, Czechoslovak Mathematical Journal, 56, No. 4, 1301–1322, 2006.

[4] F. Colombo, I. Sabadini, F. Sommen, D. Struppa, Analysis of Dirac Systems and Computational Algebra, Springer, 2004.

[5] S.D. Eidelman; Y. Krasnov, Operator method for solution of PDEs based on their symmetries. Alpay, Daniel (ed.) et al., Operator theory, systems theory and scattering theory: multidimensional generalizations, 107–137, Oper. Theory Adv. Appl., 157, Birkhäuser, Basel, 2005.

[6] K. Gürlebeck; U. Kähler; M. Shapiro, On the Π-Operator in Hyperholomorphic Function Theory. Adv. Appl. Clifford Algebras., 9, No. 1, 23–40, 1999.

[7] K. Gürlebeck; W. Sprößig, Quaternionic Analysis and Elliptic Boundary Value Problems. Int. Ser. Num. Math. (ISNM): Vol. 89. Basel: Birkhäuser Verlag 1990.

[8] K. Gürlebeck; W. Sprößig, Quaternionic and Clifford Calculus for Physicists and Engineers. John Wiley & Sons, 1997.

[9] Y. Krasnov, Symmetries of Cauchy–Riemann–Fueter equation, Complex Variables, Theory and Application: An International Journal, Vol. 41, No. 3, 279–292, 2000.

[10] Le Hung Son; Nguyen Thanh Van, Necessary and sufficient conditions for associated pairs in quaternionic analysis. Hypercomplex analysis, 207–220, Trends Math., Birkhäuser Verlag, Basel, 2009.

[11] Le Hung Son; Nguyen Canh Luong; Nguyen Quoc Hung, First-order differential operators associated to the Cauchy–Riemann operator in quaternion analysis. Le Hung Son (ed.) et al., Function spaces in complex and Clifford analysis. Proceedings of the 14th international conference on finite or infinite dimensional complex analysis and applications (ICFICCA 2006), Hue, Vietnam, August 1–5, 2007. Hanoi: National University Publishers. 269–273 (2008).

[12] W. Miller, Symmetries and separation of variables, Cambridge University Press, 1977.

[13] I.M. Mitelman; M. Shapiro, Differentiation of the Martinelli–Bochner integrals and the notion of hyperderivability. Math. Nachr. 172, 211–238, 1995.

[14] K. Nouno (1986), On the quaternion linearization of Laplacian Δ. Bull. Fukuoka Univ. Ed. III, 35 (1985), 5–10.

[15] P.J. Olver, Symmetry and explicit solutions of partial differential equations. Applied Numerical Mathematics 10, 307–324, 1992.

[16] N. Taghizadeh; A. Neirameh, Generalization of first-order differential operators associated to the Cauchy–Riemann operator in the $\mathbb{R}^n$. Australian Journal of Basic and Applied Sciences, 4(8): 3895–3899.

[17] Nguyen T.V. (2011), Differential Operators Associated to the Cauchy–Fueter Operator in Quaternion Algebra. Adv. Appl. Clifford Algebras 21, 591–605, 2010.

[18] M. Shapiro, Some remarks on generalizations of the one dimensional complex analysis: Hypercomplex approach. Functional Analytic Methods in Complex Analysis and Applications to Partial Differential Equations (Trieste, 1993). World Scienti. Publ., 379–401, 1995.

[19] M. Shapiro; N.L. Vasilevski, Quaternionic ψ-hyperholomorphic functions, singular integral operators and boundary value problems. I. ψ-hyperholomorphic function theory. Complex Var. Theory Appl., 1995, vol. 27, 17–46.

[20] N.L. Vasilevsky; M.V. Shapiro, On the Bergman Kernel Function in Hyperholomorphic Analysis. Acta Appl. Math., vol. 46, pp. 1–27. 1997.

[21] F. Sommen; N. van Acker, Differential Operators on Clifford Algebras, Foundation of Physics 23 (1993) 11, 1491–1519.

[22] F. Sommen; N. Van Acker, Monogenic differential operators, Results in Mathematics 22, 781–798 (1992).

[23] V. Soucek, Representation theory in Clifford Analysis, Springer References, to appear.

[24] N.L. Vasilevsky; M.V. Shapiro, Some questions of hypercomplex analysis. Complex analysis and applications '87 (Varna, 1987), 523–531, Publ. House Bulgar. Acad. Sci., Sofia, 1998.

Ricardo Abreu Blaya
Facultad de Informática y Matemática
Universidad de Holguín
Holguín, 80100, Cuba
e-mail: rabreu@facinf.uho.edu.cu

Juan Bory Reyes and Alí Guzmán Adán
Departamento de Matemática
Universidad de Oriente
Santiago de Cuba 90500, Cuba
e-mail: jbory@rect.uo.edu.cu
ali.guzman@csd.uo.edu.cu

Uwe Kähler
Departamento de Matemática
Universidade de Aveiro
P-3810-159 Aveiro, Portugal
e-mail: ukaehler@ua.pt

Hypercomplex Analysis: New Perspectives and Applications
Trends in Mathematics, 19–41

Generalized Quaternionic Schur Functions in the Ball and Half-space and Krein–Langer Factorization

Daniel Alpay, Fabrizio Colombo and Irene Sabadini

Abstract. In this paper we prove a new version of Krein–Langer factorization theorem in the slice hyperholomorphic setting which is more general than the one proved in [9]. We treat both the case of functions with κ negative squares defined on subsets of the quaternionic unit ball or on subsets of the half-space of quaternions with positive real part. A crucial tool in the proof of our results is the Schauder–Tychonoff theorem and an invariant subspace theorem for contractions in a Pontryagin space.

Mathematics Subject Classification (2010). 47A48, 30G35, 30D50.

Keywords. Schur functions, realization, reproducing kernels, slice hyperholomorphic functions, Blaschke products.

1. Introduction

1.1. Some history

Functions analytic and contractive in the open unit disk $\mathbb{D}$ play an important role in various fields of mathematics, in electrical engineering and digital signal processing. They bear various names, and in particular are called Schur functions. We refer to [33] for reprints of some of the original works. André Bloch's 1926 memoir [20] contains also valuable historical background.

Schur functions admit a number of generalizations, within function theory of one complex variable and outside. A $\mathbb{C}^{r\times s}$-valued function S analytic in $\mathbb{D}$ is a

D. Alpay thanks the Earl Katz family for endowing the chair which supported his research, and the Binational Science Foundation Grant number 2010117. I. Sabadini was partially supported by the FIRB Project *Geometria Differenziale Complessa e Dinamica Olomorfa*.

Schur function if and only if the kernel

$$K_S(z,w)=\frac{I_r-S(z)S(w)^*}{1-z\bar{w}}$$

is positive definite in $\mathbb{D}$. In fact, much more is true. It is enough to assume that the kernel $K_S(z,w)$ is positive definite in some subset Ω of $\mathbb{D}$ to insure that S is the restriction to Ω of a (not necessarily unique) function analytic and contractive in $\mathbb{D}$; see [3, 29].

A Schur function has no poles inside the open unit disk. Motivated by lowering a lower bound given by Carathéodory and Fejér in an interpolation problem, Takagi considered in [46, 47] rational functions bounded by 1 in modulus on the unit circle and with poles inside $\mathbb{D}$. These are the first instances of what is called a generalized Schur function. Later studies of such functions include Chamfy [23], Dufresnoy [32] (these authors being motivated by the study of Pisot numbers), Delsarte, Genin and Kamp [26, 28] and Krein and Langer [36, 37, 38], to mention a few names. The precise definition of a generalized Schur function was given (in the setting of operator-valued functions) by Krein and Langer [36]:

Definition 1.1. A $\mathbb{C}^{r\times s}$-valued function analytic in an open subset Ω of the unit disc is called a generalized Schur function if the kernel K_S has a finite number (say κ) of negative squares in Ω, meaning that for every choice of $N\in\mathbb{N}$, $c_1,\ldots,c_N\in\mathbb{C}^r$ and $w_1,\ldots,w_N\in\Omega$, the $N\times N$ Hermitian matrix with (ℓ,j) entry $c_\ell^*K_S(w_\ell,w_j)c_j$ has at most κ strictly negative eigenvalues, and exactly κ strictly negative eigenvalues for some choice of $N,c_1,\ldots,c_N,w_1,\ldots,w_N$.

In the setting of matrix-valued functions, the result of Krein and Langer states that S is a generalized Schur function if and only if it is the restriction to Ω of a function of the form

$$B_0(z)^{-1}S_0(z),$$

where S_0 is a $\mathbb{C}^{r\times s}$-valued Schur function and B_0 is $\mathbb{C}^{r\times r}$-valued Blaschke product of degree κ. Besides [36], there exist various proofs of this result; see for instance [10, p. 141], [22].

That one cannot remove the analyticity condition in the result of Krein and Langer when $\kappa>0$ is illustrated by the well-known counterexample $S(z)=\delta_0(z)$, where $\delta_0(z)=0$ if $z\neq 0$ and $\delta_0(0)=1$ (see for instance [10, p. 82]). Taking into account that $z^n\delta_0(z)\equiv 0$ for $n>0$ we have

$$\begin{aligned}\frac{1-S(z)S(w)^*}{1-z\bar{w}}&=\frac{1}{1-z\bar{w}}-\sum_{n=0}^{\infty}z^n\delta_0(z)\bar{w}^n\delta_0(w)\\&=\frac{1}{1-z\bar{w}}-\delta_0(z)\delta_0(w).\end{aligned}$$

The reproducing kernel Hilbert space associated with $\delta_0(z)\delta_0(w)$ is $\mathbb{C}\delta_0$ and has a zero intersection with $\mathbf{H}_2(\mathbb{D})$, and hence the kernel K_S has one negative square.

1.2. The slice hyperholomorphic case

Schur functions have been extended to numerous settings, and we mention in particular the setting of several complex variables [2, 19], compact Riemann surfaces [16] and hypercomplex functions [13, 14]. Generalized Schur functions do not exist necessarily in all these settings.

In [7] we began a study of Schur analysis in the framework of slice hyperholomorphic functions. The purpose of this paper is to prove the theorem of Krein and Langer (we considered a particular case in [9]) and we treat both the unit ball and half-space cases in the quaternionic setting. To that purpose we need in particular the following:

(i) The notion of negative squares and of reproducing kernel Pontryagin spaces in the quaternionic setting. This was done in [15].
(ii) The notion of generalized Schur functions and of Blaschke products, see [8].
(iii) A result on invariant subspaces of contractions in quaternionic Pontryagin spaces.
(iv) The notion of realization in the slice-hyperholomorphic setting, in particular when the state space is a one-sided (as opposed to two-sided) Pontryagin space.

The paper contains 6 sections, besides the Introduction. Section 2 contains a quick survey of the Krein–Langer result in the classical case. Section 3 introduces slice hyperholomorphic functions and discusses Blaschke products. Section 4 contains some useful results in quaternionic functional analysis, among which Schauder–Tychonoff theorem. In Section 5 we present generalized Schur functions and their realizations. Finally, in Section 6 we prove the Krein–Langer factorization for generalized Schur functions defined in a subset of the unit ball and finally, in Section 7, we state the analogous result in the case of the half-space.

2. A survey of the classical case

The celebrated one-to-one correspondence between positive definite functions and reproducing kernel Hilbert spaces (see [17]) extends to the indefinite case, when one considers functions with a finite number of negative squares and reproducing kernel Pontryagin spaces; see [11, 43, 44]. We recall the definition of the latter for the convenience of the reader.

A complex vector space $\mathcal{V}$ endowed with a sesquilinear form $[\cdot,\cdot]$ is called an indefinite inner product space (which we will also denote by the pair $(\mathcal{V},[\cdot,\cdot])$). The form $[\cdot,\cdot]$ defines an orthogonality: two vectors $v,w\in\mathcal{V}$ are orthogonal if $[v,w]=0$, and two linear subspaces $\mathcal{V}_1$ and $\mathcal{V}_2$ of $\mathcal{V}$ are orthogonal if every vector of $\mathcal{V}_1$ is orthogonal to every vector of $\mathcal{V}_2$. Orthogonal sums will be denoted by the symbol $[+]$. Note that two orthogonal spaces may intersect. We will denote by the symbol $[\oplus]$ a direct orthogonal sum. A complex vector space $\mathcal{V}$ is a Krein space if it can be written (in general in a non-unique way) as

$$\mathcal{V} = \mathcal{V}_+[\oplus]\mathcal{V}_-, \tag{2.1}$$

where $(\mathcal{V}_+, [\cdot,\cdot])$ and $(\mathcal{V}_-, -[\cdot,\cdot])$ are Hilbert spaces. When the space $\mathcal{V}_-$ (or, as in [35], the space $\mathcal{V}_+$) is finite dimensional (note that this property does not depend on the decomposition), $\mathcal{V}$ is called a Pontryagin space. The space $\mathcal{V}$ endowed with the form

$$\langle h, g\rangle = [h_+, g_+] - [h_-, g_-],$$

where $h = h_+ + h_-$ and $g = g_+ + g_-$ are the decompositions of $f, g \in \mathcal{V}$ along (2.1), is a Hilbert space. One endows $\mathcal{V}$ with the corresponding topology. This topology is independent of the decomposition (2.1) (the latter is not unique, but it is so in the definite case).

Let now T be a linear densely defined map from a Pontryagin space $(\mathcal{P}_1, [\cdot,\cdot]_1)$ into a Pontryagin space $(\mathcal{P}_2, [\cdot,\cdot]_2)$. Its adjoint is the operator T^* with domain $\mathrm{Dom}\,(T^*)$ defined by:

$$\{g \in \mathcal{P}_2 \ : \ h \mapsto \ [Th, g]_2 \text{ is continuous}\}\,.$$

One then defines by T^*g the unique element in $\mathcal{P}_1$ which satisfies

$$[Th, g]_2 = [h, T^*g]_1.$$

Such an element exists by the Riesz representation theorem.

The operator T is called a contraction if

$$[Th, Th]_2 \le [h, h]_1, \quad \forall\, h \in \mathrm{Dom}\,(T),$$

while it is said to be a coisometry if $TT^* = I$.

Theorem 2.1. *A densely defined contraction between Pontryagin spaces of the same index has a unique contractive extension and its adjoint is also a contraction.*

We refer to [18, 21, 31] for the theory of Pontryagin and Krein spaces, and of their operators.

With these definitions, we can state the following theorem, which gathers the main properties of generalized Schur functions.

Theorem 2.2. *Let S be a $\mathbb{C}^{r\times s}$-valued function analytic in a neighborhood Ω of the origin. Then the following are equivalent:*

(1) *The kernel $K_S(z, w)$ has a finite number of negative squares in Ω.*
(2) *There is a Pontryagin space $\mathcal{P}$ and a coisometric operator matrix*

$$\begin{pmatrix} A & B\\ C & D\end{pmatrix} \ : \ \mathcal{P}\oplus\mathbb{C}^s \to \mathcal{P}\oplus\mathbb{C}^r$$

such that

$$S(z) = D + zC(I - zA)^{-1}B, \quad z \in \Omega. \tag{2.2}$$

(3) *There exists a* $\mathbb{C}^{r\times s}$*-valued Schur function* S_0 *and a* $\mathbb{C}^{r\times r}$*-valued Blaschke product* B_0 *such that*

$$S(z) = B_0(z)^{-1}S_0(z), \quad z \in \Omega.$$

As a corollary we note that S can be extended to a function of bounded type in $\mathbb{D}$, with boundary limits almost everywhere of norm less than or equal 1.

We note the following:

(a) When the pair (C, A) is observable, meaning

$$\cap_{n=0}^{\infty} \ker CA^n = \{0\}, \tag{2.3}$$

the realization (2.2) is unique, up to an isomorphism of Pontryagin spaces.

(b) One can take for $\mathcal{P}$ the reproducing kernel Pontryagin space $\mathcal{P}(S)$ with reproducing kernel K_S. When $0 \in \Omega$ we have the backward shift realization

$$\begin{aligned} Af &= R_0 f, \\ Bc &= R_0 Sc, \\ Cf &= f(0), \\ Dc &= S(0)c, \end{aligned}$$

where $f \in \mathcal{P}(S)$, $c \in \mathbb{C}^s$ and where R_0 denotes the backward shift operator

$$R_0 f(z) = \begin{cases} \dfrac{f(z) - f(0)}{z}, & z \neq 0, \\ f'(0), & z = 0. \end{cases}$$

See [10] for more details on this construction, and on the related isometric and unitary realizations.

3. Slice hyperholomorphic functions and Blaschke products

Let $\mathbb{H}$ be the real associative algebra of quaternions, where a quaternion p is denoted by $p = x_0 + ix_1 + jx_2 + kx_3$, $x_i \in \mathbb{R}$, and the elements $\{1, i, j, k\}$ satisfy the relations $i^2 = j^2 = k^2 = -1$, $ij = -ji = k$, $jk = -kj = i$, $ki = -ik = j$. As is customary, $\bar{p} = x_0 - ix_1 - jx_2 - kx_3$ is called the conjugate of p, the real part $x_0 = \frac{1}{2}(p + \bar{p})$ of a quaternion is also denoted by $\mathrm{Re}(p)$, while $|p|^2 = p\bar{p}$. The symbol $\mathbb{S}$ denotes the 2-sphere of purely imaginary unit quaternions, i.e.,

$$\mathbb{S} = \{p = ix_1 + jx_2 + kx_3 \mid x_1^2 + x_2^2 + x_3^2 = 1\}.$$

If $I \in \mathbb{S}$ then $I^2 = -1$ and any nonreal quaternion $p = x_0 + ix_1 + jx_2 + kx_3$ uniquely determines an element $I_p = (ix_1 + jx_2 + kx_3)/|ix_1 + jx_2 + kx_3| \in \mathbb{S}$. (We note that later i, j, k may also denote some indices, but the context will make clear the use of the notation.)

Let $\mathbb{C}_I$ be the complex plane $\mathbb{R} + I\mathbb{R}$ passing through 1 and I and let $x + Iy$ be an element on $\mathbb{C}_I$. Any $p = x + Iy$ defines a 2-sphere $[p] = \{x + Jy \ : \ J \in \mathbb{S}\}$.

We now recall the notion of slice hyperholomorphic function:

Definition 3.1. Let $\Omega \subseteq \mathbb{H}$ be an open set and let $f : \Omega \to \mathbb{H}$ be a real differentiable function. Let $I \in \mathbb{S}$ and let f_I be the restriction of f to the complex plane $\mathbb{C}_I$. We say that f is a (left) slice hyperholomorphic function in Ω if, for every $I \in \mathbb{S}$, f_I satisfies

$$\frac{1}{2}\left(\frac{\partial}{\partial x}+I\frac{\partial}{\partial y}\right) f_I(x+Iy)=0.$$

We say that f is a right slice hyperholomorphic function in Ω if, for every $I \in \mathbb{S}$, f_I satisfies

$$\frac{1}{2}\left(\frac{\partial}{\partial x}f_I(x+Iy)+\frac{\partial}{\partial y}f_I(x+Iy)I\right)=0.$$

The set of slice hyperholomorphic functions on Ω will be denoted by $\mathcal{R}(\Omega)$. It is a right linear space on $\mathbb{H}$.
Slice hyperholomorphic functions possess good properties when they are defined on the so-called axially symmetric slice domains defined below.

Definition 3.2. Let Ω be a domain in $\mathbb{H}$. We say that Ω is a slice domain (s-domain for short) if $\Omega \cap \mathbb{R}$ is non empty and if $\Omega \cap \mathbb{C}_I$ is a domain in $\mathbb{C}_I$ for all $I \in \mathbb{S}$. We say that Ω is axially symmetric if, for all $q \in \Omega$, the sphere $[q]$ is contained in Ω.

A function f slice hyperholomorphic on an axially symmetric s-domain Ω is determined by its restriction to any complex plane $\mathbb{C}_I$, see [25, Theorem 4.3.2].

Theorem 3.3 (Structure formula). *Let $\Omega \subseteq \mathbb{H}$ be an axially symmetric s-domain, and let $f \in \mathcal{R}(\Omega)$. Then for any $x+Jy \in \Omega$ the following formula holds*

$$f(x+Jy)=\frac{1}{2}\left[f(x+Iy)+f(x-Iy)+JI(f(x-Iy)-f(x+Iy))\right]. \tag{3.1}$$

As a consequence of this result, we have the following definition:

Definition 3.4. Let Ω be an axially symmetric s-domain. Let $h : \Omega\cap\mathbb{C}_I \to \mathbb{H}$ be a holomorphic map. Then it admits a (unique) left slice hyperholomorphic extension $\mathrm{ext}(h) : \Omega \to \mathbb{H}$ defined by:

$$\mathrm{ext}(h)(x+Jy)=\frac{1}{2}\left[h(x+Iy)+h(x-Iy)+JI(h(x-Iy)-h(x+Iy))\right]. \tag{3.2}$$

Remark 3.5. Let $\Omega \subseteq \mathbb{H}$ be an axially symmetric s-domain and let $f, g \in \mathcal{R}(\Omega)$. We can define a suitable product, called the $\star$-product, such that the resulting function $f \star g$ is slice hyperholomorphic. We first define a product between the restrictions f_I, g_I of f, g to $\Omega \cap \mathbb{C}_I$. This product can be extended to the whole Ω using formula (3.2). Outside the spheres associated with the zeroes of $f \in \mathcal{R}(\Omega)$ we can consider its slice regular inverse $f^{-\star}$. Note also that $(f\star g)^{-\star} = g^{-\star}\star f^{-\star}$ where it is defined. We refer the reader to [25, p. 125–129] for the details on the $\star$-product and $\star$-inverse.

The $\star$-product can be related to the pointwise product as described in the following result, [25, Proposition 4.3.22]:

Proposition 3.6. *Let $\Omega \subseteq \mathbb{H}$ be an axially symmetric s-domain, $f, g : \Omega \to \mathbb{H}$ be slice hyperholomorphic functions. Then*

$$(f \star g)(p) = f(p)g(f(p)^{-1}pf(p)), \tag{3.3}$$

for all $p \in \Omega$, $f(p) \neq 0$, while $(f \star g)(p) = 0$ when $p \in \Omega$, $f(p) = 0$.

An immediate consequence is the following:

Corollary 3.7. *If $(f \star g)(p) = 0$ then either $f(p) = 0$ or $f(p) \neq 0$ and $g(f(p)^{-1}pf(p)) = 0$.*

Remark 3.8. Corollary 3.7 applies in particular to polynomials, allowing to recover a well-known result, see [39]: if a polynomial $\mathcal{Q}(p)$ factors as

$$\mathcal{Q}(p) = (p - \alpha_1) \star \cdots \star (p - \alpha_n), \quad \alpha_{j+1} \neq \bar{\alpha}_j, j = 1, \dots, n-1 \tag{3.4}$$

then α_1 is a root of $\mathcal{Q}(p)$ while all the other zeroes $\tilde{\alpha}_j$, $j = 2, \dots, n$ belong to the spheres $[\alpha_j]$, $j = 2, \dots, n$. The decomposition of the polynomial $\mathcal{Q}$, in general, is not unique.
Note that when $\alpha_{j+1} = \bar{\alpha}_j$ then $\mathcal{Q}(p)$ contains the second degree factor $p^2 + 2\mathrm{Re}(\alpha_j)p + |\alpha_j|^2$ and the zero set of $\mathcal{Q}(p)$ contains the whole sphere $[\alpha_j]$. We will say that $[\alpha_j]$ is a spherical zero of the polynomial $\mathcal{Q}$.

Remark 3.9. Assume that $\mathcal{Q}(p)$ factors as in (3.4) and assume that $\alpha_j \in [\alpha_1]$ for all $j = 2, \dots, n$. Then the only root of $\mathcal{Q}(p)$ is $p = \alpha_1$, see [40, Lemma 2.2.11], [41, p. 519] the decomposition in linear factors is unique, and α_1 is the only root of $\mathcal{Q}$.
Assume that $[\alpha_j]$ is a spherical zero. Then, for any $a_j \in [\alpha_j]$ we have

$$p^2 + 2\mathrm{Re}(\alpha_j)p + |\alpha_j|^2 = (p - a_j) \star (p - \bar{a}_j) = (p - \bar{a}_j) \star (p - a_j)$$

thus showing that both a_j and $\bar{a}_j$ are zeroes of multiplicity 1. So we can say that the (points of the) sphere $[\alpha_j]$ have multiplicity 1. Thus the multiplicity of a spherical zero $[\alpha_j]$ equals the exponent of $p^2 + 2\mathrm{Re}(\alpha_j)p + |\alpha_j|^2$ in a factorization of $\mathcal{Q}(p)$.

The discussion in the previous remark justifies the following:

Definition 3.10. Let

$$\mathcal{Q}(p) = (p - \alpha_1) \star \cdots \star (p - \alpha_n), \quad \alpha_{j+1} \neq \bar{\alpha}_j, \quad j = 1, \dots, n-1.$$

We say that α_1 is a zero of $\mathcal{Q}$ of *multiplicity* 1 if $\alpha_j \notin [\alpha_1]$ for $j = 2, \dots, n$.
We say that α_1 is a zero of $\mathcal{Q}$ of *multiplicity* $n \geq 2$ if $\alpha_j \in [\alpha_1]$ for all $j = 2, \dots, n$.
Assume now that $\mathcal{Q}(p)$ contains the factor $(p^2 + 2\mathrm{Re}(\alpha_j)p + |\alpha_j|^2)$ and $[\alpha_j]$ is a zero of $\mathcal{Q}(p)$. We say that the *multiplicity of the spherical zero* $[\alpha_j]$ is m_j if m_j is the maximum of the integers m such that $(p^2 + 2\mathrm{Re}(\alpha_j)p + |\alpha_j|^2)^m$ divides $\mathcal{Q}(p)$.

Note that the notion of multiplicity of a spherical zero given in [34] is different since, under the same conditions described in Definition 3.10, it would be $2m_j$.

Remark 3.11. The polynomial $\mathcal{Q}(p)$ can be factored as follows, see, e.g., [34, Theorem 2.1]:

$$\mathcal{Q}(p) = \prod_{j=1}^{r}(p^2 + 2\mathrm{Re}(\alpha_j)p + |\alpha_j|^2)^{m_j} \left(\prod_{i=1}^{\star s}\prod_{j=1}^{\star n_i} (p - \alpha_{ij})\right) a,$$

where $\prod^{\star}$ denotes the $\star$-product of the factors, $[\alpha_i] \neq [\alpha_j]$ for $i \neq j$, $\alpha_{ij} \in [a_i]$ for all $j = 1, \ldots, n_i$ and $[a_i] \neq [a_\ell]$ for $i \neq \ell$. Note that

$$\deg(\mathcal{Q}) = \sum_{j=1}^{r} 2m_j + \sum_{i=1}^{s} n_i.$$

Definition 3.12. Let $a \in \mathbb{H}$, $|a| < 1$. The function

$$B_a(p) = (1 - p\bar{a})^{-\star} \star (a - p)\frac{\bar{a}}{|a|} \tag{3.5}$$

is called a Blaschke factor at a.

Remark 3.13. Using Proposition 3.6, $B_a(p)$ can be rewritten as

$$B_a(p) = (1 - \tilde{p}\bar{a})^{-1}(a - \tilde{p})\frac{\bar{a}}{|a|}$$

where $\tilde{p} = (1 - pa)^{-1}p(1 - pa)$.

The following result is immediate, see [8]:

Proposition 3.14. *Let $a \in \mathbb{H}$, $|a| < 1$. The Blaschke factor B_a is a slice hyperholomorphic function in $\mathbb{B}$.*

As one expects, $B_a(p)$ has only one zero at $p = a$ and analogously to what happens in the case of the zeroes of a function, the product of two Blaschke factors of the form $B_a(p) \star B_{\bar{a}}(p)$ gives the Blaschke factor with zeroes at the sphere $[a]$. Thus we give the following definition:

Definition 3.15. Let $a \in \mathbb{H}$, $|a| < 1$. The function

$$B_{[a]}(p) = (1 - 2\mathrm{Re}(a)p + p^2|a|^2)^{-1}(|a|^2 - 2\mathrm{Re}(a)p + p^2) \tag{3.6}$$

is called Blaschke factor at the sphere $[a]$.

Theorem 5.16 in [8] assigns a Blaschke product having zeroes at a given set of points a_j with multiplicities n_j, $j \geq 1$ and at spheres $[c_i]$ with multiplicities m_i, $i \geq 1$, where the multiplicities are meant as exponents of the factors $(p - a_j)$ and $(p^2 - \mathrm{Re}(a_j)p + |a_j|^2)$, respectively. In view of Definition 3.10, the polynomial $(p - a_j)^{\star n_j}$ is not the unique polynomial having a zero at a_j with the given multiplicity n_j, thus the Blaschke product $\prod_{j=1}^{\star n_j} B_{a_j}$ is not the unique Blaschke product having zero at a_j with multiplicity n_j.
We give below a form of Theorem 5.16 in [8] in which we use the notion of multiplicity in Definition 3.10:

Theorem 3.16. *A Blaschke product having zeroes at the set*

$$Z = \{(a_1, n_1), \dots, ([c_1], m_1), \dots\}$$

where $a_j \in \mathbb{B}$, a_j *have respective multiplicities* $n_j \geq 1$, $a_j \neq 0$ *for* $j = 1, 2, \dots$, $[a_i] \neq [a_j]$ *if* $i \neq j$, $c_i \in \mathbb{B}$, *the spheres* $[c_j]$ *have respective multiplicities* $m_j \geq 1$, $j = 1, 2, \dots$, $[c_i] \neq [c_j]$ *if* $i \neq j$ *and*

$$\sum_{i,j \geq 1} \Big(n_i(1 - |a_i|) + 2m_j(1 - |c_j|) \Big) < \infty \tag{3.7}$$

is of the form

$$\prod_{i \geq 1} (B_{[c_i]}(p))^{m_i} \prod_{i \geq 1}^{\star} \prod_{j=1}^{\star n_i} (B_{\alpha_{ij}}(p)),$$

where $n_j \geq 1$, $\alpha_{11} = a_1$ *and* α_{ij} *are suitable elements in* $[a_i]$, $\alpha_{i\,j+1} \neq \overline{\alpha_{ij}}$, *for* $j = 2, 3, \dots$.

Proof. The fact that (3.7) ensure the convergence of the product follows from [8, Theorem 5.6]. The zeroes of the pointwise product $\prod_{i \geq 1} (B_{[c_i]}(p))^{m_i}$ correspond to the given spheres with their multiplicities. Let us consider the product:

$$\prod_{i=1}^{\star n_1} (B_{\alpha_{i1}}(p)) = B_{\alpha_{11}}(p) \star B_{\alpha_{12}}(p) \star \cdots \star B_{\alpha_{1n_1}}(p).$$

As we already observed in the proof of Proposition 5.10 in [8] this product admits a zero at the point $\alpha_{11} = a_1$ and it is a zero of multiplicity 1 if $n_1 = 1$; if $n_1 \geq 2$, the other zeroes are $\tilde{\alpha}_{12}, \dots, \tilde{\alpha}_{1n_1}$ where $\tilde{\alpha}_{1j}$ belong to the sphere $[\alpha_{1j}] = [a_1]$. This fact can be seen directly using formula (3.3). Thus, according to Remark 3.8, a_1 is a zero of multiplicity n_1. Let us now consider $r \geq 2$ and

$$\prod_{j=1}^{\star n_r} (B_{\alpha_{rj}}(p)) = B_{\alpha_{r1}}(p) \star \cdots \star B_{\alpha_{rn_r}}(p), \tag{3.8}$$

and set

$$B_{r-1}(p) := \prod_{i \geq 1}^{\star (r-1)} \prod_{j=1}^{\star n_i} (B_{\alpha_{ij}}(p)).$$

Then

$$B_{r-1}(p) \star B_{\alpha_{r1}}(p) = B_{r-1}(p) B_{\alpha_{r1}}(B_{r-1}(p)^{-1} p B_{r-1}(p))$$

has a zero at a_r if and only if $B_{\alpha_{r1}}(B_{r-1}(a_r)^{-1} a_r B_{r-1}(a_r)) = 0$, i.e., if and only if $\alpha_{r1} = B_{r-1}(a_r)^{-1} a_r B_{r-1}(a_r)$. If $n_r = 1$ then a_r is a zero of multiplicity 1 while if $n_r \geq 2$, all the other zeroes of the product (3.8) belongs to the sphere $[a_r]$ thus, by Remark 3.8, the zero a_r has multiplicity n_r. This completes the proof. □

Remark 3.17. In the case in which one has to construct a Blaschke product having a zero at a_i with multiplicity n_i by prescribing the factors $(p - a_{i1}) \star \cdots \star (p - a_{in_i})$, $a_{ij} \in [a_i]$ for all $j = 1, \dots, n_i$, the factors in the Blaschke product must be chosen accordingly (see the proof of Theorem 3.16).

Proposition 3.18. *The $\star$-inverse of B_a and $B_{[a]}$ are $B_{\bar{a}^{-1}}$, $B_{[a^{-1}]}$ respectively.*

Proof. It follows from straightforward computations, by verifying that the products $B_a \star B_{\bar{a}^{-1}}$ and $B_{[a]} \star B_{[a^{-1}]}$ equal 1. □

Definition 3.19. A Blaschke product of the form

$$B(p) = \prod_{i=1}^{r} (B_{[c_i]}(p))^{m_i} \prod_{i=1}^{\star s} \prod_{j=1}^{\star n_i} (B_{\alpha_{ij}}(p)), \tag{3.9}$$

is said to have degree $d = \sum_{i=1}^{r} 2m_i + \sum_{j=1}^{s} n_j$.

Proposition 3.20. *Let $B(p)$ be a Blaschke product as in* (3.9). *Then* $\dim(\mathcal{H}(B)) = \deg B$.

Proof. Let us rewrite $B(p)$ as

$$B(p) = \prod_{i=1}^{r} (B_{c_i}(p) \star B_{\bar{c}_i}(p))^{m_i} \prod_{i=1}^{\star s} \prod_{j=1}^{\star n_i} (B_{\alpha_{ij}}(p)) = \prod_{j=1}^{\star d} B_{\beta_j}(p),$$

$d = \deg B$. Let us first observe that in the case in which the factors B_{β_j} are such that no three of the quaternions β_j belong to the same sphere, then the statement follows from the fact that $\mathcal{H}(B)$ is the span of $(1 - p\bar{\beta}_j)^{-\star}$. Moreover $(1 - p\bar{\beta}_1)^{-\star}, \ldots, (1 - p\bar{\beta}_d)^{-\star}$ are linearly independent in the Hardy space $\mathbf{H}_2(\mathbb{B})$, see [5, Remark 3.1]. So we now assume that $d \geq 3$ and at least three among the β_j's belong the same sphere. We proceed by induction. Assume that $d = 3$ and $\beta_1, \beta_2, \beta_3$ belong to the same sphere. Since

$$\begin{aligned} K_B(p,q) &= \sum_n p^n (1 - B(p)B(q)^*)\bar{q}^n = \sum_n p^n (1 - B_{\beta_1}(p)B_{\beta_1}(q)^*)\bar{q}^n \\ &\quad + B_{\beta_1}(p) \star \sum_n p^n (1 - B_{\beta_2}(p)B_{\beta_2}(q)^*)\bar{q}^n \star_r B_{\beta_1}(q)^* \\ &\quad + B_{\beta_1}(p) \star B_{\beta_2}(p) \star \sum_n p^n (1 - B_{\beta_3}(p)B_{\beta_3}(q)^*)\bar{q}^n \star_r B_{\beta_1}(q)^* \star_r B_{\beta_1}(q)^* \end{aligned}$$

we have

$$\mathcal{H}(B_\beta) = \mathcal{H}(B_{\beta_1}) + B_{\beta_1} \star \mathcal{H}(B_{\beta_2}) + B_{\beta_1} \star B_{\beta_2} \star \mathcal{H}(B_{\beta_3}). \tag{3.10}$$

Now note that $\mathcal{H}(B_{\beta_1})$ is spanned by $f_1(p) = (1 - p\bar{\beta}_1)^{-\star}$, $B_{\beta_1} \star \mathcal{H}(B_{\beta_2})$ is spanned by $f_2(p) = B_{\beta_1}(p) \star (1 - p\bar{\beta}_2)^{-\star}$ and, finally, $B_{\beta_1} \star B_{\beta_2} \star \mathcal{H}(B_{\beta_3})$ is spanned by $f_3(p) = B_{\beta_1}(p) \star B_{\beta_2}(p) \star (1 - p\bar{\beta}_3)^{-\star}$. By using the reproducing property of f_1 we have $[f_1, f_2] = 0$ and $[f_1, f_3] = 0$ (here $[\cdot, \cdot]$ denotes the inner product in $\mathbf{H}_2(\mathbb{B})$). Observe that

$$[f_2, f_3] = [(1 - p\bar{\beta}_2)^{-\star}, B_{\beta_2}(p) \star (1 - p\bar{\beta}_3)^{-\star}] = 0$$

since the left multiplication by $B_{\beta_1}(p)$ is an isometry in $\mathbf{H}_2(\mathbb{B})$ and by the reproducing property of $(1 - p\bar{\beta}_2)^{-\star}$. So f_1, f_2, f_3 are orthogonal in $\mathbf{H}_2(\mathbb{B})$ and so they are linearly independent. We conclude that the sum (3.10) is direct and has dimension 3. Now assume that the assertion holds when $d = n$ and there in $B(p)$

are at least three Blaschke factors at points on the same sphere. We show that the assertion holds for $d = n+1$. We generalize the above discussion by considering

$$\begin{aligned}(\mathcal{H}(B_{\beta_1}) + B_{\beta_1} \star \mathcal{H}(B_{\beta_2}) + \cdots + B_{\beta_1} \star \cdots \star B_{\beta_{n-1}} \star \mathcal{H}(B_{\beta_n}) + \cdots + \\ + B_{\beta_1} \star \cdots \star B_{\beta_n} \star \mathcal{H}(B_{\beta_{n+1}}).\end{aligned} \tag{3.11}$$

Let us denote, as before, by $f_1(p) = (1 - p\bar{\beta}_1)^{-\star}$ a generator of $\mathcal{H}(B_{\beta_1})$ and by $f_j(p) = B_{\beta_1} \star \cdots \star B_{\beta_{j-1}} \star (1 - p\bar{\beta}_j)^{-\star}$ a generator of $B_{\beta_1} \star \cdots \star B_{\beta_{j-1}} \star \mathcal{H}(B_{\beta_j})$, $j = 1, \ldots, n+1$. By the induction hypothesis, the sum of the first n terms is direct and orthogonal so we show that $[f_j, f_{n+1}] = 0$ for $j = 1, \ldots, n$. This follows, as before, from the fact that the multiplication by a Blaschke factor is an isometry and by the reproducing property. The statement follows. □

We now introduce the Blaschke factors in the half-space

$$\mathbb{H}_+ = \{p \in \mathbb{H} \ : \ \mathrm{Re}(p) > 0\}.$$

Definition 3.21. For $a \in \mathbb{H}_+$ set

$$b_a(p) = (p + \bar{a})^{-\star} \star (p - a).$$

The function $b_a(p)$ is called Blaschke factor at a in the half-space $\mathbb{H}_+$.

Remark 3.22. The function $b_a(p)$ is defined outside the sphere $[-a]$ and it has a zero at $p = a$. A Blaschke factor b_a is slice hyperholomorphic in $\mathbb{H}^+$.

As before, we can also introduce Blaschke factors at spheres:

Definition 3.23. For $a \in \mathbb{H}_+$ set

$$b_{[a]}(p) = (p^2 + 2\mathrm{Re}(a)p + |a|^2)^{-1}(p^2 - 2\mathrm{Re}(a)p + |a|^2).$$

The function $b_a(p)$ is called Blaschke factor at the sphere $[a]$ in the half-space $\mathbb{H}_+$.

We now state the following result whose proof mimics the lines of the proof of Theorem 3.16 with obvious changes. Note that an analog of Remark 3.17 holds also in this case.

Theorem 3.24. *A Blaschke product having zeroes at the set*

$$Z = \{(a_1, n_1), \ldots, ([c_1], m_1), \ldots\}$$

where $a_j \in \mathbb{H}_+$, a_j *have respective multiplicities* $n_j \geq 1$, $[a_i] \neq [a_j]$ *if* $i \neq j$, $c_i \in \mathbb{H}_+$, *the spheres* $[c_j]$ *have respective multiplicities* $m_j \geq 1$, $j = 1, 2, \ldots$, $[c_i] \neq [c_j]$ *if* $i \neq j$ *and*

$$\sum_{i,j \geq 1} \Big(n_i(1 - |a_i|) + 2m_j(1 - |c_j|) \Big) < \infty$$

is given by

$$\prod_{i \geq 1} (b_{[c_i]}(p))^{m_i} \prod_{i \geq 1}^{\star} \prod_{j=1}^{\star n_i} (b_{\alpha_{ij}}(p)),$$

where $\alpha_{11} = a_1$ *and* α_{ij} *are suitable elements in* $[a_i]$ *for* $i = 2, 3, \ldots$.

Let $f(p) = \sum_{n=-\infty}^{+\infty}(p - p_0)^{\star n} a_n$ where $a_n \in \mathbb{H}$. Following the standard nomenclature and [45] we now give the definition of singularity of a slice regular function:

Definition 3.25. A function f has a pole at the point p_0 if there exists $m \geq 0$ such that $a_{-k} = 0$ for $k > m$. The minimum of such m is called the order of the pole; If p is not a pole then we call it an essential singularity for f;
f has a removable singularity at p_0 if it can be extended in a neighborhood of p_0 as a slice hyperholomorphic function.

A function f has a pole at p_0 if and only if its restriction to a complex plane has a pole. In this framework there can be poles of order 0. To give an example, let $I \in \mathbb{S}$; then the function $(p + I)^{-\star} = (p^2 + 1)^{-1}(p - I)$ has a pole of order 0 at the point $-I$ which, however, is not a removable singularity, see [25, p. 55].

Definition 3.26. Let Ω be an axially symmetric s-domain in $\mathbb{H}$. We say that a function $f : \Omega \to \mathbb{H}$ is slice hypermeromorphic in Ω if f is slice hyperholomorphic in $\Omega' \subset \Omega$ such that every point in $\Omega \setminus \Omega'$ is a pole and $(\Omega \setminus \Omega') \cap \mathbb{C}_I$ has no limit point in $\Omega \cap \mathbb{C}_I$ for all $I \in \mathbb{S}$.

4. Some results from quaternionic functional analysis

The tools from quaternionic functional analysis needed in the present paper are of two kinds. On one hand, we need some results from the theory of quaternionic Pontryagin spaces, taken essentially from [15]. On the other hand, we also need the quaternionic version of the Schauder–Tychonoff theorem in order to prove an invariant subspace theorem for contractions in Pontryagin spaces. More generally we note that in our on-going project on Schur analysis in the slice hyperholomorphic setting we were lead to prove a number of results in quaternionic functional analysis not readily available in the literature.

Operator theory in (quaternionic) Pontryagin spaces plays an important role in (quaternionic) Schur analysis, and we here recall some definitions and results needed in the sequel. We refer to [15] for more information.

Definition 4.1. Let $\mathcal{V}$ be a right quaternionic vector space. The map

$$[\cdot,\cdot] : \mathcal{V} \times \mathcal{V} \longrightarrow \mathbb{H}$$

is called an inner product if it is a (right) sesquilinear form:

$$[v_1 c_1, v_2 c_2] = \overline{c_2}[v_1, v_2]c_1, \quad \forall v_1, v_2 \in \mathcal{V}, \ \textit{and}\ c_1, c_2 \in \mathbb{H},$$

which is Hermitian in the sense that:

$$[v, w] = \overline{[w, v]}, \quad \forall v, w \in \mathcal{V}.$$

A quaternionic inner product space $\mathcal{V}$ is called a Pontryagin space if it can be written as a direct and orthogonal sum

$$\mathcal{V} = \mathcal{V}_+[\oplus]\mathcal{V}_-, \tag{4.1}$$

where $(\mathcal{V}_+, [\cdot,\cdot])$ is a Hilbert space, and $(\mathcal{V}_-, -[\cdot,\cdot])$ is a finite-dimensional Hilbert space. As in the complex case, the space $\mathcal{V}$ endowed with the form

$$\langle h, g\rangle = [h_+, g_+] - [h_-, g_-], \tag{4.2}$$

where $h = h_+ + h_-$ and $g = g_+ + g_-$ are the decompositions of $f, g \in \mathcal{V}$ along (4.1), is a Hilbert space and the norms associated with the inner products (4.2) are equivalent, and hence define the same topology. The notions of adjoint and contraction are defined as in the complex case, and Theorem 2.1 still holds in the quaternionic setting:

Theorem 4.2 ([9, Theorem 7.2]). *A densely defined contraction between quaternionic Pontryagin spaces of the same index has a unique contractive extension and its adjoint is also a contraction.*

A key result used in the proof of the Krein–Langer factorization is the following invariant subspace theorem.

Theorem 4.3 ([6, Theorem 4.6]). *A contraction in a quaternionic Pontryagin space has a unique maximal invariant negative subspace, and it is one-to-one on it.*

The arguments there follow the ones given in the complex case in [30], and require in particular to prove first a quaternionic version of the Schauder–Tychonoff theorem, and an associated lemma. We recall these for completeness:

Lemma 4.4 ([6, Lemma 4.4]). *Let $\mathcal{K}$ be a compact convex subset of a locally convex linear quaternionic space $\mathcal{V}$ and let $T : \mathcal{K} \to \mathcal{K}$ be continuous. If $\mathcal{K}$ contains at least two points, then there exists a proper closed convex subset $\mathcal{K}_1 \subset \mathcal{K}$ such that $T(\mathcal{K}_1) \subseteq \mathcal{K}_1$.*

Theorem 4.5 (Schauder–Tychonoff [6, Theorem 4.5]). *A compact convex subset of a locally convex quaternionic linear space has the fixed point property.*

5. Generalized Schur functions and their realizations

The definition of negative squares makes sense in the quaternionic setting since an Hermitian quaternionic matrix H is diagonalizable: it can be written as $H = UDU^*$, where U is unitary and D is unique and with real entries. The number of strictly negative eigenvalues of H is exactly the number of strictly negative elements of D, see [48]. The one-to-one correspondence between reproducing kernel Pontryagin spaces and functions with a finite number of negative squares, proved in the classical case by [43, 44], extends to the Pontryagin space setting, see [15].

We first recall a definition. A quaternionic matrix J is called a signature matrix if it is both self-adjoint and unitary. The index of J is the number of strictly negative eigenvalues of J, and the latter is well defined because of the spectral theorem for quaternionic matrices. See, e.g., [48].

Definition 5.1. Let Ω be an axially symmetric s-domain contained in the unit ball, let $J_1 \in \mathbb{H}^{s\times s}$ and $J_2 \in \mathbb{H}^{r\times r}$ be two signature matrix of the same index, and let S be a $\mathbb{H}^{r\times s}$-valued function, slice hyperholomorphic in Ω. Then S is called a generalized Schur function if the kernel

$$K_S(p,q) = \sum_{n=0}^{\infty} p^n(J_2 - S(p)J_1S(q)^*)\overline{q}^n$$

has a finite number of negative squares, say κ, in Ω. We set $\kappa = \operatorname{ind} S$ and call it the index of S.

We will denote by $\mathscr{S}_\kappa(J_1, J_2)$ the family of generalized Schur functions of index κ.

Lemma 5.2. *In the notation of Definition* 5.1, *let* $x_0 \in \Omega \cap \mathbb{R}$. *Let* $b(p) = \frac{p+x_0}{1+px_0}$. *Then the function* $S \circ b$ *is a generalized Schur function slice hyperholomorphic at the origin and with the same index as* S.

Proof. First of all, we note that $(1+px_0)^{-\star} = (1+px_0)^{-1}$ since $x_0 \in \mathbb{R}$, and that $(1+px_0)^{-1}$ commute with $p+x_0$ thus the rational function $b(p)$ is well defined. The result then follows from the formula

$$\begin{aligned}\sum_{n=0}^{\infty} p^n(J_2 - S(b(p))J_1S(b(q))^*)\overline{q}^n &= (1-x_0^2)\\ &\times (1+px_0)^{-1}\left(\sum_{n=0}^{\infty} b(p)^n(J_2 - S(b(p))J_1S(b(q))^*)\overline{b(q)}^n\right)(1+\overline{q}x_0)^{-1}.\end{aligned} \tag{5.1}$$

To show the validity of (5.1) we use [6, Proposition 2.22] to compute the left-hand side which gives

$$\sum_{n=0}^{\infty} p^n(J_2 - S(b(p))J_1S(b(q))^*)\overline{q}^n = (J_2 - S(b(p))J_1S(b(q))^*) \star (1-p\overline{q})^{-\star}, \tag{5.2}$$

where the $\star$-product is the left one and it is computed with respect to p. The right-hand side of (5.1) can be computed in a similar was and gives

$$\begin{aligned}&(1-x_0^2)(1+px_0)^{-1}\left(\sum_{n=0}^{\infty} b(p)^n(J_2 - S(b(p))J_1S(b(q))^*)\overline{b(q)}^n\right)(1+\overline{q}x_0)^{-1}\\ &= (1-x_0^2)(1+px_0)^{-1}(J_2 - S(b(p))J_1S(b(q))^*) \star (1-b(p)\overline{b(q)})^{-\star}(1+\overline{q}x_0)^{-1}.\end{aligned} \tag{5.3}$$

We now note that

$$(1-b(p)\overline{b(q)})^{-\star} = \left(1 - \frac{p+x_0}{1+px_0}\frac{\overline{q}+x_0}{1+\overline{q}x_0}\right)^{-\star} = \frac{1}{1-x_0^2}(1+px_0)(1-p\overline{q})^{-\star}(1+\overline{q}x_0)$$

and substituting this expression in (5.3), and using the property that

$$(J_2 - S(b(p))J_1S(b(q))^*) \star (1 + px_0) = (1 + px_0)(J_2 - S(b(p))J_1S(b(q))^*)$$

since $x_0 \in \mathbb{R}$, we obtain

$$\begin{aligned}
(1 - x_0^2)(1 + px_0)^{-1}(J_2 - S(b(p))J_1S(b(q))^*) \\
\star \frac{1}{1 - x_0^2}(1 + px_0)(1 - p\bar{q})^{-\star}(1 + \bar{q}x_0)(1 + \overline{q}x_0)^{-1} \\
(1 + px_0)^{-1}(1 + px_0)(J_2 - S(b(p))J_1S(b(q))^*)(1 - p\bar{q})^{-\star} \\
= (J_2 - S(b(p))J_1S(b(q))^*)(1 - p\bar{q})^{-\star}
\end{aligned}$$

and the statement follows. □

The reproducing kernel Pontryagin space $\mathcal{P}(S)$ associated with a generalized Schur function S, namely the space with reproducing kernel K_S, is a right quaternionic vector space, with functions taking values in a two-sided quaternionic vector space. To present the counterpart of (2.2) with $\mathcal{P}(S)$ as a state space we first recall the following result, see [6, Proposition 2.22].

Proposition 5.3. *Let A be a bounded linear operator from a right-sided quaternionic Hilbert $\mathcal{P}$ space into itself, and let C be a bounded linear operator from $\mathcal{P}$ into $\mathcal{C}$, where $\mathcal{C}$ is a two-sided quaternionic Hilbert space. The slice hyperholomorphic extension of $C(I - xA)^{-1}$, $1/x \in \rho_S(A) \cap \mathbb{R}$, is*

$$(C - \overline{p}CA)(I - 2\mathrm{Re}(p)\, A + |p|^2A^2)^{-1}.$$

We will use the notation

$$C \star (I - pA)^{-\star} \stackrel{\text{def.}}{=} (C - \overline{p}CA)(I - 2\mathrm{Re}(p)\, A + |p|^2A^2)^{-1}. \tag{5.4}$$

For the following result see [7, 8]. First two remarks: in the statement, an observable pair is defined, as in the complex case, by (2.3). Next, we denote by M^* the adjoint of a quaternionic bounded linear operator from a Pontryagin space $\mathcal{P}_1$ into a Pontryagin space $\mathcal{P}_2$:

$$[Mp_1\,,\, p_2]_{\mathcal{P}_2} = [p_1\,,\, M^*p_2]_{\mathcal{P}_1}, \quad p_1 \in \mathcal{P}_1 \quad \text{and} \quad p_2 \in \mathcal{P}_2.$$

Theorem 5.4. *Let $J_1 \in \mathbb{H}^{s\times s}$ and $J_2 \in \mathbb{H}^{r\times r}$ be two signature matrices of the same index, and let S be slice hyperholomorphic in a neighborhood of the origin. Then, S is in $\mathscr{S}_\kappa(J_1, J_2)$ if and only if it can written in the form*

$$S(p) = D + pC \star (I_{\mathcal{P}} - pA)^{-\star}B, \tag{5.5}$$

where $\mathcal{P}$ is a right quaternionic Pontryagin space of index κ, the pair (C, A) is observable, and the operator matrix

$$M = \begin{pmatrix} A & B \\ C & D \end{pmatrix} : \ \mathcal{P} \oplus \mathbb{H}^s \ \longrightarrow \ \mathcal{P} \oplus \mathbb{H}^r \tag{5.6}$$

satisfies

$$\begin{pmatrix} A & B \\ C & D \end{pmatrix} \begin{pmatrix} I_{\mathcal{P}} & 0 \\ 0 & J_1 \end{pmatrix} \begin{pmatrix} A & B \\ C & D \end{pmatrix}^* = \begin{pmatrix} I_{\mathcal{P}} & 0 \\ 0 & J_2 \end{pmatrix}. \tag{5.7}$$

The space $\mathcal{P}$ can be chosen to be the reproducing kernel Pontryagin space $\mathcal{P}(S)$ with reproducing kernel $K_S(p,q)$. Then the coisometric colligation (5.6) is given by:

$$
\begin{aligned}
(Af)(p) &= \begin{cases} p^{-1}(f(p)-f(0)), & p\neq 0,\\ f_1, & p=0,\end{cases}\\
(Bv)(p) &= \begin{cases} p^{-1}(S(p)-S(0))v, & p\neq 0,\\ s_1 v, & p=0,\end{cases}\\
Cf &= f(0),\\
Dv &= S(0)v,
\end{aligned}
\tag{5.8}
$$

where $v\in\mathbb{H}^s$, $S(p)=\sum_{n=0}^{\infty}p^n s_n$ and $f\in\mathcal{P}$ with power series $f(p)=\sum_{n=0}^{\infty}p^n f_n$ at the origin.

Assume now in the previous theorem that $r=s$, $J_1=J_2=J$, and that $\dim\mathcal{P}(S)$ is finite. Then, equation (5.7) is an equality in finite-dimensional spaces (or as matrices) and the function S is called J-unitary. The function S is moreover rational and its McMillan degree, denoted by $\deg S$, is the dimension of the space $\mathcal{P}(S)$ (we refer to [7] for the notion of rational slice-hyperholomorphic functions. Suffices here to say that the restriction of S to the real axis is an $\mathbb{H}^{r\times r}$-valued rational function of a real variable).

The $\star$-factorization $S=S_1\star S_2$ of S as a $\star$-product of two $\mathbb{H}^{r\times r}$-valued J-unitary functions is called minimal if $\deg S=\deg S_1+\deg S_2$. When $\kappa=0$, S is a minimal product of elements of three types, called Blaschke–Potapov factors, and was first introduced by V. Potapov in [42] in the complex case. We give now a formal definition of the Blaschke–Potapov factors:

Definition 5.5. A $\mathbb{H}^{r\times r}$-valued Blaschke–Potapov factor of the first kind (resp. second kind) is defined as:

$$B_a(p,P)=I_r+(B_a(p)-I_r)P$$

where $|a|<1$ (resp. $|a|>1$) and $J,P\in\mathbb{H}^{r\times r}$, J being a signature matrix, and P a matrix such that $P^2=P$ and $JP\geq 0$.
A $\mathbb{H}^{r\times r}$-valued Blaschke–Potapov factor of the third kind is defined as:

$$I_r-ku\star(p+w_0)\star(p-w_0)^{-\star}u^*J$$

where $u\in\mathbb{H}^r$ is J-neutral (meaning $uJu^*=0$), $|w_0|=1$ and $k>0$.

Remark 5.6. In the setting of circuit theory, Blaschke–Potapov factors of the third kind are also called Brune sections, see, e.g., [27], [4].

In the sequel, by Blaschke product we mean the product of Blaschke–Potapov factors.

When $\kappa>0$ there need not exist minimal factorizations. We refer to [11, 12] for examples in the complex-valued case. On the other hand, still when $\kappa>0$ but for $J = I_r$, a special factorization exists, as a $\star$-quotient of two Blaschke products. This is a special case of the factorization of Krein–Langer. The following result plays a key role in the proof of this factorization. It is specific of the case $J_1 = I_s$ and $J_2 = I_r$, which allows us to use the fact that the adjoint of a contraction between quaternionic Pontryagin spaces of the same index is still a contraction.

Proposition 5.7. *In the notation of Theorem* 5.4, *assume* $J_1 = I_s$ *and* $J_2 = I_r$. *Then the operator* A *is a Pontryagin contraction.*

Proof. Equation (5.7) expresses that the operator matrix M (defined by (5.6)) is a coisometry, and in particular a contraction, between Pontryagin spaces of same index. Its adjoint is a Pontryagin space contraction (see [15]) and we have

$$\begin{pmatrix} A & B \\ C & D \end{pmatrix}^* \begin{pmatrix} I_{\mathcal{P}} & 0 \\ 0 & I_r \end{pmatrix} \begin{pmatrix} A & B \\ C & D \end{pmatrix} \leq \begin{pmatrix} I_{\mathcal{P}} & 0 \\ 0 & I_s \end{pmatrix}.$$

It follows from this inequality that

$$A^*A + C^*C \leq I_s. \tag{5.9}$$

Since the range of C is inside the Hilbert space $\mathbb{H}^r$ we have that A^* is a contraction from $\mathcal{P}$ into itself, and so is its adjoint $A = (A^*)^*$. □

6. The factorization theorem

Below we prove a version of the Krein–Langer factorization theorem in the slice hyperholomorphic setting which generalizes [9, Theorem 9.2]. The role of the Blaschke factors B_a in the scalar case is played here by the Blaschke–Potapov factors with $J = I$.

Theorem 6.1. *Let* $J_1 = I_s$ *and* $J_2 = I_r$, *and let* S *be a* $\mathbb{H}^{r\times s}$*-valued generalized Schur function of index* κ. *Then there exists a* $\mathbb{H}^{r\times r}$*-valued Blaschke product* B_0 *of degree* κ *and a* $\mathbb{H}^{r\times s}$*-valued Schur function* S_0 *such that*

$$S(p) = (B_0^{-\star} \star S_0)(p).$$

Proof. We proceed in a number of steps:

STEP 1: *One can assume that* S *is slice hyperholomorphic at the origin.*

To check this, we note that whenever $f = g\star h$, we have $f\circ b = (g\circ b)\star(h\circ b)$ where $b(p) = \frac{p+x_0}{1+px_0}$, $x_0 \in \mathbb{R}$. This equality is true on $\Omega\cap\mathbb{R}_+$, and extends to Ω by slice hyperholomorphic extension. Thus, taking into account Lemma 5.2, we now assume $0 \in \Omega$.

STEP 2: *Let* (5.5) *be a coisometric realization of* S. *Then* A *has a unique maximal strictly negative invariant subspace* $\mathcal{M}$.

Indeed, A is a contraction as proved in Proposition 5.7. The result then follows from Theorem 4.3.

The rest of the proof is as in [9], and is as follows. Let $\mathcal{M}$ be the space defined in STEP 2, and let $A_{\mathcal{M}}$, $C_{\mathcal{M}}$ denote the matrix representations of A and

C, respectively, in a basis of $\mathcal{M}$, and let $G_{\mathcal{M}}$ be the corresponding Gram matrix. It follows from (5.9) that

$$A_{\mathcal{M}}^* G_{\mathcal{M}} A_{\mathcal{M}} \leq G_{\mathcal{M}} - C_{\mathcal{M}}^* C_{\mathcal{M}}.$$

STEP 3: *The equation*

$$A_{\mathcal{M}}^* P_{\mathcal{M}} A_{\mathcal{M}} = P_{\mathcal{M}} - C_{\mathcal{M}}^* C_{\mathcal{M}}$$

has a unique solution. It is strictly negative and $\mathcal{M}$ endowed with this metric is contractively included in $\mathcal{P}(S)$.

Recall that the S-spectrum of an operator A is defined as the set of quaternions p such that $A^2 - 2\mathrm{Re}(p)A + |p|^2 I$ is not invertible, see [25]. Then, the first two claims follow from the fact that the S-spectrum of $A_{\mathcal{M}}$, which coincides with the right spectrum of $A_{\mathcal{M}}$, is outside the closed unit ball. Moreover, the matrix $G_{\mathcal{M}} - P_{\mathcal{M}}$ satisfies

$$A_{\mathcal{M}}^* (G_{\mathcal{M}} - P_{\mathcal{M}}) A_{\mathcal{M}} \leq G_{\mathcal{M}} - P_{\mathcal{M}},$$

or equivalently (since A is invertible)

$$G_{\mathcal{M}} - P_{\mathcal{M}} \leq A_{\mathcal{M}}^{-*} (G_{\mathcal{M}} - P_{\mathcal{M}}) A_{\mathcal{M}}^{-1}$$

and so, for every $n \in \mathbb{N}$,

$$G_{\mathcal{M}} - P_{\mathcal{M}} \leq (A_{\mathcal{M}}^{-*})^n (G_{\mathcal{M}} - P_{\mathcal{M}}) A_{\mathcal{M}}^{-n}. \tag{6.1}$$

By the spectral theorem (see [24, Theorem 3.10, p. 616] and [25, Theorem 4.12.6, p. 155] for the spectral radius theorem) we have:

$$\lim_{n \to \infty} \|A_{\mathcal{M}}^{-n}\|^{1/n} = 0,$$

and so $\lim_{n \to \infty} \|(A_{\mathcal{M}}^{-*})^n (P_{\mathcal{M}} - G_{\mathcal{M}}) A_{\mathcal{M}}^{-n}\| = 0$. Thus entry-wise

$$\lim_{n \to \infty} (A_{\mathcal{M}}^{-*})^n (P_{\mathcal{M}} - G_{\mathcal{M}}) A_{\mathcal{M}}^{-n} = 0$$

and it follows from (6.1) that $G_{\mathcal{M}} - P_{\mathcal{M}} \leq 0$.

By [9, Proposition 8.8]

$$\mathcal{M} = \mathcal{P}(B),$$

when $\mathcal{P}$ is endowed with the $P_{\mathcal{M}}$ metric and where B is a rational function with associated de Branges–Rovnyak space which is finite dimensional and an anti Hilbert space. Such functions B are (inverses of) inner functions, and can be seen, as in [1], to be a finite product of Blaschke–Potapov factors of the second kind. Furthermore:

STEP 4: *The kernel $K_S - K_B$ is positive.*

Let $k_{\mathcal{M}}$ denote the reproducing kernel of $\mathcal{M}$ when endowed with the $\mathcal{P}(S)$ metric. Then

$$k_{\mathcal{M}}(p, q) - K_B(p, q) \geq 0$$

and

$$K_S(p, q) - k_{\mathcal{M}}(p, q) \geq 0.$$

Moreover

$$K_S(p,q) - K_B(p,q) = K_S(p,q) - k_{\mathcal{M}}(p,q) + k_{\mathcal{M}}(p,q) - K_B(p,q)$$

and so it is positive definite.

Finally we apply [9, Proposition 5.1] to

$$K_S(p,q) - K_B(p,q) = B(p) \star (I_r - S_0(p)S_0(q)^*) \star_r B(q)^*$$

where where $S_0(p) = B(p) \star S(p)$, to conclude that S_0 is a Schur function. □

7. The case of the half-space

Since the map (where $x_0 \in \mathbb{R}_+$)

$$p \mapsto (p - x_0)(p + x_0)^{-1}$$

sends the right half-space onto the open unit ball, one can translate the previous results to the case of the half-space $\mathbb{H}_+$. In particular the Blaschke–Potapov factors are of the form

$$B_a(p, P) = I_r + (b_a(p) - 1)P$$

where P is a matrix such that $P^2 = P$ and $JP \geq 0$ where, in general, J is signature matrix, and $a \in \mathbb{H}^+$. The factors of the third type are now functions of the form

$$I_r - ku \star (p + w_0)^{-\star} u^* J$$

where $u \in \mathbb{H}^r$ is such that $uJu^* = 0$, and $w_0 + \overline{w_0} = 0$, $k > 0$. The various definitions and considerations on rational J-unitary functions introduced in Section 5 have counterparts here. We will not explicit them, but restrict ourselves to the case $J_1 = I_s$ and $J_2 = I_r$, and only mention the counterpart of the Krein–Langer factorization in the half-space setting. We outline the results and leave the proofs to the reader.

In the setting of slice hyperholomorphic functions in $\mathbb{H}_+$ the counterpart of the kernel $\sum_{n=0}^{\infty} p^n \overline{q}^n$ is

$$k(p,q) = (\bar{p} + \bar{q})(|p|^2 + 2\mathrm{Re}(p)\bar{q} + \bar{q}^2)^{-1}. \tag{7.1}$$

Definition 7.1. The $\mathbb{H}^{r\times s}$-valued function S slice hypermeromorphic in an axially symmetric s-domain Ω which intersects the positive real line belongs to the class $\mathscr{S}_\kappa(\mathbb{H}_+)$ if the kernel

$$K_S(p,q) = I_r k(p,q) - S(p) \star k(p,q) \star_r S(q)^*$$

has κ negative squares in Ω, where $k(p,q)$ is defined in (7.1).

The following realization theorem has been proved in [6, Theorem 6.2].

Theorem 7.2. *Let x_0 be a strictly positive real number. A $\mathbb{H}^{r\times s}$-valued function S slice hyperholomorphic in an axially symmetric s-domain Ω containing x_0 is the restriction to Ω of an element of $S_\kappa(\mathbb{H}_+)$ if and only if it can be written as*

$$S(p) = H - (p - x_0)\left(G - (\overline{p} - x_0)(\overline{p} + x_0)^{-1}GA\right) \\ \times \left(\frac{|p - x_0|^2}{|p + x_0|^2}A^2 - 2\mathrm{Re}\left(\frac{p - x_0}{p + x_0}\right)A + I\right)^{-1} F, \tag{7.2}$$

where A is a linear bounded operator in a right-sided quaternionic Pontryagin space Π_κ of index κ, and, with $B = -(I + x_0A)$, the operator matrix

$$\begin{pmatrix} B & F \\ G & H \end{pmatrix} : \begin{pmatrix} \Pi_k \\ \mathbb{H}^s \end{pmatrix} \longrightarrow \begin{pmatrix} \Pi_k \\ \mathbb{H}^r \end{pmatrix}$$

is co-isometric. In particular S has a unique slice hypermeromorphic extension to $\mathbb{H}_+$. Furthermore, when the pair (G, A) is observable, the realization is unique up to a unitary isomorphism of Pontryagin right quaternionic spaces.

By an abuse of notation, we write

$$S(p) = H - (p - x_0)G \star ((x_0 + p)I + (p - x_0)B)^{-\star}F$$

rather than (7.2).

In the following statement, the degree of the Blaschke product B_0 is the dimension of the associated reproducing kernel Hilbert space with reproducing kernel K_{B_0}.

Theorem 7.3. *Let S be a $\mathbb{H}^{r\times s}$-valued function slice hypermeromorphic in an axially symmetric s-domain Ω which intersects the positive real line. Then, $S \in \mathscr{S}_\kappa(\mathbb{H}_+)$ if and only if it can be written as $S = B_0^{-\star} \star S_0$, where B_0 is a $\mathbb{H}^{r\times r}$-valued finite Blaschke product of degree κ, and $S_0 \in \mathscr{S}_0(\mathbb{H}_+)$.*

References

[1] K. Abu-Ghanem, D. Alpay, F. Colombo, D.P. Kimsey, I. Sabadini. Boundary interpolation for slice hyperholomorphic Schur functions, preprint arxiv:1404.3352.

[2] J. Agler. *On the representation of certain holomorphic functions defined on a polydisk*, volume 48 of *Operator Theory: Advances and Applications*, pages 47–66. Birkhäuser Verlag, Basel, 1990.

[3] D. Alpay. *The Schur algorithm, reproducing kernel spaces and system theory.* American Mathematical Society, Providence, RI, 2001. Translated from the 1998 French original by Stephen S. Wilson, Panoramas et Synthèses.

[4] D. Alpay, V. Bolotnikov, P. Dewilde, A. Dijksma. Brune sections in the nonstationary case. *Linear Algebra Appl.*, 343/344, 389–418, 2002.

[5] D. Alpay, V. Bolotnikov, F. Colombo, and I. Sabadini. Self-mappings of the quaternionic unit ball: multiplier properties, Schwarz–Pick inequality, and Nevanlinna–Pick interpolation problem. To appear in Indiana Univ. Math. J. (2014).

[6] D. Alpay, F. Colombo, I. Lewkowicz, and I. Sabadini. Realizations of slice hyperholomorphic generalized contractive and positive functions. *ArXiv e-prints*, October 2013.

[7] D. Alpay, F. Colombo, and I. Sabadini. Schur functions and their realizations in the slice hyperholomorphic setting. *Integral Equations and Operator Theory*, 72:253–289, 2012.

[8] D. Alpay, F. Colombo, and I. Sabadini. Pontryagin–de Branges–Rovnyak spaces of slice hyperholomorphic functions. *J. Anal. Math.*, 121:87–125, 2013.

[9] D. Alpay, F. Colombo, and I. Sabadini. Krein–Langer factorization and related topics in the slice hyperholomorphic setting. *J. Geom. Anal.* 24(2): 843–872, 2014.

[10] D. Alpay, A. Dijksma, J. Rovnyak, and H. de Snoo. *Schur functions, operator colligations, and reproducing kernel Pontryagin spaces*, volume 96 of *Operator theory: Advances and Applications*. Birkhäuser Verlag, Basel, 1997.

[11] D. Alpay and H. Dym. On applications of reproducing kernel spaces to the Schur algorithm and rational J-unitary factorization. In I. Gohberg, editor, *I. Schur methods in operator theory and signal processing*, volume 18 of *Operator Theory: Advances and Applications*, pages 89–159. Birkhäuser Verlag, Basel, 1986.

[12] D. Alpay and I. Gohberg. Unitary rational matrix functions. In I. Gohberg, editor, *Topics in interpolation theory of rational matrix-valued functions*, volume 33 of *Operator Theory: Advances and Applications*, pages 175–222. Birkhäuser Verlag, Basel, 1988.

[13] D. Alpay, M.E. Luna Elizarraras, M. Shapiro, and D.C. Struppa. Basics of functional analysis with bicomplex scalars, and bicomplex Schur analysis. Springer Briefs in Mathematics, Springer, 2014.

[14] D. Alpay, B. Schneider, M. Shapiro, and D. Volok. Fonctions rationnelles et théorie de la réalisation: le cas hyper-analytique. *Comptes Rendus Mathématiques*, 336:975–980, 2003.

[15] D. Alpay and M. Shapiro. Reproducing kernel quaternionic Pontryagin spaces. *Integral Equations and Operator Theory*, 50:431–476, 2004.

[16] D. Alpay and V. Vinnikov. Finite-dimensional de Branges spaces on Riemann surfaces. *J. Funct. Anal.*, 189(2):283–324, 2002.

[17] N. Aronszajn. Theory of reproducing kernels. *Trans. Amer. Math. Soc.*, 68:337–404, 1950.

[18] T.Ya. Azizov and I.S. Iohvidov. *Foundations of the theory of linear operators in spaces with indefinite metric.* Nauka, Moscow, 1986 (Russian). English translation: *Linear operators in spaces with an indefinite metric.* John Wiley, New York, 1989.

[19] J. Ball, T. Trent, and V. Vinnikov. Interpolation and commutant lifting for multipliers on reproducing kernel Hilbert spaces. In *Proceedings of Conference in honor of the* 60*th birthday of M.A. Kaashoek*, volume 122 of *Operator Theory: Advances and Applications*, pages 89–138. Birkhäuser, 2001.

[20] A. Bloch. Les fonctions holomorphes et méromorphes dans le cercle unité. *Mémorial des sciences mathématiques*, pages 1–61, 1926. Fascicule 20.

[21] J. Bognár. *Indefinite inner product spaces.* Ergebnisse der Mathematik und ihrer Grenzgebiete, Band 78. Springer-Verlag, Berlin, 1974.

[22] V. Bolotnikov and L. Rodman. Krein–Langer factorizations via pole triples. *Integral Equations and Operator Theory*, 47(2):169–195, 2003.

[23] C. Chamfy. Fonctions méromorphes sur le cercle unité et leurs séries de Taylor. *Ann. Inst. Fourier*, 8:211–251, 1958.

[24] F. Colombo and I. Sabadini. On some properties of the quaternionic functional calculus. *J. Geom. Anal.*, 19(3):601–627, 2009.

[25] F. Colombo, I. Sabadini, and D.C. Struppa. *Noncommutative functional calculus. Theory and applications of slice hyperholomorphic functions*, volume 289 of *Progress in Mathematics*. Birkhäuser/Springer Basel AG, Basel, 2011.

[26] P. Delsarte, Y. Genin, and Y. Kamp. Pseudo-Carathéodory functions and Hermitian Toeplitz matrices. *Philips J. Res.*, 41(1):1–54, 1986.

[27] P. Dewilde and H. Dym. Lossless inverse scattering, digital filters, and estimation theory. *IEEE Trans. Inform. Theory*, 30 (4): 644–662, 1984.

[28] B. Dickinson, Ph. Delsarte, Y. Genin, and Y. Kamp. Minimal realizations of pseudo-positive and pseudo-bounded rational matrices. *IEEE Transactions on Circuits and Systems*, 32:603–605, 1985.

[29] W.F. Donoghue. *Monotone matrix functions and analytic continuation*, volume 207 of *Die Grundlehren der mathematischen Wissenschaften*. Springer-Verlag, 1974.

[30] M.A. Dritschel and J. Rovnyak. *Extension theorems for contractions on Kreĭn spaces*, volume 47 of *Operator theory: Advances and Applications*, pages 221–305. Birkhäuser Verlag, Basel, 1990.

[31] M.A. Dritschel and J. Rovnyak. Operators on indefinite inner product spaces. In P. Lancaster, editor, *Lectures on operator theory and its applications* (*Waterloo, ON,* 1994), volume 3 of *Fields Inst. Monogr.*, pages 141–232. Amer. Math. Soc., Providence, RI, 1996.

[32] J. Dufresnoy. Le problème des coefficients pour certaines fonctions méromorphes dans le cercle unité. *Ann. Acad. Sci. Fenn. Ser. A. I, no.*, 250/9:7, 1958.

[33] B. Fritzsche and B. Kirstein, editors. *Ausgewählte Arbeiten zu den Ursprüngen der Schur-Analysis*, volume 16 of *Teubner-Archiv zur Mathematik*. B.G. Teubner Verlagsgesellschaft, Stuttgart-Leipzig, 1991.

[34] G. Gentili and D.C. Struppa. On the multiplicity of zeroes of polynomials with quaternionic coefficients. *Milan J. Math.*, 76:15–25, 2008.

[35] I.S. Iohvidov, M.G. Kreĭn, and H. Langer. *Introduction to the spectral theory of operators in spaces with an indefinite metric*. Akademie-Verlag, Berlin, 1982.

[36] M.G. Kreĭn and H. Langer. Über die verallgemeinerten Resolventen und die charakteristische Funktion eines isometrischen Operators im Raume Π_k. In *Hilbert space operators and operator algebras* (*Proc. Int. Conf. Tihany,* 1970), pages 353–399. North-Holland, Amsterdam, 1972. Colloquia Math. Soc. János Bolyai.

[37] M.G. Kreĭn and H. Langer. Über einige Fortsetzungsprobleme, die eng mit der Theorie hermitescher Operatoren im Raume π_k zusammenhängen. I. Einige Funktionenklassen und ihre Darstellungen. *Math. Nachrichten*, 77:187–236, 1977.

[38] M.G. Kreĭn and H. Langer. Some propositions on analytic matrix functions related to the theory of operators in the space π_k. *Acta Sci. Math.*, 43:181–205, 1981.

[39] T.Y. Lam. *A first course in noncommutative rings*, volume 131 of *Graduate Texts in Mathematics*. Springer-Verlag, New York, second edition, 2001.

[40] R. Pereira. *Quaternionic Polynomials and Behavioral Systems*. PhD thesis, University of Aveiro, 2006.

[41] R. Pereira and P. Vettori. Stability of quaternionic linear systems. *IEEE Trans. Automat. Control*, 51(3):518–523, 2006.

[42] V.P. Potapov. The multiplicative structure of J-contractive matrix functions. *Amer. Math. Soc. Transl.*, 15: 131–243, 1960.

[43] L. Schwartz. Sous espaces hilbertiens d'espaces vectoriels topologiques et noyaux associés (noyaux reproduisants). *J. Analyse Math.*, 13:115–256, 1964.

[44] P. Sorjonen. Pontryagin Räume mit einem reproduzierenden Kern. *Ann. Acad. Fenn. Ser. A*, 1:1–30, 1973.

[45] C. Stoppato. Singularities of slice regular functions. *Math. Nachr.*, 285(10):1274–1293, 2012.

[46] T. Takagi. On an algebraic problem related to an analytic theorem of Carathéodory and Fejér and on an allied theorem of Landau. *Japanese journal of mathematics*, 1:83–93, 1924.

[47] T. Takagi. Remarks on an algebraic problem. *Japanese journal of mathematics*, 2:13–17, 1925.

[48] F. Zhang. Quaternions and matrices of quaternions. *Linear Algebra Appl.*, 251:21–57, 1997.

Daniel Alpay
Department of Mathematics
Ben-Gurion University of the Negev
Beer-Sheva 84105 Israel
e-mail: `dany@math.bgu.ac.il`

Fabrizio Colombo and Irene Sabadini
(FC) Politecnico di Milano
Dipartimento di Matematica
Via E. Bonardi, 9
I-20133 Milano, Italy
e-mail: `fabrizio.colombo@polimi.it`
`irene.sabadini@polimi.it`

Hypercomplex Analysis: New Perspectives and Applications
Trends in Mathematics, 43–59

The Fock Space in the Slice Hyperholomorphic Setting

Daniel Alpay, Fabrizio Colombo, Irene Sabadini and Guy Salomon

Abstract. In this paper we introduce and study some basic properties of the Fock space (also known as Segal–Bargmann space) in the slice hyperholomorphic setting. We discuss both the case of slice regular functions over quaternions and the case of slice monogenic functions with values in a Clifford algebra. In the specific setting of quaternions, we also introduce the full Fock space. This paper can be seen as the beginning of the study of infinite-dimensional analysis in the quaternionic setting.

Mathematics Subject Classification (2010). MSC: 30G35, 30H20.

Keywords. Fock space, slice hyperholomorphic functions, quaternions, Clifford algebras.

1. Introduction

Fock spaces are a very important tool in quantum mechanics, and also in its quaternionic formulation; see the book of Adler [1] and the paper [31]. Roughly speaking, they can be seen as the completion of the direct sum of the symmetric or anti-symmetric, or full tensor powers of a Hilbert space which, from the point of view of Physics, represents a single particle. There is an alternative description of the Fock spaces in the holomorphic setting which, in this framework, are also known as Segal–Bargmann spaces.

In this note we work first in the setting of slice hyperholomorphic functions, namely either we work with slice regular functions (these are functions defined on subsets of the quaternions with values in the quaternions) or with slice monogenic functions (these functions are defined on the Euclidean space $\mathbb{R}^{n+1}$ and have values in the Clifford algebra $\mathbb{R}_n$), see the book [21].

D. Alpay thanks the Earl Katz family for endowing the chair which supported his research, and the Binational Science Foundation Grant number 2010117. F. Colombo and I. Sabadini acknowledge the Center for Advanced Studies of the Mathematical Department of the Ben-Gurion University of the Negev for the support and the kind hospitality during the period in which part of this paper has been written.

Slice hyperholomorphic functions have been introduced quite recently but they have already several applications, for example in Schur analysis and to define some functional calculi. The application to Schur analysis started with the paper [6] and it is rapidly growing, see for example [3, 4, 5, 7, 8].

The applications to the functional calculus range from the so-called S-functional calculus, which works for n-tuples non necessarily commuting operators, to a quaternionic version of the classical Riesz–Dunford functional calculus, see [23]. The literature on slice hyperholomorphic functions and the related functional calculi is wide, and we refer the reader to the book [21] and the references therein.

We note that Fock spaces have been treated in the more classical setting of monogenic functions, see for example the book [22]. In the treatment in [22] no tensor products of Hilbert-Clifford modules are involved. In the framework of slice hyperholomorphic analysis we have already introduced and studied the Hardy spaces (see [7, 3, 4]), and Bergman spaces (see [18, 20, 19]). Here we begin the study of the main properties of the quaternionic Fock spaces.

We start by recalling the definition of the Fock space in the classical complex analysis case (for the origins of the theory see [24]).
For $n \in \mathbb{N}$ let $z = (z_1, \dots, z_n) \in \mathbb{C}^n$ where $z_j = x_j + iy_j$, x_j, $y_j \in \mathbb{R}$ $(j = 1, \dots, n)$ and denote by

$$d\mu(z) := \pi^{-n}\Pi_{j=1}^n dx_j dy_j$$

the normalized Lebesgue measure on $\mathbb{C}^n$. The Fock space of holomorphic functions f defined on $\mathbb{C}^n$ is

$$F_n := \left\{ f : \mathbb{C}^n \mapsto \mathbb{C} \text{ such that } \int_{\mathbb{C}^n} |f(z)|^2 e^{-|z|^2} d\mu(z) < \infty \right\}. \tag{1.1}$$

The space F_n with the scalar product

$$\langle f, g\rangle_{F_n} = \int_{\mathbb{C}^n} f(z)\overline{g(z)}e^{-|z|^2} d\mu(z)$$

becomes a Hilbert space and the norm is

$$\|f\|^2_{F_n} = \int_{\mathbb{C}^n} |f(z)|^2 e^{-|z|^2} d\mu(z), \quad f \in F_n.$$

The space F_n is called boson Fock space and since we will treat this case in the sequel we will refer to it simply as Fock space. One of its most important properties is that it is a reproducing kernel Hilbert space. If we denote by $\langle \cdot, \cdot \rangle_{\mathbb{C}^n}$ the natural scalar product in $\mathbb{C}^n$ defined by $\langle u, v\rangle_{\mathbb{C}^n} := \sum_{j=1}^n u_j \overline{v}_j$, for every u, $v \in \mathbb{C}^n$ we define the function

$$\psi_u(z) = e^{\langle u, v\rangle_{\mathbb{C}^n}} = e^{\sum_{j=1}^n u_j \overline{v}_j}. \tag{1.2}$$

We have the reproducing property

$$\langle f, \psi_u\rangle_{F_n} = f(u), \qquad \text{for all} \quad f \in F_n.$$

So there are two equivalent characterizations of the Fock space F_n; one geometric, in terms of integrals (see (1.1)), and one analytic, obtained by the reproducing kernel property (or, directly from (1.1)): an entire function $f(z) = \sum_{m\in\mathbb{N}_0^n} a_m z^m$ of n complex variables $z = (z_1, \ldots, z_n)$ is in F_n if and only if its Taylor coefficients satisfy

$$\sum_{m\in\mathbb{N}_0^n} m!|a_m|^2 < \infty,$$

where we have used the multi-index notation. A third characterization is of importance, namely (with appropriate identification, and with $\circ$ denoting the symmetric tensor product)

$$F_n = \oplus_{k=0}^{\infty}(\mathbb{C}^n)^{\circ k}.$$

In this paper we will address some aspects of these three characterizations in the quaternionic and Clifford algebras settings.

The paper consists of four sections besides the introduction. In Section 2 we give a brief survey of infinite-dimensional analysis. In Section 3 we study the quaternionic Fock space in one quaternionic variable. We then discuss, in Section 4, the full Fock space. In order to define it, we need to study tensor products of quaternionic two-sided Hilbert spaces. Tensor product of quaternionic vector spaces have been treated in the literature at various level, see, e.g., [11], [31, 30]. This section in particular opens the way to study a quaternionic infinite-dimensional analysis. The last section considers the case of slice monogenic functions.

2. A brief survey of infinite-dimensional analysis

There are various ways to introduce infinite-dimensional analysis. We mention here four related approaches:

1. The white noise space and the Bochner–Minlos theorem: The formula

$$e^{-\frac{t^2}{2}} = \frac{1}{\sqrt{2\pi}}\int_{\mathbb{R}} e^{-\frac{u^2}{2}} e^{-itu} du \tag{2.1}$$

is an illustration of Bochner's theorem. It is well known that there is no such formula when $\mathbb{R}$ is replaced by an infinite-dimensional Hilbert space. On the other hand, the Bochner–Minlos theorem asserts that there exists a probability measure P on the space $\mathcal{S}'$ of real tempered distributions such that

$$e^{-\frac{\|s\|_2^2}{2}} = \int_{\mathcal{S}'} e^{i\langle s', s\rangle} dP(s'). \tag{2.2}$$

In this expression, s belongs to the space $\mathcal{S}$ of real-valued Schwartz function, the duality between $\mathcal{S}$ and $\mathcal{S}'$ is denoted by $\langle s', s\rangle$ and $\|\cdot\|_2$ denotes the $\mathbf{L}_2(\mathbb{R}, dx)$ norm.

The probability space $\mathbf{L}_2(\mathcal{S}', P)$ is called the white noise space, and is denoted by $\mathcal{W}$. Denoting by Q_s the map $s' \mapsto \langle s', s\rangle$ we see that (2.2) induces an isometry, which we denote Q_f, from the Lebesgue space $\mathbf{L}_2(\mathbb{R}, dx)$ into the white noise space. We now give an important family of orthogonal basis $(H_\alpha,\ \alpha \in \ell)$ of the white noise space, indexed by the set ℓ of sequences $(\alpha_1, \alpha_2, \ldots)$, with entries in

$$\mathbb{N}_0 = \{0, 1, 2, 3, \ldots\},$$

where $\alpha_k \neq 0$ for only a finite number of indices k. Let $h_0, h_1, \ldots$ denote the Hermite polynomials, and let $\xi_1, \xi_2, \ldots$ be an orthonormal basis of $\mathbf{L}_2(\mathbb{R}, dx)$ (typically, the Hermite functions, but other choices are possible). Then

$$H_\alpha = \prod_{k=1}^{\infty} h_{\alpha_k}(Q_{\xi_k}), \tag{2.3}$$

and, with the multi-index notation

$$\alpha! = \alpha_1!\alpha_2!\cdots,$$

we have

$$\|H_\alpha\|_{\mathcal{W}}^2 = \alpha!. \tag{2.4}$$

The decomposition of an element $f \in \mathcal{W}$ along the basis $(H_\alpha)_{\alpha\in\ell}$ is called the chaos expansion.

2. The Bargmann space in infinitely many variables: When in (1.2), $\mathbb{C}^n$ is replaced by $\ell_2(\mathbb{N})$, we have the function

$$\psi_u(z) = e^{\langle u,v\rangle_{\ell_2(\mathbb{N})}} = e^{\sum_{j=1}^{\infty} u_j \overline{v}_j}. \tag{2.5}$$

The map $H_\alpha \mapsto z^\alpha$ is called the Hermite transform, and is unitary from the white noise space onto the reproducing kernel Hilbert space with reproducing kernel (2.5).

3. The Fock space: We denote by $\circ$ the symmetrized tensor product and by

$$\Gamma^\circ(\mathcal{H}) = \oplus_{n=0}^{\infty} \mathcal{H}^{\circ n},$$

the symmetric Fock space associated to a Hilbert space $\mathcal{H}$. Then, $\Gamma^\circ(\mathbf{L}_2(\mathbb{R}, dx))$ can be identified with the white noise space via the Wiener–Itô–Segal transform defined as follows (see [37, p. 165]):

$$\xi_\alpha = \xi_{i_1}^{\circ\alpha_{i_1}} \circ \cdots \circ \xi_{i_m}^{\circ\alpha_{i_m}} \in \mathcal{H}^{\circ n} \quad \mapsto H_\alpha$$

This is the starting point of our approach to quaternionic infinite-dimensional analysis; see Section 4.

4. The free setting. The full Fock space: It is defined by

$$\Gamma(\mathcal{H}) = \oplus_{n=0}^{\infty} \mathcal{H}^{\otimes n},$$

and allows to develop the free analog of the white noise space theory. See [36, 35] for background for the free setting. See [12] for recent applications to the theory of non commutative stochastic distributions.

We refer in particular to the papers [33, 34, 13, 14] and the books [26, 27, 28, 29, 37, 25] for more information on these various aspects.

3. The Fock space in the slice regular case

The algebra of quaternions is indicated by the symbol $\mathbb{H}$. The imaginary units in $\mathbb{H}$ are denoted by i, j and k, respectively, and an element in $\mathbb{H}$ is of the form $q = x_0 + ix_1 + jx_2 + kx_3$, for $x_\ell \in \mathbb{R}$. The real part, the imaginary part and the modulus of a quaternion are defined as $\mathrm{Re}(q) = x_0$, $\mathrm{Im}(q) = ix_1 + jx_2 + kx_3$, $|q|^2 = x_0^2 + x_1^2 + x_2^2 + x_3^2$, respectively. The conjugate of the quaternion $q = x_0 + ix_1 + jx_2 + kx_3$ is defined by $\bar{q} = \mathrm{Re}(q) - \mathrm{Im}(q) = x_0 - ix_1 - jx_2 - kx_3$ and it satisfies

$$|q|^2 = q\bar{q} = \bar{q}q.$$

The unit sphere of purely imaginary quaternions is

$$\mathbb{S} = \{q = ix_1 + jx_2 + kx_3 \text{ such that } x_1^2 + x_2^2 + x_3^2 = 1\}.$$

Notice that if $I \in \mathbb{S}$, then $I^2 = -1$; for this reason the elements of $\mathbb{S}$ are also called imaginary units. Note that $\mathbb{S}$ is a two-dimensional sphere in $\mathbb{R}^4$. Given a nonreal quaternion $q = x_0 + \mathrm{Im}(q) = x_0 + I|\mathrm{Im}(q)|$, $I = \mathrm{Im}(q)/|\mathrm{Im}(q)| \in \mathbb{S}$, we can associate to it the two-dimensional sphere defined by

$$[q] = \{x_0 + I\mathrm{Im}(q)| \ : \ I \in \mathbb{S}\}.$$

This sphere has center at the real point x_0 and radius $|\mathrm{Im}(q)|$. An element in the complex plane $\mathbb{C}_I = \mathbb{R} + I\mathbb{R}$ is denoted by $x + Iy$.

Definition 3.1 (Slice regular (or slice hyperholomorphic) functions). Let U be an open set in $\mathbb{H}$ and consider a real differentiable function $f : U \to \mathbb{H}$. Denote by f_I the restriction of f to the complex plane $\mathbb{C}_I$.
The function f is (left) slice regular (or (left) slice hyperholomorphic) if, for every $I \in \mathbb{S}$, it satisfies:

$$\overline{\partial}_I f_I(x + Iy) := \frac{1}{2}\left(\frac{\partial}{\partial x} + I\frac{\partial}{\partial y}\right) f_I(x + Iy) = 0,$$

on $U \cap \mathbb{C}_I$. The set of (left) slice regular functions on U will be denoted by $\mathcal{R}(U)$. The function f is right slice regular (or right slice hyperholomorphic) if, for every $I \in \mathbb{S}$, it satisfies:

$$(f_I\overline{\partial}_I)(x + Iy) := \frac{1}{2}\left(\frac{\partial}{\partial x} f_I(x + Iy) + \frac{\partial}{\partial y} f_I(x + Iy)I\right) = 0,$$

on $U \cap \mathbb{C}_I$.

The class of slice hyperholomorphic quaternionic-valued functions is important since power series centered at real points are slice hyperholomorphic: if $B = B(y_0, R)$ is the open ball centered at the real point y_0 and radius $R > 0$ and if $f : B \to \mathbb{H}$ is a left slice regular function then f admits the power series expansion

$$f(q) = \sum_{m=0}^{+\infty} (q - y_0)^m \frac{1}{m!} \frac{\partial^m f}{\partial x^m}(y_0),$$

converging on B.

A main property of the slice hyperholomorphic functions is the so-called Representation Formula (or Structure Formula). It holds on a particular class of open sets which are described below.

Definition 3.2 (Axially symmetric domain). Let $U \subseteq \mathbb{H}$. We say that U is axially symmetric if, for all $x + Iy \in U$, the whole 2-sphere $[x + Iy]$ is contained in U.

Definition 3.3 (Slice domain). Let $U \subseteq \mathbb{H}$ be a domain in $\mathbb{H}$. We say that U is a slice domain (s-domain for short) if $U \cap \mathbb{R}$ is non empty and if $U \cap \mathbb{C}_I$ is a domain in $\mathbb{C}_I$ for all $I \in \mathbb{S}$.

Theorem 3.4 (Representation Formula). *Let U be an axially symmetric s-domain $U \subseteq \mathbb{H}$.*

Let f be a (left) slice regular function on U. Choose any $J \in \mathbb{S}$. Then the following equality holds for all $q = x + yI \in U$:

$$f(x+Iy) = \frac{1}{2}\Big[f(x+Jy) + f(x-Jy)\Big] + I\frac{1}{2}\Big[J[f(x-Jy) - f(x+Jy)]\Big]. \quad (3.1)$$

Remark 3.5. One of the applications of the Representation Formula is the fact that any function defined on an open set Ω_I of a complex plane $\mathbb{C}_I$ which belongs to the kernel of the Cauchy–Riemann operator can be uniquely extended to a slice hyperholomorphic function defined on the axially symmetric completion of Ω_I (see [21]).

We now define the Fock space in this framework.

Definition 3.6 (Slice hyperholomorphic quaternionic Fock space). Let I be any element in $\mathbb{S}$. Consider the set

$$\mathcal{F}(\mathbb{H}) = \left\{ f \in \mathcal{R}(\mathbb{H}) \mid \int_{\mathbb{C}_I} e^{-|p|^2} |f_I(p)|^2 d\sigma(x,y) < \infty \right\}$$

where $p = x + Iy$, $d\sigma(x,y) := \frac{1}{\pi} dxdy$. We will call $\mathcal{F}(\mathbb{H})$ (slice hyperholomorphic) quaternionic Fock space.

We endow $\mathcal{F}(\mathbb{H})$ with the inner product

$$\langle f, g \rangle := \int_{\mathbb{C}_I} e^{-|p|^2} \overline{g_I(p)} f_I(p) d\sigma(x,y); \quad (3.2)$$

we will show below that this definition, as well as the definition of Fock space, do not depend on the imaginary unit $I \in \mathbb{S}$.
The norm induced by the inner product is then

$$\|f\|^2 = \int_{\mathbb{C}_I} e^{-|p|^2} |f_I(p)|^2 d\sigma(x,y).$$

We have the following result:

Proposition 3.7. *The quaternionic Fock space $\mathcal{F}(\mathbb{H})$ contains the monomials p^n, $n \in \mathbb{N}$ which form an orthogonal basis.*

Proof. Let us choose an imaginary unit $I \in \mathbb{S}$ and, for $n, m \in \mathbb{N}$, compute

$$\langle p^n, p^m \rangle = \int_{\mathbb{C}_I} e^{-|p|^2} \overline{p^m} p^n \, d\sigma(x,y).$$

By using polar coordinates, we write $p = \rho e^{I\theta}$ and we have

$$\begin{aligned}
\langle p^n, p^m \rangle &= \frac{1}{\pi} \int_0^{2\pi} \int_0^{+\infty} e^{-\rho^2} \rho^m e^{-Im\theta} \rho^n e^{In\theta} \rho \, d\rho \, d\theta \\
&= \frac{1}{\pi} \int_0^{2\pi} \int_0^{+\infty} e^{-\rho^2} \rho^{m+n+1} e^{I(n-m)\theta} d\rho \, d\theta \\
&= \frac{1}{2\pi} \int_0^{2\pi} e^{I(n-m)\theta} d\theta \int_0^{+\infty} e^{-\rho^2} \rho^{m+n} d\rho^2.
\end{aligned}$$

Since $\int_0^{2\pi} e^{I(n-m)\theta} d\theta$ vanishes for $n \neq m$ and equals 2π for $n = m$, we have $\langle p^n, p^m \rangle = 0$ for $n \neq m$. For $n = m$, standard computations give

$$\langle p^n, p^n \rangle = \int_0^{+\infty} e^{-\rho^2} \rho^{2n} d\rho^2 = n!.$$

Thus the monomials p^n belong to $\mathcal{F}(\mathbb{H})$ and any two of them are orthogonal. We now show that these monomials form a basis for $\mathcal{F}(\mathbb{H})$. A function $f \in \mathcal{F}(\mathbb{H})$ is entire so it admits series expansion of the form $f(p) = \sum_{m=0}^{+\infty} p^m a_m$ and thus the monomials p^n are generators. To show that they are independent, we show that if $\langle f, p^n \rangle = 0$ for all $n \in \mathbb{N}$ then f is identically zero. We have:

$$\begin{aligned}
\langle f, p^n \rangle &= \left\langle \sum_{m=0}^{+\infty} p^m a_m, p^n \right\rangle \\
&= \int_{\mathbb{C}_I} e^{-|p|^2} \overline{p^n} \left(\sum_{m=0}^{+\infty} p^m a_m \right) d\sigma(x,y)
\end{aligned}$$

and so

$$\begin{aligned}
\langle f, p^n \rangle &= \lim_{r \to +\infty} \int_{|p|<r,\, p \in \mathbb{C}_I} e^{-|p|^2} \overline{p^n} \left(\sum_{m=0}^{+\infty} p^m a_m \right) d\sigma(x,y) \\
&= \lim_{r \to +\infty} \sum_{m=0}^{+\infty} \left(\int_{|p|<r,\, p \in \mathbb{C}_I} e^{-|p|^2} \overline{p^n} p^m \, d\sigma(x,y) \right) a_m \\
&= \lim_{r \to +\infty} \int_0^r \rho^{2n} e^{-r^2} dr^2 a_n = n! a_n,
\end{aligned}$$

where we used the fact that the series expansion converges uniformly on $|p| < r$, thus we can exchange the series with the integration where needed. Thus $\langle f, p^n \rangle = 0$ for all n if and only if $a_n = 0$, i.e., $f \equiv 0$. □

Proposition 3.8. *The definition of inner product* (3.2) *does not depend on the imaginary unit* $I \in \mathbb{S}$.

Proof. Let $f(p) = \sum_{m=0}^{+\infty} p^m a_m$, $g(p) = \sum_{m=0}^{+\infty} p^m b_m \in \mathcal{F}(\mathbb{H})$ and let $I \in \mathbb{S}$. We have

$$\begin{aligned}\langle f, g\rangle &= \left\langle \sum_{m=0}^{+\infty} p^m a_m, \sum_{m=0}^{+\infty} p^m b_m \right\rangle \\ &= \int_{\mathbb{C}_I} e^{-|p|^2} \overline{\left(\sum_{m=0}^{+\infty} p^m a_m\right)} \left(\sum_{n=0}^{+\infty} p^n b_n\right) d\sigma(x,y) \\ &= \int_{\mathbb{C}_I} e^{-|p|^2} \left(\sum_{m=0}^{+\infty} \bar{a}_m \bar{p}^m\right) \left(\sum_{n=0}^{+\infty} p^n b_n\right) d\sigma(x,y)\end{aligned}$$

so that

$$\begin{aligned}\langle f, g\rangle &= \sum_{n=0}^{+\infty} \int_{\mathbb{C}_I} e^{-|p|^2} \bar{a}_n \bar{p}^n p^n b_n d\sigma(x,y) \\ &= \sum_{n=0}^{+\infty} \bar{a}_n \left(\int_{\mathbb{C}_I} e^{-|p|^2} \bar{p}^n p^n d\sigma(x,y)\right) b_n \\ &= \sum_{n=0}^{+\infty} n! \bar{a}_n b_n,\end{aligned}$$

which shows that the computation does not depend on the chosen imaginary unit I. □

Let us recall that the slice regular exponential function is defined by

$$e^p := \sum_{n=0}^{+\infty} \frac{p^n}{n!}.$$

We need to generalize the definition of the function $e^{zw} = \sum_{m=0}^{+\infty} \frac{(zw)^m}{m!}$, $z, w \in \mathbb{C}$ to the slice hyperholomorphic setting.

We first observe that if we set $e^{pq} = \sum_{m=0}^{+\infty} \frac{(pq)^m}{m!}$ then the function e^{pq} does not have any good property of regularity: it is not slice regular neither in p nor in q (while e^{zw} is holomorphic in both the variables). Let us consider p as a variable and q as a parameter and set:

$$e_\star^{pq} = \sum_{n=0}^{+\infty} \frac{(pq)^{\star n}}{n!} = \sum_{n=0}^{+\infty} \frac{p^n q^n}{n!} \tag{3.3}$$

where the $\star$-product (see [21]) is computed with respect to the variable p. It is immediate that $e_\star^{pq}$ is a function left slice regular in p and right regular in q.

Remark 3.9. The definition (3.3) is consistent with the fact that we are looking for a slice regular extension of e^{zw}. In fact, we start from the function $e^{zw} =$

$\sum_{n=0}^{+\infty} \frac{z^n w^n}{n!}$, which is holomorphic in z seen as an element on the complex plane $\mathbb{C}_I$; we then use the Representation Formula to get the extension to $\mathbb{H}$:

$$\text{ext}(e^{zw}) = \frac{1}{2}(1 - I_q I) \sum_{n=0}^{+\infty} \frac{z^n w^n}{n!} + \frac{1}{2}(1 + I_q I) \sum_{n=0}^{+\infty} \frac{\bar{z}^n w^n}{n!} = e^{qw}$$

and since w is arbitrary, we get the statement.

We now set $k_q(p) := e_\star^{p\bar{q}}$ and we discuss the reproducing property in the Fock space.

Theorem 3.10. *For every $f \in \mathcal{F}(\mathbb{H})$ we have*

$$\langle f, k_q \rangle = f(q).$$

Moreover, $\langle k_q, k_s \rangle = e_\star^{s\bar{q}}$.

Proof. We have

$$\begin{aligned}
\langle f, k_q \rangle &= \int_{\mathbb{C}_I} e^{-|p|^2} \overline{e_\star^{p\bar{q}}} f(p) d\sigma(x,y) \\
&= \int_{\mathbb{C}_I} e^{-|p|^2} \left(\sum_{n=0}^{+\infty} \frac{q^n \bar{p}^n}{n!} \right) \left(\sum_{m=0}^{+\infty} p^m a_m \right) d\sigma(x,y) \\
&= \sum_{n=0}^{+\infty} \sum_{m=0}^{+\infty} \frac{q^n}{n!} \langle p^m, p^n \rangle a_m \\
&= \sum_{n=0}^{+\infty} q^n a_n \\
&= f(q).
\end{aligned}$$

Similarly, we have

$$\begin{aligned}
\langle k_q, k_s \rangle &= \int_{\mathbb{C}_I} e^{-|p|^2} \overline{e_\star^{p\bar{s}}} e_\star^{p\bar{q}} d\sigma(x,y) \\
&= \int_{\mathbb{C}_I} e^{-|p|^2} \left(\sum_{n=0}^{+\infty} \frac{s^n \bar{p}^n}{n!} \right) \left(\sum_{m=0}^{+\infty} p^m \bar{q}^m \right) d\sigma(x,y) \\
&= \sum_{n=0}^{+\infty} \sum_{m=0}^{+\infty} \frac{s^n}{n!} \langle p^m, p^n \rangle \frac{\bar{q}^m}{m!} \\
&= \sum_{n=0}^{+\infty} \frac{s^n \bar{q}^n}{n!} \\
&= e_\star^{s\bar{q}}.
\end{aligned}$$

□

Proposition 3.11. *A function $f(p) = \sum_{m=0}^{+\infty} p^m a_m$ belongs to $\mathcal{F}(\mathbb{H})$ if and only if $\sum_{m=0}^{+\infty} |a_m|^2 m! < \infty$.*

Proof. Let us use polar coordinates; with computations similar to those in the proof of Proposition 3.8 and using the Parseval identity, we have

$$\begin{aligned}\int_{\mathbb{C}_I} e^{-|p|^2}|f(p)|^2 d\sigma(x,y) &= \lim_{r\to+\infty}\frac{1}{\pi}\int_0^r e^{-\rho^2}\int_0^{2\pi}|f(\rho e^{I\theta})|^2\rho\, d\theta\, d\rho \\ &= 2\lim_{r\to+\infty}\int_0^r e^{-\rho^2}\left(\sum_{m=0}^{+\infty}\rho^{2m}|a_m|^2\right)\rho\, d\rho \\ &= 2\lim_{r\to+\infty}\sum_{m=0}^{+\infty}\int_0^r e^{-\rho^2}\rho^{2m+1}|a_m|^2 d\rho \\ &= \sum_{m=0}^{+\infty}|a_m|^2 m!\end{aligned}$$

and the statement follows. □

4. Quaternion full Fock space and symmetric Fock space

Let V be a right vector space over $\mathbb{H}$. Recall that a quaternionic inner product on V is a map $\langle\cdot,\cdot\rangle : V\times V\to\mathbb{H}$ satisfying the same properties of a complex inner product, with the exception of the homogeneity requirement which is replaced by

$$\langle u\alpha, v\beta\rangle = \overline{\beta}\langle u,v\rangle\alpha,$$

and that if V is complete with respect to the norm induced by the inner product, it is called a right quaternionic Hilbert space. A similar definition can be given in the case of a quaternionic vector space on the left or two-sided.

Let $\mathcal{H}$ be a two-sided quaternionic Hilbert space. Then one may consider the quaternionic n-fold Hilbert space tensor power $\mathcal{H}^{\otimes n}$ defined by

$$\mathcal{H}^{\otimes n} = \mathcal{H}\otimes\mathcal{H}\otimes\cdots\otimes\mathcal{H}\quad (n \text{ times}),$$

where all tensor products are over $\mathbb{H}$.

Remark 4.1. A convenient way of constructing $\mathcal{H}^{\otimes n}$ is inductively. Recall that if M is a left R-module and N is a right R-module, then a tensor product of them $M_R\otimes{}_RN$ is an abelian group together with a bilinear map $\delta : M\times N\to M_R\otimes{}_RN$ which is universal in the sense that for any abelian group A and a bilinear map $f : M\times N\to A$, there is a unique group homomorphism $\tilde{f} : M_R\otimes{}_RN\to A$ such that $\tilde{f}\otimes\delta = f$. If furthermore, M is a right S-module and N is a left T-module, then ${}_SM_R\otimes{}_RN_T$ is a (S,T)-bi-module if one defines $sz = (s\otimes 1)z$ and $zt = z(1\otimes t)$ for $z\in {}_SM_R\otimes{}_RN_T$. Since it holds that

$$({}_RM_S\otimes{}_SN_T)\otimes{}_TP_U \cong {}_RM_S\otimes({}_SN_T\otimes{}_TP_U),$$

one can define inductively the tensor product of $M_1,\dots,M_n$, where M_i is a (R_{i-1},R_i)-bi-module, and obtain a (R_0,R_n)-bi-module,

$${}_{R_0}{M_1}_{R_1}\otimes{}_{R_1}{M_2}_{R_2}\otimes\cdots\otimes{}_{R_{n-1}}{M_n}_{R_n}.$$

For more details see [32, pp. 133–135]. One can also define it non-inductively (see [17, p. 264]).

We make the convention $\mathcal{H}^{\otimes 0} = \mathbb{H}$, and the element $1 \in \mathbb{H}$ is called the vacuum vector and denoted by $\mathbf{1}$. For the case of two Hilbert spaces in the next proposition, see also [11, equation (3)].

Proposition 4.2. *Let $\langle \cdot, \cdot \rangle$ be the inner product of $\mathcal{H}$, and assume that it satisfies also the additional property*

$$\langle u, \lambda v \rangle = \langle \overline{\lambda} u, v \rangle.$$

Then, it induces an inner product on $\mathcal{H}^{\otimes n}$,

$$\langle u_1 \otimes \cdots \otimes u_n, v_1 \otimes \cdots \otimes v_n \rangle = \langle \langle \cdots \langle \langle \langle u_1, v_1 \rangle u_2, v_2 \rangle u_3, v_3 \rangle \cdots \rangle u_n, v_n \rangle,$$

with the same additional property.

Proof. The statement clearly holds for $n = 1$. By induction,

$$\begin{aligned}\langle u_1 \otimes \cdots \otimes u_n \alpha, v_1 \otimes \cdots \otimes v_n \beta \rangle &= \langle \langle u_1 \otimes \cdots \otimes u_{n-1}, v_1 \otimes \cdots \otimes v_{n-1} \rangle u_n \alpha, v_n \beta \rangle \\ &= \overline{\beta} \langle \langle u_1 \otimes \cdots \otimes u_{n-1}, v_1 \otimes \cdots \otimes v_{n-1} \rangle u_n, v_n \rangle \alpha \\ &= \overline{\beta} \langle u_1 \otimes \cdots \otimes u_n, v_1 \otimes \cdots \otimes v_n \rangle \alpha,\end{aligned}$$

and

$$\begin{aligned}\overline{\langle v_1 \otimes \cdots \otimes v_n, u_1 \otimes \cdots \otimes u_n \rangle} &= \overline{\langle \langle v_1 \otimes \cdots \otimes v_{n-1}, u_1 \otimes \cdots \otimes u_{n-1} \rangle v_n, u_n \rangle} \\ &= \langle u_n, \langle v_1 \otimes \cdots \otimes v_{n-1}, u_1 \otimes \cdots \otimes u_{n-1} \rangle v_n \rangle \\ &= \langle \overline{\langle v_1 \otimes \cdots \otimes v_{n-1}, u_1 \otimes \cdots \otimes u_{n-1} \rangle} u_n, v_n \rangle \\ &= \langle \langle u_1 \otimes \cdots \otimes u_{n-1}, v_1 \otimes \cdots \otimes v_{n-1} \rangle u_n, v_n \rangle \\ &= \langle u_1 \otimes \cdots \otimes u_n, v_1 \otimes \cdots \otimes v_n \rangle.\end{aligned}$$

For the additional property, we obtain

$$\begin{aligned}\langle u_1 \otimes \cdots \otimes u_n, \lambda v_1 \otimes \cdots \otimes v_n \rangle &= \langle \langle u_1 \otimes \cdots \otimes u_{n-1}, \lambda v_1 \otimes \cdots \otimes v_{n-1} \rangle u_n, v_n \rangle \\ &= \langle \langle \overline{\lambda} u_1 \otimes \cdots \otimes u_{n-1}, v_1 \otimes \cdots \otimes v_{n-1} \rangle u_n, v_n \rangle \\ &= \langle \overline{\lambda} u_1 \otimes \cdots \otimes u_n, v_1 \otimes \cdots \otimes v_n \rangle.\end{aligned}$$

Additivity and positivity are obvious. □

Definition 4.3. The quaternionic full Fock module over a Hilbert space $\mathcal{H}$ is the space

$$\mathcal{F}(\mathcal{H}) = \oplus_{n=0}^{\infty} \mathcal{H}^{\otimes n},$$

with the corresponding inner product.

Definition 4.4. Let $u \in \mathcal{H}$. The right-linear map $T_u : \mathcal{F}(\mathcal{H}) \to \mathcal{F}(\mathcal{H})$ defined by

$$T_u(u_1 \otimes \cdots \otimes u_n) = u \otimes u_1 \otimes \cdots \otimes u_n,$$

is called the creation map. The right-linear map $T_u^* : \mathcal{F}(\mathcal{H}) \to \mathcal{F}(\mathcal{H})$ defined by

$$T_u^*(u_0 \otimes \cdots \otimes u_n) = \overline{\langle u, u_0 \rangle} u_1 \otimes \cdots \otimes u_n,$$

is called the annihilator map.

The following result is the quaternionic counterpart of a classical result:

Proposition 4.5. *T_u^* is the adjoint of T_u.*

Proof. The statement follows from

$$\begin{aligned}\langle T_u^*(u_0\otimes\cdots\otimes u_n), v_1\otimes\cdots\otimes v_n\rangle &= \langle \overline{\langle u,u_0\rangle} u_1\otimes\cdots\otimes u_n, v_1\otimes\cdots\otimes v_n\rangle\\ &= \langle \langle u_0,u\rangle u_1\otimes\cdots\otimes u_n, v_1\otimes\cdots\otimes v_n\rangle\\ &= \langle u_0\otimes\cdots\otimes u_n, u\otimes v_1\otimes\cdots\otimes v_n\rangle\\ &= \langle u_0\otimes\cdots\otimes u_n, T_u(v_1\otimes\cdots\otimes v_n)\rangle. \qquad \square\end{aligned}$$

Remark 4.6. Note that the isometry $u\mapsto T_u$ is both left-linear and right-linear.

The complex-valued version of the following proposition appears in [15, 16], where the free Brownian motion is defined and studied. The derivative of the function $X(t)$ is studied in [10].

Proposition 4.7 (The non-symmetric quaternionic Brownian motion). *Let $\mathcal{H} = \mathbf{L}_2(\mathbb{R}^+, dx)$, and consider $X(t) = T_{\mathbf{1}_{[0,t]}} + T^*_{\mathbf{1}_{[0,t]}}$. Then*

$$\langle X(t)\mathbf{1}, X(s)\mathbf{1}\rangle = \min\{t,s\}.$$

In particular $X(t)$ is self-adjoint, and if one considers the expectation

$$E : B(\mathcal{F}(\mathcal{H})) \to \mathbb{H}$$

defined by $E(T) = \langle T\mathbf{1},\mathbf{1}\rangle$, then

$$E(X(s)^*X(t)) = \min\{t,s\}.$$

Proof. More generally, note that

$$\langle (T_u+T_u^*)\mathbf{1}, (T_u+T_u^*)\mathbf{1}\rangle = \langle T_u\mathbf{1}, T_u\mathbf{1}\rangle = \langle u,u\rangle.$$

Since $\langle \mathbf{1}_{[0,t]}, \mathbf{1}_{[0,s]}\rangle_{\mathcal{H}} = \min\{t,s\}$, the result follows. $\square$

The symmetric product $\circ$ is defined by

$$u_1\circ\cdots\circ u_n = \frac{1}{n!}\sum_{\sigma\in S_n} u_{\sigma(1)}\otimes\cdots\otimes u_{\sigma(n)},$$

and the closed subspace of $\mathcal{H}^{\otimes n}$ generated by all vectors of this form is called the nth symmetric power of $\mathcal{H}$, and denoted by $\mathcal{H}^{\circ n}$.

Proposition 4.8. *Let $\langle\cdot,\cdot\rangle$ be the inner product of $\mathcal{H}$, and assume that it satisfies also the additional property*

$$\langle u,\lambda v\rangle = \langle \overline{\lambda}u, v\rangle.$$

Then, it induces an inner product on $\mathcal{H}^{\circ n}$

$$\begin{aligned}&\langle u_1\circ\cdots\circ u_n, v_1\circ\cdots\circ v_n\rangle\\ &\quad= \frac{1}{n!^2}\sum_{\sigma,\tau\in S_n}\langle\langle\cdots\langle\langle\langle u_{\sigma(1)}, v_{\tau(1)}\rangle u_{\sigma(2)}, v_{\tau(2)}\rangle u_{\sigma(3)}, v_{\tau(3)}\rangle\cdots\rangle u_{\sigma(n)}, v_{\tau(n)}\rangle,\end{aligned}$$

with the same additional property.

Proof. The result follows as in the proof of Proposition 4.2. □

In the classical case (where $\mathcal{H}$ is a Hilbert space over the field $\mathbb{R}$ or $\mathbb{C}$), another natural inner-product is usually being used, namely the symmetric inner product. It is defined by

$$\langle u_1 \circ \cdots \circ u_n, v_1 \circ \cdots \circ v_n \rangle = per\left(\langle u_i, v_j \rangle\right),$$

where $per(A)$ is called the permanent of A and has the same definition as a determinant, with the exception that the factor $sgn(\sigma)$ is omitted. An easy computation implies that when restricted to the n-fold symmetric tensor power $\mathcal{H}^{\circ n}$, the second inner product (i.e., the symmetric inner product) is simply $n!$ times the first inner product (the one which is defined in Proposition 4.8). This gives rise to the following definition.

Definition 4.9. Let $\langle \cdot, \cdot \rangle$ be the inner product of $\mathcal{H}$, and assume that it satisfies also the additional property

$$\langle u, \lambda v \rangle = \langle \overline{\lambda} u, v \rangle.$$

Then, the symmetric inner product on $\mathcal{H}^{\circ n}$ is defined by,

$$\begin{aligned} &\langle u_1 \circ \cdots \circ u_n, v_1 \circ \cdots \circ v_n \rangle \\ &\quad = \frac{1}{n!} \sum_{\sigma, \tau \in S_n} \langle \langle \cdots \langle \langle \langle u_{\sigma(1)}, v_{\tau(1)} \rangle u_{\sigma(2)}, v_{\tau(2)} \rangle u_{\sigma(3)}, v_{\tau(3)} \rangle \cdots \rangle u_{\sigma(n)}, v_{\tau(n)} \rangle. \end{aligned}$$

We now focus on the special case of the symmetric Fock space $\mathcal{F}^{\circ}(\mathcal{H})$ where p is a quaternion variable and $\mathcal{H} = p\mathbb{H}$. When no confusion can arise, we will simply denote it by $\mathcal{F}^{\circ}(\mathbb{H})$. The following result shows the relation with the Fock space as introduced in Definition 3.6, see Proposition 3.11.

Proposition 4.10. *$\mathcal{F}^{\circ}(\mathbb{H})$ is the space of all entire functions*

$$\sum_{n=0}^{\infty} p^n a_n$$

satisfying $\sum_{n=0}^{\infty} |a_n|^2 n! < \infty$, under an identification of $p^{\circ n}$ with p^n.

Proof. Clearly, any element in the nth level $\mathcal{H}^{\circ n}$ can be written as $p^{\circ n} a$ for some $a \in \mathbb{H}$, and

$$\langle p^{\circ n}, p^{\circ n} \rangle = \frac{1}{n!} \sum_{\sigma, \tau \in S_n} \langle \langle \cdots \langle \langle \langle p, p \rangle p, p \rangle p, p \rangle \cdots \rangle p, p \rangle = n!$$ □

5. The slice monogenic case

In this section we recall just the definition and some properties of slice monogenic functions and we show how the results obtained in Section 2 can be reformulated in this case. We work with the real Clifford algebra $\mathbb{R}_n$ over n imaginary units $e_1, \ldots, e_n$ satisfying the relations $e_i e_j + e_j e_i = -2\delta_{ij}$. An element in the Clifford algebra $\mathbb{R}_n$ is of the form $\sum_A e_A x_A$ where $A = i_1 \ldots i_r$, $i_\ell \in \{1, 2, \ldots, n\}$, $i_1 < \cdots < i_r$ is a multi-index, $e_A = e_{i_1} e_{i_2} \ldots e_{i_r}$ and $e_\emptyset = 1$. We set $|A| = i_1 + \cdots + i_r$ and we call k-vectors the elements of the form $\sum_{A, |A|=k} e_A x_A$, if $k > 0$. In the Clifford algebra $\mathbb{R}_n$, we can identify some specific elements with the vectors in the Euclidean space $\mathbb{R}^{n+1}$: an element $(x_0, x_1, \ldots, x_n) \in \mathbb{R}^{n+1}$ will be identified with the element $x = x_0 + \underline{x} = x_0 + \sum_{j=1}^n x_j e_j$ called, in short, paravector. The norm of $x \in \mathbb{R}^{n+1}$ is defined as $|x|^2 = x_0^2 + x_1^2 + \cdots + x_n^2$. The real part x_0 of x will be also denoted by $\mathrm{Re}(x)$. Using the above identification, a function $f : U \subseteq \mathbb{R}^{n+1} \to \mathbb{R}_n$ is seen as a function $f(x)$ of the paravector x. We will denote by $\mathbb{S}$ the $(n-1)$-dimensional sphere of unit 1-vectors in $\mathbb{R}^n$, i.e.,

$$\mathbb{S} = \{e_1 x_1 + \cdots + e_n x_n \ : \ x_1^2 + \cdots + x_n^2 = 1\}.$$

Note that to any nonreal paravector $x = x_0 + e_1 x_1 + \cdots + e_n x_n$ we can associate a $(n-1)$-dimensional sphere defined as the set, denoted by $[x]$, of elements of the form $x_0 + I|e_1 x_1 + \cdots + e_n x_n|$ when I varies in $\mathbb{S}$.
As it is well known, for $n \geq 3$ the Clifford algebra $\mathbb{R}_n$ contains zero divisors. Thus, in general, the result which hold in the quaternionic setting do not necessarily hold in Clifford algebra. For this reasons, we quickly revise the definitions and results given for the quaternionic Fock space. We omit the proofs since, as the reader may easily check, the proofs given in the quaternionic case are valid also in this setting. We begin by giving the definition of slice monogenic functions (see [21]).

Definition 5.1. Let $U \subseteq \mathbb{R}^{n+1}$ be an open set and let $f : \ U \to \mathbb{R}_n$ be a real differentiable function. Let $I \in \mathbb{S}$ and let f_I be the restriction of f to the complex plane $\mathbb{C}_I$. We say that f is a (left) slice monogenic function if for every $I \in \mathbb{S}$, we have

$$\frac{1}{2}\left(\frac{\partial}{\partial u} + I\frac{\partial}{\partial v}\right) f_I(u + Iv) = 0,$$

on $U \cap \mathbb{C}_I$. The set of (left) slice monogenic functions on U will be denoted by $\mathcal{SM}(U)$.

The slice monogenic Fock spaces and their properties are as follows.

Definition 5.2 (Slice hyperholomorphic Clifford–Fock space). Let I be any element in $\mathbb{S}$. Consider the set

$$\mathcal{F}(\mathbb{R}^{n+1}) = \left\{ f \in \mathcal{SM}(\mathbb{R}^{n+1}) \mid \int_{\mathbb{C}_I} e^{-|x|^2} |f_I(x)|^2 d\sigma(u, v) < \infty \right\}$$

where $x = u + Iv$, $d\sigma(u, v) := \frac{1}{\pi} du dv$. We will call $\mathcal{F}(\mathbb{R}^{n+1})$ (slice hyperholomorphic) Clifford–Fock space.

We endow $\mathcal{F}(\mathbb{R}^{n+1})$ with the inner product (which does not depend on the choice of the imaginary unit $I \in \mathbb{S}$):

$$\langle f, g \rangle := \int_{\mathbb{C}_I} e^{-|x|^2} \overline{g_I(x)} f_I(x) d\sigma(u, v). \tag{5.1}$$

Proposition 5.3. *The Clifford–Fock space $\mathcal{F}(\mathbb{R}^{n+1})$ contains the monomials x^m, $m \in \mathbb{N}$ which form an orthogonal basis, where x is a paravector in $\mathbb{R}^{n+1}$.*

Starting from the function $e^{zy} = \sum_{m=0}^{+\infty} \frac{z^m y^m}{m!}$, holomorphic in z that can be interpreted as an element on a complex plane $\mathbb{C}_I$ we can extend it to a slice monogenic function as

$$\mathrm{ext}(e^{zy}) = \frac{1}{2}(1 - I_x I) \sum_{m=0}^{+\infty} \frac{z^m y^m}{m!} + \frac{1}{2}(1 + I_x I) \sum_{m=0}^{+\infty} \frac{\bar{z}^m y^m}{m!} = e_\star^{xy}$$

and since y is arbitrary, we get the function we need. We now consider the function $k_y(x) := e_\star^{x\bar{y}}$ and we have the reproducing property in the Clifford–Fock space.

Theorem 5.4. *For every $f \in \mathcal{F}(\mathbb{R}^{n+1})$ we have*

$$\langle f, k_x \rangle = f(x).$$

Moreover, $\langle k_x, k_y \rangle = e_\star^{y\bar{x}}$.

Proposition 5.5. *A function $f(x) = \sum_{m=0}^{+\infty} x^m a_m$, $a_m \in \mathbb{R}_n$ for $m \in \mathbb{N}$, belongs to $\mathcal{F}(\mathbb{R}^{n+1})$ if and only if $\sum_{m=0}^{+\infty} |a_m|^2 m! < \infty$.*

In the case of modules over $\mathbb{R}_n$ the full Fock module is still under investigation.

References

[1] S. Adler, *Quaternionic Quantum Field Theory*, Oxford University Press, 1995.

[2] D. Alpay, *The Schur algorithm, reproducing kernel spaces and system theory*, American Mathematical Society, Providence, RI, 2001. Translated from the 1998 French original by Stephen S. Wilson, Panoramas et Synthèses.

[3] D. Alpay, V. Bolotnikov, F. Colombo, I. Sabadini, *Self-mappings of the quaternionic unit ball: multiplier properties, Schwarz–Pick inequality, and Nevanlinna–Pick interpolation problem*, to appear in Indiana Univ. Math. J., arXiv:1308.2658.

[4] D. Alpay, F. Colombo, I. Lewkowicz, I. Sabadini, *Realizations of slice hyperholomorphic generalized contractive and positive functions*, preprint 2013. Submitted.

[5] D. Alpay, F. Colombo, I. Sabadini, *Inner product spaces and Krein spaces in the quaternionic setting*, in *Recent advances in inverse scattering, Schur analysis and stochastic processes. A collection of papers dedicated to Lev Sakhnovich*, Operator Theory Advances and Applications. Linear Operators and Linear Systems, 2014.

[6] D. Alpay, F. Colombo, I. Sabadini, *Schur functions and their realizations in the slice hyperholomorphic setting*, Integral Equations Operator Theory, **72** (2012), 253–289.

[7] D. Alpay, F. Colombo, I. Sabadini, *Pontryagin De Branges Rovnyak spaces of slice hyperholomorphic functions*, Journal d'Analyse Mathématique vol. 121 (2013), no. 1, 87–125.

[8] D. Alpay, F. Colombo, I. Sabadini, *Krein–Langer factorization and related topics in the slice hyperholomorphic setting*, J. Geom. Anal., **24** (2014), no. 2, 843–872.

[9] D. Alpay, A. Dijksma, J. Rovnyak, and H. de Snoo, *Schur functions, operator colligations, and reproducing kernel Pontryagin spaces*, volume 96 of *Operator theory: Advances and Applications*. Birkhäuser Verlag, Basel, 1997.

[10] D. Alpay, P. Jorgensen, and G. Salomon. On free stochastic processes and their derivatives. *ArXiv e-prints*, November 2013.

[11] D. Alpay, H.T. Kaptanoglu, *Quaternionic Hilbert spaces and a von Neumann inequality*, Complex Var. Elliptic Equ., **57** (2012), 667–675.

[12] D. Alpay and G. Salomon. Non-commutative stochastic distributions and applications to linear systems theory. *Stochastic Process. Appl.*, 123(6):2303–2322, 2013.

[13] V. Bargmann. On a Hilbert space of analytic functions and an associated integral transform. *Comm. Pure Appl. Math.*, 14:187–214, 1961.

[14] V. Bargmann. Remarks on a Hilbert space of analytic functions. *Proceedings of the National Academy of Arts*, 48:199–204, 1962.

[15] M. Bożejko and E. Lytvynov. Meixner class of non-commutative generalized stochastic processes with freely independent values. I. A characterization. *Comm. Math. Phys.*, 292(1):99–129, 2009.

[16] M. Bożejko and E. Lytvynov. Meixner class of non-commutative generalized stochastic processes with freely independent values II. The generating function. *Comm. Math. Phys.*, 302(2):425–451, 2011.

[17] N. Bourbaki. *Algebra I. Chapters* 1–3. Translated from the French. Reprint of the 1989 English translation. Elements of Mathematics (Berlin). Springer-Verlag, Berlin, 1998.

[18] F. Colombo, J.O. Gonzalez-Cervantes, I. Sabadini, *On slice biregular functions and isomorphisms of Bergman spaces*, Complex Variables and Elliptic Equations, **57** (2012), 825–839.

[19] F. Colombo, J.O. Gonzalez-Cervantes, I. Sabadini, *The C-property for slice regular functions and applications to the Bergman space*, Complex Variables and Elliptic Equations, **58** (2013), 1355–1372.

[20] F. Colombo, J.O. Gonzalez-Cervantes, M.E. Luna-Elizarraras, I. Sabadini, M.V. Shapiro, *On two approaches to the Bergman theory for slice regular functions*, Advances in Hypercomplex Analysis; Springer INdAM Series 1, (2013), 39–54.

[21] F. Colombo, I. Sabadini, D.C. Struppa, *Noncommutative Functional Calculus. Theory and Applications of Slice Hyperholomorphic Functions*, Progress in Mathematics, Vol. 289, Birkhäuser, 2011.

[22] R. Delanghe, F. Sommen, V. Soucek, *Clifford algebra and spinor-valued functions. A function theory for the Dirac operator*, Mathematics and its Applications, 53. Kluwer Academic Publishers Group, Dordrecht, 1992.

[23] N. Dunford, J. Schwartz, *Linear operators, part I: general theory* , J. Wiley and Sons (1988).

[24] V. Fock. Verallgemeinerung und Lösung der Diracschen statistischen Gleichung, *Zeitschrift für Physik.* 49: 339–357, 1928.

[25] F. Hiai and D. Petz. *The semicircle law, free random variables and entropy*, volume 77 of *Mathematical Surveys and Monographs*. American Mathematical Society, Providence, RI, 2000.

[26] T. Hida, H. Kuo, J. Potthoff, and L. Streit. *White noise*, volume 253 of *Mathematics and its Applications.* Kluwer Academic Publishers Group, Dordrecht, 1993. An infinite-dimensional calculus.

[27] T. Hida and Si Si. *Lectures on white noise functionals.* World Scientific Publishing Co. Pte. Ltd., Hackensack, NJ, 2008.

[28] T. Hida and M. Hitsuda. *Gaussian processes*, volume 120 of *Translations of Mathematical Monographs.* American Mathematical Society, Providence, RI, 1993. Translated from the 1976 Japanese original by the authors.

[29] H. Holden, B. Øksendal, J. Ubøe, and T. Zhang. *Stochastic partial differential equations.* Probability and its Applications. Birkhäuser Boston Inc., Boston, MA, 1996.

[30] L.P. Horwitz and L.C. Biedenharn. Quaternion quantum mechanics: second quantization and gauge fields. *Ann. Physics*, 157(2):432–488, 1984.

[31] L.P. Horwitz and A. Razon. Tensor product of quaternion Hilbert modules. In *Classical and quantum systems (Goslar,* 1991), pages 266–268. World Sci. Publishing, River Edge, NJ, 1993.

[32] N. Jacobson. *Basic algebra. II.* Second edition. W.H. Freeman and Company, New York, 1989.

[33] I.E. Segal. Tensor algebras over Hilbert spaces. I. *Trans. Amer. Math. Soc.*, 81:106–134, 1956.

[34] I.E. Segal. Tensor algebras over Hilbert spaces. II. *Ann. of Math.* (2), 63:160–175, 1956.

[35] D. Voiculescu. Aspects of free probability. In *XIVth International Congress on Mathematical Physics*, pages 145–157. World Sci. Publ., Hackensack, NJ, 2005.

[36] D.V. Voiculescu, K.J. Dykema, and A. Nica. *Free random variables*, volume 1 of *CRM Monograph Series.* American Mathematical Society, Providence, RI, 1992. A noncommutative probability approach to free products with applications to random matrices, operator algebras and harmonic analysis on free groups.

[37] Zhi-yuan Huang and Jia-an Yan. *Introduction to infinite dimensional stochastic analysis*, volume 502 of *Mathematics and its Applications.* Kluwer Academic Publishers, Dordrecht, Chinese edition, 2000.

Daniel Alpay and Guy Salomon
Department of Mathematics
Ben-Gurion University of the Negev
Beer-Sheva 84105 Israel
e-mail: `dany@math.bgu.ac.il`
`guysal@math.bgu.ac.il`

Fabrizio Colombo and Irene Sabadini
(FC) Politecnico di Milano
Dipartimento di Matematica
Via E. Bonardi, 9
I-20133 Milano, Italy
e-mail: `fabrizio.colombo@polimi.it`
`irene.sabadini@polimi.it`

Hypercomplex Analysis: New Perspectives and Applications
Trends in Mathematics, 61–73

Multi Mq-monogenic Function in Different Dimension

Eusebio Ariza and Antonio Di Teodoro

Abstract. A metamonogenic of first-order function or simply metamonogenic function is a function that satisfies the differential equation $(D-\lambda)u=0$, where D is the Cauchy–Riemann operator and λ can be real or Clifford-valued constant (see [4]). Using this definition we can say that a multi-metamonogenic function u is separately metamonogenic in several variables $x^{(j)}$, $j=1,\dots,n$ with $n\geq 2$, if $x^{(j)}=(x_1^{(j)},\dots,x_{m_j}^{(j)})$ runs in the Euclidean space $\mathbb{R}^{m_j}$ and $(D_j-\lambda)u=0$, for each $j=1,\dots,n$,, where D_j is the corresponding Cauchy–Riemann operator in the space $\mathbb{R}^{m_j}$. Using the theory of algebras of Clifford type depending on parameters (see [11, 12]), the present proposal discusses the properties of u in case the dimensions m_j are different from each other for multi Mq-monogenic functions, following the ideas exhibited in [9, 10].

Mathematics Subject Classification (2010). 30A05; 15A66; 30G35.

Keywords. Monogenic function, metamonogenic function, multi-metamonogenic function, multi Mq-monogenic functions, Clifford algebras, Clifford type depending on parameters.

1. Introduction

In [2] are defined the n-order meta-analytic functions as the solutions of the equation $(\partial_{\bar{z}}-\lambda)^n u=0$. Replacing the operator $\partial_{\bar{z}}$ by D, a new set of functions are introduced which we will call n-order meta-monogenic functions. See [1, 4]. In this sense, we can see the meta-monogenic functions as a generalization of meta-analytic functions. For $n=1$, a continuously differentiable Clifford algebra-valued function, u, is said to be meta-monogenic of first order if it satisfies the equation $(D-\lambda)u=0$. See [1, 4, 14]. We will denote the operator $D-\lambda$ by D_λ.
If we replace the Cauchy–Riemann operator, D, by the more general modified q-Cauchy–Riemann operator, $D_q=\sum_{i=0}^{n} q_i e_i \partial_i$, for q_i real-valued functions defined in

$\mathbb{R}^{n+1}$, in the meta of first-order operator, we have a more general operator called *Meta-q of first-order operator* or simply *Mq* operator:

$$D_{q,\lambda} := D_q - \lambda = \sum_{i=0}^{n} q_i e_i \partial_i - \lambda, \tag{1.1}$$

where λ can be real or Clifford valued. Solutions of the equation $D_{q,\lambda}u = 0$ are called *Mq-monogenic functions.* If $\lambda = \sum_{i=0}^{n} \lambda_i e_i$ and $\lambda_i \in \mathbb{R}$ for each $i = 0, \dots, n$, the operator (1.1) can be rewritten as:

$$D_{q,\lambda} = D_q - \lambda = (q_0\partial_0 - \lambda_0)e_0 + (q_1\partial_1 - \lambda_1)e_1 + \cdots + (q_n\partial_n - \lambda_n)e_n. \tag{1.2}$$

This kind of operator is useful in physics (quantum mechanics and electromagnetism) and was studied recently by Kravchenko in [7] in the context of quaternionic analysis and is called the electromagnetic Dirac operator.

Example. When $n = 2$ and the q_i, $i = 0, 1, 2$ are constants or real-valued nonzero functions, the Mq-monogenic system or Mq system, $D_{q,\lambda}u = 0$ for a Clifford-algebra-valued function

$$\begin{aligned} u(x_0, x_1, x_2) = u_0(x_0, x_1, x_2) &+ u_1(x_0, x_1, x_2)e_1 \\ &+ u_1(x_0, x_1, x_2)e_2 + u_{12}(x_0, x_1, x_2)e_{12} \end{aligned}$$

gives

$$\begin{aligned} (q_0\partial_0 - \lambda_0)u_0 - (q_1\partial_1 - \lambda_1)u_1 - (q_2\partial_2 - \lambda_2)u_2 &= 0, \\ (q_1\partial_1 - \lambda_1)u_0 + (q_0\partial_0 - \lambda_0)u_1 + (q_2\partial_2 - \lambda_2)u_{12} &= 0, \\ (q_2\partial_2 - \lambda_2)u_0 + (q_0\partial_0 - \lambda_0)u_2 - (q_1\partial_1 - \lambda_1)u_{12} &= 0, \\ -(q_2\partial_2 - \lambda_2)u_1 + (q_1\partial_1 - \lambda_1)u_2 + (q_0\partial_0 - \lambda_0)u_{12} &= 0. \end{aligned}$$

The theory of multi-monogenic functions generalizes the theory of holomorphic functions in several complex variables to the case of monogenic functions. In the holomorphic case the desired function is complex valued. In the monogenic case it is necessary to choose a suitable space in which the values of the function are running.

Consider the function u depending on n variables

$$x^{(j)} = (x_0^{(j)}, \dots, x_{m_j}^{(j)}), \quad j = 1, \dots, n.$$

Thus, u is defined in the (real) Euclidean space

$$\mathbf{R} = \mathbb{R}^{m_1+1} \times \mathbb{R}^{m_2+1} \times \cdots \times \mathbb{R}^{m_n+1}$$

whose dimension is equal to

$$\sum_{j=1}^{n} m_j + n.$$

A function u is a *Multi-Monogenic Function* in several variables $x^{(j)}$, for $j = 1, \ldots, n$, if

$$D_j u = \left(\partial_{x_0^{(j)}} + \sum_{i=1}^{m_j} e_i \partial_i^{(j)} \right) u = 0, \tag{1.3}$$

where $\partial_i^{(j)} = \frac{\partial}{\partial x_i^{(j)}}$, and D_j is the Cauchy–Riemann operator in $\mathbb{R}^{m_j}$.

Similarly, a function u is *Multi Mq-monogenic* in several variables $x^{(j)}$, for $j = 1, \ldots, n$, if

$$D_{j,q,\lambda} u = 0, \tag{1.4}$$

where

$$D_{j,q,\lambda} := q_0 \partial_{x_0^{(j)}} + \sum_{i=1}^{m_j} q_i e_i \partial_i^{(j)} - \lambda,$$

$q = (q_0,\ q_1, \ldots, q_{m_j})$ and q_i are real valued functions defined in $\mathbb{R}^{m_j}$ and λ can be real or Clifford valued.

If $n = 1$ these functions are the Mq-monogenic functions.

Remark 1.1. Since $D_{j,1,0} = D_j$, we can say that the multi Mq-monogenic functions are a generalization of the multi-monogenic functions in the same way the Mq-monogenic functions are a generalization of the monogenic functions.

When we consider $n \geq 2$, in the simplest case we have the same dimension $m_j = m + 1$ for all the Euclidean spaces $\mathbb{R}^{m_j}$, and so it is possible to assume that the desired function u has values belonging to the Clifford algebra $\mathcal{A}_m$ which is the usual extension of $\mathbb{R}^{m+1}$. Since the dimension of the $\mathcal{A}_m$ equals 2^m, the Mq-monogenic equations consist of 2^m real equations for the 2^m real-valued components of u.

If, however, the numbers m_j associated to the variables $x^{(j)}$ are different from each other, then $m + 1$ can be chosen by

$$m + 1 \geq \max_j m_j \tag{1.5}$$

Thus the number 2^m of real-valued components of u is greater than the number $2^{m_j - 1}$ of real-value components of Mq-monogenic functions in $\mathbb{R}^{m_j}$ for those j for which $m > m_j$.

In the specific (1.5) case a theory of multi-monogenic functions (Cauchy's Integral Formula, Hartog's extension theorem, Cousin problem and so on) can be found in [6] as an extension of the works [5, 8] to the case of holomorphic functions.

On the other hand in [10] Tutschke and Hung Son discuss a theory of multi-monogenic functions in the case that the dimension 2^m of the corresponding algebra of Clifford type depending on parameters will be defined by

$$m + 1 = \sum_{j=1}^{n} m_j.$$

Finally in [9] using the theory of algebras of Clifford type depending on parameters, the authors consider the same real part for all of the factor spaces $\mathbb{R}^{m_1+1}$. The corresponding algebraic structure of Clifford type belongs to the product

$$\mathbb{R} \times \mathbb{R}^{m_1} \times \mathbb{R}^{m_2} \times \cdots \times \mathbb{R}^{m_n}.$$

Following the same ideas presented [9, 10], in this article we develop a theory of multi Mq-monogenic functions in case the dimensions m_j are different from each other using an algebraic structure of Clifford type depending on parameters.

2. Separately holomorphic and monogenic functions

Definition 2.1. A **separately holomorphic function** is a continuous complex-valued function which depends holomorphically on a finite number of complex variables.

Definition 2.2. A **separately monogenic function** takes its values in the 2^m-dimensional Clifford algebra, whereas it depends o a finite number of independent variables all of which run in the **same** $m+1$-dimensional Euclidean space $\mathbb{R}^{m+1}$.

In case the independent variables of a separately monogenic function run in **Euclidean space of different dimensions**, then usually one assumes that the values belong to the Clifford algebra with *maximal dimension* 2^m.

This has the disadvantage that the function has too much components for a variable for which m is not maximal.

Example. Consider the function u holomorphic in the (x_0, x_1)-plane (u must have two components, real and imaginary parts) and monogenic in the (y_0, y_1, y_2)-space (u must have four components in $\mathbb{R}^3$). Then the desired function u must have at least four components depending on $x = (x_0, x_1)$ and $y = (y_0, y_1, y_2)$, that is

$$u(x,y) = u_0(x,y) + u_1(x,y)e_1 + u_2(x,y)e_2 + u_{12}(x,y)e_{12}$$

where $e_1^2 = e_2^2 = -1$ and $e_1e_2 + e_2e_1 = 0$.

In case u is a (left-)monogenic function in the y-space, the Cauchy–Riemann system

$$D_y u = \sum_{i=0}^{2} e_i \partial_{y_i} u = 0,$$

has to be satisfied.

Since at the same time u is holomorphic in the x-plane, if we put $\partial_i = \partial_{x_i}$ for $i = 0, 1$, then

$$D_x u = (\partial_0 + e_1\partial_1)u = 0.$$

Splitting up $D_x u$ in its components, one gets:

$$\partial_0 u_0 - \partial_1 u_1 = 0, \quad \partial_0 u_1 + \partial_1 u_0 = 0,$$
$$\partial_0 u_2 - \partial_1 u_{12} = 0, \quad \partial_0 u_{12} + \partial_1 u_2 = 0.$$

In view of this, the function u, having four components, is holomorphic in the x-plane if and only if the two functions

$$u_0 + u_1 e_1 \quad \text{and} \quad u_2 e_2 + u_{12} e_{12}$$

are holomorphic in the x-plane. Moreover, other possible combinations for u are as follows.
Considering the operator

$$\partial_0 + e_2 \partial_1,$$

then the holomorphy of u is equivalent to the holomorphy of

$$u_0 + u_2 e_2 \quad \text{and} \quad u_1 e_1 + u_{12} e_{12}.$$

Finally, the holomorphy of u, considering the operator

$$\partial_0 + e_{12} \partial_1$$

is equivalent to the holomorphy of

$$u_0 + u_{12} e_{12} \quad \text{and} \quad u_1 e_1 + u_2 e_2.$$

This example shows that, if we have the case that the number of real components is smaller as the maximal values (which determines the dimension of the Clifford algebra), then there exist different possibilities for the choice of the Cauchy–Riemann operator to an independent variable.

In order to overcome these differences of the influence of the independent variables, for each independent variable is assigned its own range.

This means that if the variable $x = (x_0, \dots, x_m)$ is $(m+1)$-dimensional, then (for fixed other independent variables) the range of the function is 2^m-dimensional Clifford algebra defined by $\mathbb{R}^{m+1}$.

3. Clifford-algebra-valued functions in several variables

If the basis vector of $\mathbb{R}^{m+1}$ is $\{e_0, \dots, e_m\}$, then

$$D = \partial_0 + \sum_{i=1}^{m} e_i \partial_i$$

is the uniquely defined Cauchy–Riemann operator which corresponds to the $(m+1)$-dimensional variable $x = (x_0, \dots, x_m)$, with $\partial_i = \partial_{x_i}$ for $i = 0, 1, \dots, m$.

Now consider the function u depending on n variables

$$x^{(j)} = (x_0^{(j)}, \dots, x_{m_j}^{(j)}) \quad j = 1, \dots, n.$$

Then, u is defined in the (real) Euclidean space

$$\mathbf{R} = \mathbb{R}^{m_1+1} \times \mathbb{R}^{m_2+1} \times \cdots \times \mathbb{R}^{m_n+1}$$

with dimension equals

$$\sum_{j=1}^{n} m_j + n.$$

Thus we have the following possibilities:

(a) One possibility is to introduce, for each partial space $\mathbb{R}^{m_j+1}$, an own real part and m_j imaginary units.

Let $e_1, \ldots, e_{m_1}$ be the m_1 imaginary unit vectors of $\mathbb{R}^{m_1}$. Analogously, the m_j unit vectors of $\mathbb{R}^{m_j}$ are denoted by

$$e_i, \; i = m_1 + \cdots + m_{j-1} + 1, \ldots, \; m_1 + \cdots + m_{j-1} + m_j.$$

In this case the Cauchy–Riemann operator acting in $\mathbb{R}^{m_j+1}$ is giving by

$$\mathcal{D}_j = \partial_{x_0^{(j)}} + \sum_{i=m_1+\cdots+m_{j-1}+1}^{m_1+\cdots+m_j} e_i \partial_{x_i} \tag{3.1}$$

where $x_{m_1+\cdots+m_{j-1}+1} = x_1^{(j)}, \ldots, x_{m_j}^{(j)} = x_{m_1+\cdots+m_{j-1}+m_j}$.

(b) Another simpler possibility is, for instance, to use the same real part for all of the factor spaces $\mathbb{R}^{m_1+1}$. In other words, the corresponding algebraic structure of Clifford type belongs to the product

$$\mathbb{R} \times \mathbb{R}^{m_1} \times \mathbb{R}^{m_2} \times \cdots \times \mathbb{R}^{m_n}.$$

In this case, the Cauchy–Riemann operator acting in $\mathbb{R}^{m_j+1}$ is given by

$$\mathfrak{D}_j = \partial_{x_0} + \sum_{i=m_1+\cdots+m_{j-1}+1}^{m_1+\cdots+m_j-1} e_i \partial_{x_i} \tag{3.2}$$

for $j = 1, \ldots, n$, where ∂_0 correspond to the derivative with respect to the common real part x_0, $m_0 = 0$ and

$$x_{m_1+\cdots+m_{j-1}+1} = x_1^{(j)}, \ldots, x_{m_1+\cdots+m_{j-1}+m_j} = x_{m_j}^{(j)}.$$

Remark 3.1. Note that the derivatives with respect to the variables x_i where $i = m_1, m_1 + m_2, \ldots, m_1 + m_2 + \cdots + m_n$ does not appear in the operator $\mathfrak{D}_j$.

Note, additionally, that in the operator D_j we have the unit vectors to the corresponding Clifford algebra $\mathcal{A}_{m_j}$. On the other hand, if we consider the operators given by (3.1) and (3.2), the unit vectors are those of the Clifford algebra $\mathcal{A}_m$, where m depends on the m_j for $j = 1, \ldots, n$. See Sections 4 and 6.

In order to multiply vectors, it is necessary to introduce an algebraic structure of Clifford type.

4. Associated algebra of Clifford type 1

Clifford algebras over $\mathbb{R}^{m+1}$ can be constructed as equivalence classes in the ring $R[X_1, \ldots, X_m]$ of polynomials in m variables $X_1, \ldots, X_m$ with real coefficients, where two polynomials are said to be equivalent if their difference is a polynomial for which each term contains at least one of the factors

$$X_j^2 + 1 \quad \text{and} \quad X_i X_j + X_j X_i \, , \tag{4.1}$$

where $i, j = 1, \dots, m$ and $i \neq j$. Denoting X_j by e_j, $j = 1, \dots, m$, one obtains the usual Clifford algebra $\mathcal{A}_m$, which extends the Euclidean space $\mathbb{R}^{m+1}$ whose basis is $e_0 = 1$, $e_1, \dots, e_m$. The structure polynomials (4.1) yield the well-known rules $e_j^2 = -1$ and $e_i e_j + e_j e_i = 0$ for its basis elements, $i \neq j$. See [3].

In 2008 the authors of [11] introduced the Clifford algebras depending on parameters in order to describe more general systems of partial differential equations and this is possible in the framework of classical Clifford analysis. These algebras can be obtained if the structure polynomials (4.1) are replaced by

$$X_j^{k_j} + \alpha_j \text{ and } X_i X_j + X_j X_i - 2\gamma_{ij}, \tag{4.2}$$

where $i, j = 1, \dots, n$, $i \neq j$, and the $k_j \geq 2$ are natural numbers. The parameters α_j and $\gamma_{ij} = \gamma_{ji}$ have to be real and may depend also on further variables such as the variable x in $\mathbb{R}^{n+1}$. If the parameters do not depend on further variables and if $n \geq 3$, the Clifford type algebra generated by the structure polynomials (4.2) is denoted by $\mathcal{A}_n(k_j, \alpha_j, \gamma_{ij})$. For $n = 1$ we write $\mathcal{A}_1(k, \alpha)$. If in $\mathcal{A}_n(k_j, \alpha_j, \gamma_{ij})$ all of the k_j are equal to 2 and α_j and γ_{ij} are constant we denote this algebra by $A_{n,2}^*$. This algebra has the dimension 2^n. We denoted by $\mathcal{A}_n$ the classical Clifford algebra $\mathcal{A}_n(2, 1, 0)$.

In this part we will develop a theory of multi Mq-monogenic functions in the case that the dimension 2^m of the corresponding algebra of Clifford type will be defined by

$$m + 1 = \sum_{j=1}^{n} m_j \tag{4.3}$$

and we will consider $\mathbb{R}^{m+1}$.

Remark 4.1. The choice of m in this way has the advantage that the dimensions m_j of the n given Euclidean spaces $\mathbb{R}^{m_j}$ have the same influence on the choice of the dimension m. In other terms, no space $\mathbb{R}^{m_j}$ is preferred in comparison with the other spaces.

Denote the basis vectors of

$$\mathbb{R}^{m_1} \times \mathbb{R}^{m_2} \times \cdots \times \mathbb{R}^{m_n}.$$

by

$$\{e_0 = 1, \dots, e_{m_1 - 1}; e_{m_1}, \dots, e_{m_1 + m_2 - 1}; \dots, e_{m - m_n}, \dots, e_m\}.$$

Since, in this case (4.3), the uniform use of the n Euclidean spaces $\mathbb{R}^{m_j}$ requires the introduction of n real axes, the construction of the related algebra has to be modified. This construction can be realized in the framework of algebras of Clifford type depending on parameters. This algebra allows us to repeat the arguments of case (1.5) also in case (4.3).

In order to multiply vectors, we introduce the following algebraic structure of Clifford type. Consider the (non-commutative) ring of polynomials in $X_1, \dots, X_m$.

Let j be one of the indices

$$j = m_1, m_1 + m_2, \dots, m_1 + \cdots + m_{n-1},$$

whereas k and l are indices (between 1 and m) which are different from these $n-1$ indices j. Then the algebra $\mathcal{A}_m(\sigma_1)$ is defined by the structure relations

$$\begin{aligned} e_j^2 &= 1, & e_k^2 &= -1, \\ e_k e_j &= e_j e_k, & e_k e_l &= -e_l e_k. \end{aligned} \tag{4.4}$$

The $\mathfrak{D}_{j,q,\lambda}$ operator in the space $\mathbb{R}^{m_j}$ is given by

$$\mathfrak{D}_{j,q,\lambda} = \mathfrak{D}_{j,q} - \lambda = q_0 \partial_0 + \sum_{i=m_1+\cdots+m_{j-1}+1}^{m_1+\cdots+m_j-1} q_i e_i \partial_{x_i} - \lambda \tag{4.5}$$

for $j = 1, \dots, n$, where ∂_0 correspond to the derivative with respect to the common real part x_0, $m_0 = 0$ and

$$x_{m_1+\cdots+m_{j-1}+1} = x_1^{(j)}, \dots, x_{m_j}^{(j)} = x_{m_1+\cdots+m_{j-1}+m_j}.$$

Note that, like for the operator $\mathfrak{D}_j$, the derivatives with respect to the variables x_i where

$$i = m_1,\ m_1 + m_2, \dots,\ m_1 + m_2 + \cdots + m_n,$$

does not appear in the operator $\mathfrak{D}_{j,q}$. This corresponds to the derivatives with respect to

$$x_0^{(1)}, \dots,\ x_0^{(n)}.$$

A solution of the system $\mathfrak{D}_{j,q,\lambda} u = 0$ is said to be Mq-monogenic in the space $\mathbb{R}^{m_j}$.

4.1. Decomposition of the q-Cauchy–Riemann system

Let $\mathfrak{D}_{j,q}$ the q-Cauchy–Riemann operator (3.2) of the $x^{(j)}$-space, for $j = 1, \dots, n$. Since $\mathcal{A}_m$ has dimension 2^m, the q-Cauchy–Riemann system $\mathfrak{D}_{j,q} u = 0$ can be decomposed into 2^m real equations for the 2^m real-valued components of u. On the other hand, the differential operator $\mathfrak{D}_{j,q}$ contains m_j differentiations. Note, additionally, that the usual Clifford algebra associated to $\mathbb{R}^{m_j}$ has the dimension 2^{m_j-1}. In consequence, we can decompose $\mathfrak{D}_{j,q} u = 0$ into 2^{m-m_j+1} groups of Cauchy–Riemann equations in $\mathbb{R}^{m_j}$ for 2^{m_j-1} desired real-valued components at a time.

5. Example 1

Consider a function u depending on five real variables x_0, x_1, y_0, y_1, y_2 which is Mq-holomorphic in the $x = (x_0, x_1)$-plane but at the same time a Mq-monogenic function in the $y = (y_0, y_1, y_2)$-space. Since

$$m + 1 = 5 = m_1 + m_2,$$

we have $m = 4$, whereas the corresponding algebra of Clifford type is $\mathcal{A}_4(\sigma_1)$.

Thus,

$$\mathbb{R}^{4+1} = \mathbb{R}^2 \times \mathbb{R}^3,$$

which implies

$$j = 2 \quad \text{and} \quad k,\ l \in \{1, 3, 4\}.$$

The structure relations for the five basis elements $\{e_0 = 1, e_1, e_2, e_3, e_4\}$ are given by

$$\begin{aligned}
&e_1^2 = -1, \qquad && e_2^2 = 1, \qquad && e_3^2 = -1, \quad e_4^2 = -1,\\
&e_1e_2 = e_2e_1, && e_2e_3 = e_3e_2, && e_2e_4 = e_4e_2,\\
&e_1e_3 = -e_3e_1, && e_1e_4 = -e_4e_1, && e_3e_4 = -e_4e_3.
\end{aligned}$$

Since $m = 4$, the desired function u has 16 real-valued components

$$\begin{aligned}
u = u_0 &+ u_1e_1 + u_2e_2 + u_3e_3 + u_4e_4\\
&+ u_{12}e_{12} + u_{13}e_{13} + u_{14}e_{14} + u_{23}e_{23} + u_{24}e_{24} + u_{34}e_{34}\\
&+ u_{123}e_{123} + u_{124}e_{124} + u_{134}e_{134} + u_{234}e_{234} + u_{1234}e_{1234}.
\end{aligned}$$

Let

$$\mathfrak{D}_{x,q,\lambda} = \mathfrak{D}_{1,q,\lambda} = \mathfrak{D}_{1,q} - \lambda = (q_0\partial_0 + q_1e_1\partial_1) - \lambda$$

be the Mq-monogenic operator in the (x_0, x_1)-plane. The Mq-monogenic equations $D_{x,q,\lambda}u = 0$ implies the following 8 equations for the real components of u:

$$\begin{aligned}
q_0\partial_0u_0 - q_1\partial_1u_1 - \lambda u_0 &= 0, & q_1\partial_0u_1 + q_0\partial_1u_0 - \lambda u_1 &= 0,\\
q_0\partial_0u_2 - q_1\partial_1u_{12} - \lambda u_2 &= 0, & q_0\partial_0u_{12} + q_1\partial_1u_2 - \lambda u_{12} &= 0,\\
q_0\partial_0u_3 - q_1\partial_1u_{13} - \lambda u_3 &= 0, & q_0\partial_0u_{13} + q_1\partial_1u_3 - \lambda u_{13} &= 0,\\
q_0\partial_0u_4 - q_1\partial_1u_{14} - \lambda u_4 &= 0, & q_0\partial_0u_{14} + q_1\partial_1u_4 - \lambda u_{14} &= 0,\\
q_0\partial_0u_{23} - q_1\partial_1u_{123} - \lambda u_{23} &= 0, & q_0\partial_0u_{123} + q_1\partial_1u_{23} - \lambda u_{123} &= 0,\\
q_0\partial_0u_{24} - q_1\partial_1u_{124} - \lambda u_{24} &= 0, & q_0\partial_0u_{124} + q_1\partial_1u_{24} - \lambda u_{124} &= 0,\\
q_0\partial_0u_{34} - q_1\partial_1u_{134} - \lambda u_{34} &= 0, & q_0\partial_0u_{134} + q_1\partial_1u_{34} - \lambda u_{134} &= 0,\\
q_0\partial_0u_{234} - q_1\partial_1u_{1234} - \lambda u_{234} &= 0, & q_0\partial_0u_{1234} + q_1\partial_1u_{234} - \lambda u_{1234} &= 0.
\end{aligned}$$

Therefore the following eight functions turn out to be Mq-holomorphic in the x_0, x_1-plane

$$\begin{aligned}
&u_0 + u_1e_1, && u_2 + u_{12}e_1, && u_3 + u_{13}e_1, && u_4 + u_{14}e_1,\\
&u_{23} + u_{123}e_1, && u_{24} + u_{124}e_1, && u_{34} + u_{134}e_1, && u_{234} + u_{1234}e_1.
\end{aligned}$$

This can also be seen by the following decomposition of the function u:

$$\begin{aligned}
u = (u_0 + u_1e_1) &+ (u_2 + u_{12}e_1)e_2 + (u_3 + u_{13}e_1)e_3 + (u_{23} + u_{123}e_1)e_2e_3\\
&+ (u_{24} + u_{124}e_1)e_2e_4 + (u_{34} + u_{134}e_1)e_3e_4 + (u_{234} + u_{1234}e_1)e_2e_3e_4.
\end{aligned}$$

Analogously, the Mq-monogenic equation in the (y_0, y_1, y_2)-space is

$$(\mathfrak{D}_{2,q} - \lambda)u = (q_0\partial_0 + q_3e_3\partial_{y_1} + q_4e_4\partial_{y_2} - \lambda)u = 0.$$

This shows that the four functions

$$\begin{aligned}u_0 + u_3e_3 + u_4e_4 + u_{34}e_{34},\\ u_1e_1 + u_{13}e_{13} + u_{14}e_{14} + u_{134}e_{134},\\ u_2e_2 + u_{23}e_{23} + u_{24}e_{24} + u_{234}e_{234},\\ u_{12}e_{12} + u_{123}e_{123} + u_{124}e_{124} + u_{1234}e_{1234}\end{aligned}$$

are Mq-monogenic in the (y_0, y_1, y_2)-space.

This can be seen by the following decomposition of the function u:

$$\begin{aligned}u = u_0 &+ u_3e_3 + u_4e_4 + u_{34}e_{34}\\ &+ e_1\left(u_1 + u_{13}e_3 + u_{14}e_4 + u_{134}e_{34}\right)\\ &+ e_2\left(u_2 + u_{23}e_3 + u_{24}e_4 + u_{234}e_{34}\right)\\ &+ e_{12}\left(u_{12} + u_{123}e_3 + u_{124}e_4 + u_{1234}e_{34}\right).\end{aligned}$$

6. Associated algebra of Clifford type 2

Consider the real Euclidean space $\mathbf{R}$ given by

$$\mathbf{R} = \mathbb{R}^{m_1+1} \times \mathbb{R}^{m_2+1} \times \cdots \times \mathbb{R}^{m_n+1}.$$

Define m by $m = m_1 + \cdots + m_n$ and introduce the algebraic structure of Clifford type defined by

$$\begin{aligned}e_i^2 &= -1 && \text{for each } i = 1, \ldots, m,\\ e_ke_l &= -e_le_k && \text{if } k \text{ and } l \text{ belong to the same space } \mathbb{R}^{m_j},\\ e_ke_l &= e_le_k && \text{if } k \text{ and } l \text{ belong to different spaces } \mathbb{R}^{m_j}.\end{aligned}\tag{6.1}$$

The structure relations (6.1) define the algebras of Clifford type depending on parameters $\mathcal{A}_m(\sigma_2)$.

Let $\mathcal{D}_{j,q}$ be the q-Cauchy–Riemann operator acting in $\mathbb{R}^{m_j+1}$ given by (3.1). That is,

$$\mathcal{D}_{j,q} = q_0\partial_{x_0^{(j)}} + \sum_{i=m_1+\cdots+m_{j-1}+1}^{m_1+\cdots+m_j} q_ie_i\partial_{x_i}$$

where $x_{m_1+\cdots+m_{j-1}+1} = x_1^{(j)}, \ldots, x_{m_j}^{(j)} = x_{m_1+\cdots+m_{j-1}+m_j}$.

In this case, the $\mathcal{D}_{j,q,\lambda}$ operator acting in $\mathbb{R}^{m_j+1}$ is given by $\mathcal{D}_{j,q,\lambda} = \mathcal{D}_{j,q} - \lambda$, where λ can be real or Clifford valued. A solution of the equation $\mathcal{D}_{j,q,\lambda}u = 0$ is said to be Mq-monogenic in the $\mathbb{R}^{m_j+1}$ space.

Then we can define a new class of multi Mq-monogenic functions by the solutions of the equations $\mathcal{D}_{j,q,\lambda}u = (\mathcal{D}_{j,q} - \lambda) = 0$, where the function u is a $\mathcal{A}_m$-valued function defined in $\mathbf{R}$.

Note that the commutativity of the unit vectors corresponding to different spaces implies that

$$\mathcal{D}_{j_1,q}\mathcal{D}_{j_2,q} = \mathcal{D}_{j_2,q}\mathcal{D}_{j_1,q} \quad \text{if} \quad j_1 \neq j_2. \tag{6.2}$$

Since the q-Cauchy–Riemann system in the subspace $\mathbb{R}^{m_j}$ consists of 2^{m_j} equations and the function u has $2m$ real components, the q-Cauchy–Riemann system in the subspace $\mathbb{R}^{m_j}$ can be decomposed in 2^{m-m_j} subsystems.

7. Example 2

Consider a function Mq-holomorphic in the (x_0, x_1)-plane and, at the same time, Mq-monogenic in the (y_0, y_1, y_2)-space (with respect to the operator $\mathcal{D}_{j,q,\lambda} = \mathcal{D}_{j,q} - \lambda$). We assume that the values of u belong to an algebraic structure whose basis vectors belong to

$$\mathbb{R} \times \mathbb{R}^1 \times \mathbb{R}^2.$$

The dimension is equal

$$m = m_1 + m_2 = 1 + 2.$$

The basis is

$$\{e_0, e_1, e_2, e_3, e_{12}, e_{13}, e_{23}, e_{123}\},$$

thus u has $2^3 = 8$ real components:

$$u = u_0 e_1 + u_2 e_2 + u_3 e_3 + u_{12} e_{12} + u_{13} e_{13} + u_{23} e_{23} + u_{123} e_{123}.$$

The corresponding algebra of Clifford type is $\mathcal{A}_3(\sigma_2)$, thus

$$\mathbb{R}^3 = \mathbb{R}^1 \times \mathbb{R}^2, \quad j = 1, \quad k \in \{2, 3\}.$$
$$e_1^2 = 1,\ e_2^2 = -1,\ e_3^2 = -1,$$
$$e_1 e_2 = e_2 e_1,\ e_1 e_3 = e_3 e_1,$$
$$e_2 e_3 = -e_3 e_2.$$

The Mq-holomorphic system $\mathcal{D}_{x,q,\lambda} u = \mathcal{D}_{1,q,\lambda} u = (q_0 \partial_0 + q_1 e_1 \partial_1 - \lambda) u = 0$ in the $x = (x_0, x_1)$-plane leads to the four Mq-holomorphic functions

$$u_0 + u_1 e_1, \quad u_2 + u_{12} e_1, \quad u_3 + u_{13} e_1, \quad u_{23} + u_{123} e_1.$$

This can also be seen by the following decomposition of the function u:

$$u = (u_0 + u_1 e_1) + (u_2 + u_{12} e_1) e_2 + (u_3 + u_{13} e_1) e_3 + (u_{23} + u_{123} e_1) e_2 e_3.$$

Analogously, the Mq-monogenic system in the (y_0, y_1, y_2)-space

$$\mathcal{D}_{2,q,\lambda} u = (q_0 \partial_{y_0} + q_1 e_2 \partial_{y_1} + q_2 e_3 \partial_{y_2} - \lambda) u = 0$$

show that the following two

$$u_0 + u_2 e_2 + u_3 e_3 + u_{23} e_{23},$$
$$u_1 e_1 + u_{12} e_{12} + u_{13} e_{13} + u_{123} e_{123}$$

are Mq-monogenics.

This corresponds to the following decomposition of the function u:

$$u = (u_0 + u_2 e_2 + u_3 e_3 + u_{23} e_{23}) + e_1 (u_1 + u_{12} e_2 + u_{13} e_3 + u_{123} e_{23}).$$

8. Definition of separately Mq-monogenic functions

Definition 8.1. A separately Mq-monogenic function takes its values in the 2^m-dimensional Clifford algebra defined by $(\mathcal{A}_m(\sigma_1))$ or $(\mathcal{A}_m(\sigma_2))$, whereas it depends on a finite number of independent variables all of with run in the **same** $m+1$-dimensional Euclidean space $\mathbb{R}^{m+1}$.

Definition 8.2. A Multi Mq-monogenic function u in Ω takes its values in the algebraic structure $\mathcal{A}_m(\sigma_1)$ or $\mathcal{A}_m(\sigma_2)$ of Clifford type where the structure relations are given by (4.4) (respectively, (6.1)).

Remark 8.3. The above definition of multi Mq-monogenic functions includes the case that u is separately Mq-holomorphic with respect to one or several variables $x^{(j)}$. This is the case if $m_j = 1$. In case $m_j = 1$ for each $j = 1, \ldots, n$, then one gets a new class of separately Mq-holomorphic functions in several complex variables.

9. Conclusions

In this article, we use the theory of algebras of Clifford type depending on parameters [11, 12] to construct suitable structure relations so that the separately *Meta-q of firstorder monogenic function or Mq-monogenic function* has the correct number of components with respect to each separate variables following the ideas of the papers [9, 10].

As a natural extension of this article, may be considered more general concepts of algebras of Clifford type. See [13]. Also the article leads to representations of the type of the Cauchy Integral Formula for the new classes of multi Mq-monogenic functions. Using power-series representations of the new Cauchy kernel, one also obtains power-series representations for multi Mq-monogenic functions in the new sense.

Acknowledgment

The authors would like to express their sincere gratitude to Professor Wolfgang Tutschke for his suggestions in this investigation.

References

[1] Balderrama, C., Di Teodoro, A. and Infante, A., *Some Integral Representation for Meta-Monogenic Function in Clifford Algebras Depending on Parameters*, Adv. Appl. Clifford Algebras, 23, 4, 793–813, (2013).

[2] Balk, M., *Polyanalytic functions.* Berlin: Akademie Verlag, 1991.

[3] Brackx, F. Delanghe, R and Sommen, F., *Clifford Analysis.* Pitman Research Notes, 1982.

[4] Di Teodoro, A. and Vanegas, C., *Fundamental Solutions for the First-order Meta-Monogenic Operator*, Adv. Appl. Clifford Algebras, 22, 1, 49–58, (2012).

[5] Hörmander, L., *An Introduction to Complex Analysis in Several Variables*, North-Holland Mathematical Library, 3rd ed., vol. 7, North-Holland Publishing Co., Amsterdam, 1990.

[6] Hung Son, L., *Recent Trends in Theory of Regular Functions Taking Value in Clifford Algebra*, Proceedings of the International Conference of Applied Mathematics (ICAM), Hanoi, August 25–29, 2004, SAS International Publications, Delhi, pp. 113–122, 2004.

[7] Kravchenko, V., *Applied Quaternionic Analysis*, Heldermann Verlag, 2003.

[8] Shabat, D.V., *Introduction to Complex Analysis, Part II, Functions of several variables.* (Translated from the third (1985) Russian edition by J.S. Joel), Translations of Mathematical Monographs, vol. 110. American Mathematical Society. Providence, RI, 1992.

[9] Tutschke, W. and Hung Son, L., A New Concept of Separately Holomorphic and Separately Monogenic Functions. Algebraic structures in partial differential equations related to complex and Clifford analysis, Ho Chi Minh City Univ. Educ. Press, Ho Chi Minh City, 67–78, 2010.

[10] Tutschke, W. & Hung Son, L., *Multi-monogenic functions in different dimensions.* Complex Variables and Elliptic Equations: An International Journal, vol. 58, 2, 293–298, 2013.

[11] Tutschke, W. and Vanegas, C.J., *Clifford algebras depending on parameters and their applications to partial differential equations.* Contained in Some topics on value distribution and differentiability in complex and p-adic analysis. Science Press Beijing, 430–449, 2008.

[12] Tutschke, W. and Vanegas, C.J., *Métodos del análisis complejo en dimensiones superiores.* Ediciones IVIC, Caracas 2008.

[13] Tutschke, W. and Vanegas, C.J., *General algebraic structures of Clifford type and Cauchy–Pompeiu formulae for some piecewise constant structure relations*, Adv. Appl. Clifford Algebras, 21, 4, 829–838, (2011).

[14] Xu Zhenyuan, *A function theory for the operator* D_λ, Complex Variables, Theory and Application: An International Journal, 16, 1, 27–42, (1991)

Eusebio Ariza
Departamento de matemáticas
Universidad Simón Bolívar
Valle de Sartenejas
Caracas Venezuela
e-mail: euariza@usb.ve

Antonio Di Teodoro
School of Mathematics Yachay Tech
Yachay City of Knowledge
100119-Urcuquí, Ecuador
e-mail: aditeodoro@yachay.gob.ec

Hypercomplex Analysis: New Perspectives and Applications
Trends in Mathematics, 75–88

The Fractional Monogenic Signal

Swanhild Bernstein

Abstract. The monogenic signal is a well-known generalization of the analytic signal. The advantage of such models consists in the fact that they have more parameters to characterize a signal. In this paper we study two generalizations in $\mathbb{R}^4$. Firstly, the fractional Riesz transform and secondly the fractional monogenic signal. The Riesz transform is a generalization of the Hilbert transform and builds up the monogenic signal of a scalar-valued function f. This type of fractional Riesz transform is based on rotations in $\mathbb{R}^4$ and not on a fractional Fourier transform. Rotations in $\mathbb{R}^4$ can by described by quaternions and therefore the algebraic structure we will use are quaternions.

Mathematics Subject Classification (2010). Primary 30G35; Secondary 44A15.

Keywords. Quaternions, Riesz transform, Hilbert transform, fractional Riesz transform, monogenic signals, fractional Hilbert transform.

1. Introduction

The analytic signal proposed by D. Gabor [7] is a mathematical method to construct a unique complex signal associated with a given real signal. The analytic signal can be characterized by the amplitude, phase and frequency, whereas the real signal gives just a real number at each point. Driven by applications in optics in [9] two generalizations of the Hilbert transform where proposed. In [4] the fractional Hilbert transform was employed for image processing and specifically for edge detection. Later in [3] three different generalizations of Gabor's analytic signal were constructed, all of which reduce to the analytic signal when the angle is set to be $\frac{\pi}{2}$. It was also demonstrated in this work that the fractional Hilbert transform has the semigroup property, unlike the generalized Hilbert transform in [18].

The monogenic signal proposed in [6] generalizes the analytic signal into higher dimensions based on Clifford analysis. This approach is different from the higher-dimensional analytic signal [8]. The monogenic signal has been proven to be useful in image processing too (see for example [2]). The aim of this paper is to construct a version of a fractional Riesz transform generalizing the fractional

Hilbert transform and a version of a fractional monogenic signal generalizing the fractional analytic signal in $\mathbb{R}^4$ which is embedded into the quaternions. The reason for that is that we use a rotation based approach as it is done in [16] for the analytic signal, fractional Hilbert transform and fractional analytic signal, respectively. Indeed, in the first place, it is not easy to use a fractional Fourier transform in higher dimensions and, in the second place, quaternions and rotations do fit very well together. We will also prove some simple properties of our constructions. We don't get a semigroup property because our construction does not depend solely on angles but also on axes.

The paper is organized as follows. After this introduction in Section 2 we introduce real and complex quaternions, rotations in $\mathbb{R}^3$ and $\mathbb{R}^4$, quaternionic analysis based on some kind of Dirac operator and important properties of monogenic functions, Hardy spaces and the projections onto these spaces. In Section 3 we explain the construction of the analytic signal and the fractional analytic signal as done in [16]. In Section 4 we construct a fractional Riesz operator and finally in Section 5 fractional monogenic signals. Section 6 gives some concluding remarks regarding the 2D case (images), and the situation in $\mathbb{R}^3$.

2. Preliminaries

2.1. Quaternions

2.1.1. Real quaternions. Let $\mathbb{H}$ denote the skew-field of quaternions with basis $\{\mathbf{1}, \mathbf{i}, \mathbf{j}, \mathbf{k}\}$. An arbitrary element $q \in \mathbb{H}$ is given by

$$q = q_0\mathbf{1} + q_1\mathbf{i} + q_2\mathbf{j} + q_3\mathbf{k},$$

where $q_0, q_1, q_2, q_3 \in \mathbb{R}$. Addition and multiplication by a real scalar defined in component-wise fashion. The multiplication distributes over addition; $\mathbf{1}$ is the multiplication identity; and

$$\mathbf{i}^2 = \mathbf{j}^2 = \mathbf{k}^2 = -\mathbf{1},$$
$$\mathbf{ij} = -\mathbf{ji} = \mathbf{k}, \quad \mathbf{jk} = -\mathbf{kj} = \mathbf{i}, \quad \mathbf{ki} = -\mathbf{ik} = \mathbf{j}.$$

We decompose a quaternion into two parts that are called its scalar and vector parts. If $q = q_0\mathbf{1} + q_1\mathbf{i} + q_2\mathbf{j} + q_3\mathbf{k}$, we write

$$q = q_0 + \underline{q} = \mathrm{Sc}\,(q) + Vec\,(q),$$

where $q_0 = q_0\mathbf{1}$ is the scalar part and $\underline{q} = q_1\mathbf{i} + q_2\mathbf{j} + q_3\mathbf{k}$ the vector part of q. On $\mathbb{H}$ a conjugation is defined as $\overline{q} = q_0\mathbf{1} - \underline{q}$ for all $q \in \mathbb{H}$. The length or norm $|q|$ of $q \in \mathbb{H}$ is then

$$|q| = \sqrt{q_0^2 + q_1^2 + q_2^2 + q_3^2} = \sqrt{q\overline{q}} = \sqrt{\overline{q}q}.$$

All $q \in \mathbb{H}\backslash\{0\}$ are invertible with inverse

$$q^{-1} = \frac{\overline{q}}{|q|}.$$

The product of two quaternions $p, q \in \mathbb{H}$ can also be described in terms of the usual vector products of vectors in $\mathbb{R}^3$. We have

$$pq = p_0 q_0 + q_0 \underline{p} + p_0 \underline{q} + \underline{pq}$$

and

$$\begin{aligned}\underline{pq} &= -\sum_{i=1}^{3} p_i q_i + (p_1 q_2 - p_2 q_1)\mathbf{k} + (p_2 q_3 - p_3 q_2)\mathbf{i} + (p_3 q_1 - p_1 q_3)\mathbf{j} \\ &= -\langle \underline{p}, \underline{q} \rangle + \underline{p} \times \underline{q},\end{aligned}$$

where $\langle \cdot, \cdot \rangle$ denotes the scalar product and $\times$ the vector product of vectors in $\mathbb{R}^3$.

2.1.2. Complex quaternions. We won't need much of the theory of complex quaternions. But because we are using the Fourier transform and will work in Fourier domain, we have to say something about complex quaternions. A complex quaternion is given by

$$q = q_0 \mathbf{1} + q_1 \mathbf{i} + q_2 \mathbf{j} + q_3 \mathbf{k}, \quad q_0, q_1, q_2, q_3 \in \mathbb{C}.$$

The main difference to the real case is that not all elements $\neq 0$ are invertible. There are so-called zero-divisors. A simple example are $1 + i\mathbf{k}$ and $1 - i\mathbf{k}$. It turns out that $(1 + i\mathbf{k})(1 - i\mathbf{k}) = 1 - i^2\mathbf{k}^2 = 1 - 1 = 0$. We will see that zero divisors are important for our considerations and won't create any problems.

2.2. Rotations

We identify $\mathbb{H}$ with the four-dimensional Euclidean space $\mathbb{R}^4$ by associating $q \in \mathbb{H}$ with the vector $(q_0, q_1, q_2, q_3) \in \mathbb{R}^4$.

If $q \in \mathbb{H}$ has $|q| = 1$, then we call q a unit quaternion. If $\underline{u}$ is a pure unit quaternion then $\underline{u}^{-1} = -\underline{u}$. Two pure unit quaternions $\underline{u}$ and $\underline{v}$ are orthogonal if and only if

$$\langle \underline{u}, \underline{v} \rangle = -\frac{1}{2}\left(\underline{u}\,\underline{v} + \underline{v}\,\underline{u}\right) = 0.$$

If q is a unit quaternion there is a real number φ and a pure unit quaternion $\underline{u}$ such that

$$q = \mathbf{1}\cos\varphi + \underline{u}\sin\varphi.$$

Since $\underline{u}^2 = -1$, the power series expansion leads to

$$e^{\underline{u}\varphi} = \sum_{n=0}^{\infty} \frac{(\underline{u}\varphi)^n}{n!} = \mathbf{1}\cos\varphi + \underline{u}\sin\varphi,$$

providing equivalent representations for a unit quaternion

$$q = q_0 + \underline{q} = \mathbf{1}\cos\varphi + \underline{u}\sin\varphi = e^{\underline{u}\varphi}.$$

It should be mentioned that neither $\underline{u}$ nor φ are uniquely determined. When $q \neq \pm\mathbf{1}$ then $\sin\varphi = \pm|\underline{q}|$ and $\underline{u} = \pm\frac{\underline{q}}{|\underline{q}|}$. When $q = \pm\mathbf{1}$, $\underline{u}$ can be any pure unit quaternion.

Let q be a quaternion. Then $L_q : \mathbb{H} \to \mathbb{H}$ and $R_q : \mathbb{H} \to \mathbb{H}$ are defined as follows

$$L_q(x) = qx, \quad R_q(x) = xq, \quad x \in \mathbb{H}.$$

If q is a unit quaternion, then both L_q and R_q are orthogonal transformations of $\mathbb{H}$. Thus, for unit quaternions p and q, the mapping $C_{p,q} : \mathbb{H} \to \mathbb{H}$ defined by

$$C_{p,q} = L_p \circ R_q = R_q \circ L_p$$

is also an orthogonal transform of $\mathbb{H}$ and $C_{p_1,q_1} \circ C_{p_2,q_2} = C_{p_1p_2,q_2q_1}$.

Theorem 2.1 ([17] and [10]). *If q is a unit quaternion, then there exist a pure unit quaternion $\underline{u}$ and a real scalar φ such that $q = e^{\underline{u}\varphi}$. The transform $C : \mathbb{R}^3 \to \mathbb{R}^3$ defined by $C(x) = qx\overline{q}$ is a rotation in the plane orthogonal to $\underline{u}$ through an angle 2φ.*

Theorem 2.2 ([17] and [10]). *Let $p = e^{\underline{u}\varphi}$ and $q = e^{\underline{v}\psi}$, where $\underline{u}$ and $\underline{v}$ are pure unit quaternions. The orthogonal transform $C_{p,q}$ of $\mathbb{H}$ is a product of two rotations in orthogonal planes. If $\underline{u} \neq \pm\underline{v}$, then $C_{p,q}$ rotates the plane spanned by $\underline{u}+\underline{v}$ and $\underline{u}\underline{v}-\mathbf{1}$ through the angle $|\varphi+\psi|$ and the plane spanned by $\underline{v}-\underline{u}$ and $\underline{u}\underline{v}+\mathbf{1}$ through the angle $|\varphi-\psi|$. If $\underline{u} = \pm\underline{v}$, then the invariant planes are the span of $\mathbf{1}$ and $\underline{u}$ and its orthogonal complement, and the rotation angles in appropriates planes are still $|\varphi+\psi|$ and $|\varphi-\psi|$.*

2.3. Quaternionic analysis

The quaternionic analysis presented in this section is based on the nice presentation of Clifford analysis in [5].

2.3.1. Dirac operator. The Dirac operator is defined as the first-order linear differential operator

$$D_x = \mathbf{1}\partial_{x_0} + \mathbf{i}\partial_{x_1} + \mathbf{j}\partial_{x_2} + \mathbf{k}\partial_{x_3}.$$

Definition 2.3 (Monogenic functions). Let $\Omega \subseteq \mathbb{R}^4$ be open and let f be a C^1-function in Ω which is (real or complex) quaternionic-valued. Then f is *left monogenic* or (*right monogenic*) in Ω if in Ω

$$D_x f = 0 \quad \text{or} \quad f D_x = 0.$$

The connection between monogenic and harmonic functions is due to the fact that $\Delta_x = -\partial_x^2 = \overline{D}_x D_x$.

2.3.2. Integral formulae. Let $\Omega \subset \mathbb{R}^4$ be open, let G be a compact orientable four-dimensional manifold with boundary ∂G.

Theorem 2.4 (Stokes formula). *Suppose that $f, g \in C^1(\Omega)$. Then for each $G \subset \Omega$,*

$$\int_{\partial G} f(x)n(x)g(x)\, dS(x) = \int_G [(fD_x)g + f(D_x g)]\, dx,$$

where $n(x)$ stands for the outwardly pointing unit normal and $dS(x)$ being the surface measure.

Because the Dirac operator D_x is a first-order linear differential operator with constant coefficients there exists a fundamental solution.

Lemma 2.5. *A fundamental solution is given by*

$$E(x) = \frac{1}{2\pi^2}\frac{\overline{x}}{|x|^4} = \frac{1}{2\pi^2}\frac{(x_0 - x_1\mathbf{i} - x_2\mathbf{j} - x_3\mathbf{k})}{|x|^4}.$$

E has the following properties.

1. *E is* $\mathbb{H} \sim \mathbb{R}^4$*-valued and belongs to* $L^1_{\text{loc}}(\mathbb{R}^4)$.
2. *E is left and right monogenic in* $\mathbb{R}^4\backslash\{0\}$
3. *and* $\lim_{|x|\to\infty} E(x) = 0$.
4. $D_x E = E D_x = \delta(x)$, $\delta(x)$ *being the classical* δ*-function in* $\mathbb{R}^4$.

2.3.3. Hardy spaces. Let $\Sigma \in \mathbb{R}^4$ be a graph of a Lipschitz function $g : \mathbb{R}^3 \to \mathbb{R}$ and let $\Omega^{\pm}$ be the domains in $\mathbb{R}^4$ which lie above, respectively, below Σ, $n(y)$ is the a.e. on Σ defined outward unit normal at $y \in \Sigma$ and $dS(y)$ the elementary surface element on Σ.

Definition 2.6 (Hardy spaces). Let $1 < p < \infty$. Then

$$H^p(\Omega^{\pm}) = \left\{ F \text{ monogenic in } \Omega^{\pm} : \sup_{\varepsilon > 0} \int_{\Sigma} |F(y \pm \varepsilon)|^p \, dS(y) < \infty \right\}$$

is the Hardy space of (left) monogenic functions in $\Omega^{\pm}$.

Definition 2.7 (Integral operators). For $f \in L^p(\Sigma)$ and $x \in \mathbb{R}^4\backslash\Sigma$,

$$\mathcal{C}_{\Sigma} f(x) = \int_{\Sigma} E(x-y) n(y) f(y) \, dS(y)$$

is the Cauchy transform of f.
For $f \in L^p(\Sigma)$ and a.e. $x \in \Sigma$,

$$\begin{aligned}\mathcal{R}_{\Sigma} f(x) &= 2p.v. \int_{\Sigma} E(x-y) n(y) f(y) \, dS(y) \\ &= 2 \lim_{\varepsilon \to 0+} \int_{y \in \Sigma : |x-y| > \varepsilon} E(x-y) n(y) f(y) \, dS(y)\end{aligned}$$

is the Riesz transform (or Hilbert transform) of f.

Remark 2.8. The transform $\mathcal{R}$ is in Clifford analysis and function theory usually called Hilbert transform. In signal processing this transform is called Riesz transform to distinguish between the higher-dimensional analytic signal, which is related to analytic functions in $\mathbb{C}^m$, and the monogenic signal which is built by the Riesz operators and they form the Clifford–Riesz transform. (The Riesz operators are different from the Riesz potentials!)

Theorem 2.9. *Let* $f \in L^p(\Sigma)$, $1 < p < \infty$. *Then*

1. $\mathcal{C}_{\Sigma} f \in H^p(\Omega^{\pm})$.
2. $\mathcal{C}_{\Sigma} f$ *has non-tangetial limits* $(\mathcal{C}^{\pm}_{\Sigma})$ *at almost all* $x^* \in \Sigma$.

3. *Putting*
$$\mathbb{P}_\Sigma^+ f(x^*) = (\mathcal{C}_\Sigma f)^+(x^*) \quad \textit{and} \quad \mathbb{P}_\Sigma^- f(x^*) = -(\mathcal{C}_\Sigma f)^-(x^*)$$
then $\mathbb{P}_\Sigma^\pm$ *are bounded projections in* $L^p(\Sigma)$.
4. *(Plemelj–Sokhotzki formulae). For a.e.* $x^* \in \Sigma$,
$$\mathbb{P}_\Sigma^+ f(x^*) = \tfrac{1}{2}\left(f(x^*) + \mathcal{R}_\Sigma f(x^*)\right) \quad \textit{and} \quad \mathbb{P}_\Sigma^- f(x^*) = \tfrac{1}{2}\left(f(x^*) - \mathcal{R}_\Sigma f(x^*)\right)$$
whence
$$\mathbf{1} = \mathbb{P}_\Sigma^+ + \mathbb{P}_\Sigma^- \quad \textit{and} \quad \mathcal{R}_\Sigma = \mathbb{P}_\Sigma^+ - \mathbb{P}_\Sigma^-.$$
In particular $\mathcal{R}_\Sigma$ *is a bounded linear operator on* $L^p(\Sigma)$ *and, putting* $L^{p,\pm}(\Sigma) = \mathbb{P}^\pm L^p(\Sigma)$, *leads to the decomposition into Hardy spaces*
$$L^p(\Sigma) = L^{p,+}(\Sigma) \oplus L^{p,-}(\Sigma).$$

A special situations occurs for $\Sigma = \mathbb{R}^3$. In this case if $\mathcal{C}$ denotes the Cauchy integral and $\mathcal{R}$ the Riesz transform, the boundary values of monogenic functions in Fourier domain can be characterized by the Fourier transform
$$\mathcal{F}f(\underline{\xi}) = \frac{1}{\sqrt{2\pi}^3} \int_{\mathbb{R}^3} e^{i\langle \underline{x}, \underline{\xi}\rangle} f(\underline{x})\, d\underline{x}.$$
Using that
$$\mathcal{F}\left(\frac{2x_j}{2\pi^2 |\underline{x}|^4}\right) = -i\frac{\xi_j}{|\underline{\xi}|}, \quad j = 1, \dots, 3,$$
and set [11]
$$\chi_\pm(\underline{\xi}) = \tfrac{1}{2}\left(1 \pm i\frac{\underline{\xi}}{|\underline{\xi}|}\right),$$
it should be noticed that
$$\chi_\pm^2(\underline{\xi}) = \chi_\pm(\underline{\xi}) \quad \text{and} \quad \chi_+(\underline{\xi}) + \chi_-(\underline{\xi}) = 1, \quad \chi_+(\underline{\xi})\chi_-(\underline{\xi}) = \chi_-(\underline{\xi})\chi_+(\underline{\xi}) = 0.$$
Which means that $\chi_\pm$ are projections and zero divisors. The boundary values of monogenic function in upper half-space are characterized in the next theorem.

Theorem 2.10 ([11]). *For* $f \in L^p(\mathbb{R}^m)$ *the following statements are equivalent:*
1. *The non-tangential limit of* $\mathcal{C}_\Sigma f$ *is a.e. equal to* f,
2. $\mathcal{R}f = f$,
3. $\mathcal{F}f = \chi_+ \mathcal{F}f$,

and characterizes boundary values of monogenic functions.

Specifically for $f \in L^p(\mathbb{R}^3)$ the function $f + \mathcal{R}f$ satisfies $\mathcal{R}(f + \mathcal{R}f) = \mathcal{R}f + \mathcal{R}^2 f = f + \mathcal{R}f$, i.e., $f + \mathcal{R}f \in L^{p,+}$.

The monogenic signal f_M based on the real-valued signal f is defined as
$$f_M := f + \mathcal{R}f$$
and can be described as a Fourier-integral operator with symbol
$$2\chi_+(\underline{\xi}) = \left(1 + \frac{i\underline{\xi}}{|\underline{\xi}|}\right), \quad \text{i.e.,} \quad f_M(\underline{x}) = \mathcal{F}^{-1}\left(\left(1 + \frac{i\underline{\xi}}{|\underline{\xi}|}\right)\hat{f}(\underline{\xi})\right)$$

3. The analytic signal

3.1. Hilbert transform

For a real-valued function $f(t)$, its Hilbert transform is defined by

$$(\mathcal{H}f)(t) = \text{p.v.}\,\frac{1}{\pi}\int_{-\infty}^{\infty}\frac{f(\tau)}{t-\tau}\,d\tau = (h * f)(t),$$

where $h(t) = \frac{1}{\pi t}$ is the Hilbert convolution kernel and p.v. denotes the Cauchy principal value integral. The Fourier transform of $h(t)$ is $\hat{h}(\underline{\xi}) = -i\,\mathrm{sgn}(\xi)$, where sign(.) denotes the signum function. The analytic signal by Gabor is now given by boundary values of an analytic function in the upper half-space. It is easily seen that

$$\chi_{\pm}(\underline{\xi}) = \frac{1}{2}(1 \pm \mathrm{sgn}(\xi))$$

are projections onto the Hardy spaces. The analytic signal related to $f(t)$ is defined as

$$f_A(t) = f(t) + i(\mathcal{H}f)(t)$$

and in frequency domain

$$\hat{f}_A(\xi) = (1+\mathrm{sgn}(\xi))\hat{f}(\xi) = 2\chi_+(\xi)\hat{f}(\xi).$$

The analytic signal operator $\mathcal{A}$ can be represented as

$$\mathcal{A} = \mathcal{I} + i\,\mathcal{H}.$$

3.2. Fractional Hilbert operator and analytic fractional signal

The fractional Hilbert kernel $h_\varphi(t)$ corresponding to the fractional Hilbert operator $\mathcal{H}_\varphi$ is given in Fourier domain as

$$\hat{h}_\varphi(\xi) = e^{-i\varphi}\chi_+(\xi) + e^{i\varphi}\chi_-(\xi), \ -\tfrac{\pi}{2} \le \varphi \le \tfrac{\pi}{2}.$$

This leads to the fractional Hilbert operator $\mathcal{H}_\varphi$ defined as

$$\mathcal{H}_\varphi = \cos\varphi\,\mathcal{I} + \sin\varphi\,\mathcal{H}.$$

For $\varphi = 0$ we get the identity operator and for $\varphi = \frac{\pi}{2}$ the fractional Hilbert operator coincides with the standard Hilbert operator. The analytic fractional operator is

$$f_A^\varphi(t) = (\mathcal{A}_\varphi f)(t) = f(t) + e^{i(\pi-\varphi)}(\mathcal{H}_\varphi f)(t).$$

4. The fractional Riesz operator

In what follows the function f will be considered to be scalar-valued, i.e., real- or complex-valued, and therefore the function f and its Fourier transform $\hat{f}$ will commute with all quaternions. To construct a fractional Riesz operator we start with a definition similar to that of the 1D fractional Hilbert operator $\mathcal{H}_\varphi$. We will work in Fourier domain and replace the projections for analytic functions by those of monogenic functions and rotate them in $\mathbb{R}^4$. The convolution kernel of the fractional Riesz transform is given by

$$\begin{aligned}
\hat{r}_{p,q}(\underline{\xi}) &= C_{p,q^{-1}}\chi_-(\underline{\xi}) + C_{p^{-1},q}\chi_+(\underline{\xi}) = e^{\underline{u}\frac{\varphi}{2}}\chi_-(\underline{\xi})e^{-\underline{v}\frac{\psi}{2}} + e^{-\underline{u}\frac{\varphi}{2}}\chi_+(\underline{\xi})e^{\underline{v}\frac{\psi}{2}} \\
&= (\cos\tfrac{\varphi}{2} + \underline{u}\sin\tfrac{\varphi}{2})\frac{1}{2}\left(1 - \frac{i\underline{\xi}}{|\underline{\xi}|}\right)(\cos\tfrac{\psi}{2} - \underline{v}\sin\tfrac{\psi}{2}) \\
&\quad + (\cos\tfrac{\varphi}{2} - \underline{u}\sin\tfrac{\varphi}{2})\frac{1}{2}\left(1 + \frac{i\underline{\xi}}{|\underline{\xi}|}\right)(\cos\tfrac{\psi}{2} + \underline{v}\sin\tfrac{\psi}{2}) \\
&= \frac{1}{2}\Big(\cos\tfrac{\varphi}{2}\cos\tfrac{\psi}{2} - \cos\tfrac{\varphi}{2}\sin\tfrac{\psi}{2}\,\underline{v} + \sin\tfrac{\varphi}{2}\cos\tfrac{\psi}{2}\,\underline{u} - \sin\tfrac{\varphi}{2}\sin\tfrac{\psi}{2}\,\underline{uv} \\
&\quad - \cos\tfrac{\varphi}{2}\cos\tfrac{\psi}{2}\frac{i\underline{\xi}}{|\underline{\xi}|} + \cos\tfrac{\varphi}{2}\sin\tfrac{\psi}{2}\frac{i\underline{\xi}}{|\underline{\xi}|}\underline{v} - \sin\tfrac{\varphi}{2}\cos\tfrac{\psi}{2}\,\underline{u}\frac{i\underline{\xi}}{|\underline{\xi}|} + \sin\tfrac{\varphi}{2}\sin\tfrac{\psi}{2}\,\underline{u}\frac{i\underline{\xi}}{|\underline{\xi}|}\underline{v} \\
&\quad + \cos\tfrac{\varphi}{2}\cos\tfrac{\psi}{2} + \cos\tfrac{\varphi}{2}\sin\tfrac{\psi}{2}\,\underline{v} - \sin\tfrac{\varphi}{2}\cos\tfrac{\psi}{2}\,\underline{u} - \sin\tfrac{\varphi}{2}\sin\tfrac{\psi}{2}\,\underline{uv} \\
&\quad + \cos\tfrac{\varphi}{2}\cos\tfrac{\psi}{2}\frac{i\underline{\xi}}{|\underline{\xi}|} + \cos\tfrac{\varphi}{2}\sin\tfrac{\psi}{2}\frac{i\underline{\xi}}{|\underline{\xi}|}\underline{v} - \sin\tfrac{\varphi}{2}\cos\tfrac{\psi}{2}\,\underline{u}\frac{i\underline{\xi}}{|\underline{\xi}|} - \sin\tfrac{\varphi}{2}\sin\tfrac{\psi}{2}\,\underline{u}\frac{i\underline{\xi}}{|\underline{\xi}|}\underline{v}\Big) \\
&= \cos\tfrac{\varphi}{2}\cos\tfrac{\psi}{2} - \sin\tfrac{\varphi}{2}\sin\tfrac{\psi}{2}\,\underline{uv} + \cos\tfrac{\varphi}{2}\sin\tfrac{\psi}{2}\frac{i\underline{\xi}}{|\underline{\xi}|}\underline{v} - \sin\tfrac{\varphi}{2}\cos\tfrac{\psi}{2}\,\underline{u}\frac{i\underline{\xi}}{|\underline{\xi}|} \\
&= \frac{1}{2}\Big(\cos(\tfrac{\varphi}{2}+\tfrac{\psi}{2}) + \cos(\tfrac{\varphi}{2}-\tfrac{\psi}{2}) - (\cos(\tfrac{\varphi}{2}-\tfrac{\psi}{2}) - \cos(\tfrac{\varphi}{2}+\tfrac{\psi}{2}))\underline{uv}\Big) \\
&\quad + \frac{1}{2}\left((\sin(\tfrac{\varphi}{2}+\tfrac{\psi}{2}) - \sin(\tfrac{\varphi}{2}-\tfrac{\psi}{2}))\frac{i\underline{\xi}}{|\underline{\xi}|}\underline{v} - (\sin(\tfrac{\varphi}{2}+\tfrac{\psi}{2}) + \sin(\tfrac{\varphi}{2}-\tfrac{\psi}{2}))\underline{u}\frac{i\underline{\xi}}{|\underline{\xi}|}\right) \\
&= \frac{1}{2}\cos(\tfrac{\varphi}{2}+\tfrac{\psi}{2})(1+\underline{uv}) + \frac{1}{2}\cos(\tfrac{\varphi}{2}-\tfrac{\psi}{2})(1-\underline{uv}) \\
&\quad + \frac{1}{2}\sin(\tfrac{\varphi}{2}+\tfrac{\psi}{2})\left(\frac{i\underline{\xi}}{|\underline{\xi}|}\underline{v} - \underline{u}\frac{i\underline{\xi}}{|\underline{\xi}|}\right) - \frac{1}{2}\sin(\tfrac{\varphi}{2}-\tfrac{\psi}{2})\left(\frac{i\underline{\xi}}{|\underline{\xi}|}\underline{v} + \underline{u}\frac{i\underline{\xi}}{|\underline{\xi}|}\right) \\
&= \frac{1}{2}\cos(\tfrac{\varphi}{2}+\tfrac{\psi}{2})(1+\underline{uv}) + \frac{1}{2}\cos(\tfrac{\varphi}{2}-\tfrac{\psi}{2})(1-\underline{uv}) \\
&\quad + \frac{1}{2}\sin(\tfrac{\varphi}{2}+\tfrac{\psi}{2})\left(-\left\langle\frac{i\underline{\xi}}{|\underline{\xi}|}, \underline{v}-\underline{u}\right\rangle + \frac{i\underline{\xi}}{|\underline{\xi}|}\times(\underline{v}+\underline{u})\right) \\
&\quad - \frac{1}{2}\sin(\tfrac{\varphi}{2}-\tfrac{\psi}{2})\left(-\left\langle\frac{i\underline{\xi}}{|\underline{\xi}|}, \underline{v}+\underline{u}\right\rangle + \frac{i\underline{\xi}}{|\underline{\xi}|}\times(\underline{v}-\underline{u})\right)
\end{aligned}$$

and hence

$$\begin{aligned}
\hat{r}_{p,q},\hat{f} &= \frac{1}{2}\cos(\tfrac{\varphi}{2}+\tfrac{\psi}{2})(1+\underline{uv})\hat{f} + \frac{1}{2}\cos(\tfrac{\varphi}{2}-\tfrac{\psi}{2})(1-\underline{uv})\hat{f} \\
&\quad + \frac{1}{2}\sin(\tfrac{\varphi}{2}+\tfrac{\psi}{2})\left(-\left\langle\frac{i\underline{\xi}}{|\underline{\xi}|}\hat{f}, \underline{v}-\underline{u}\right\rangle + \frac{i\underline{\xi}}{|\underline{\xi}|}\hat{f}\times(\underline{v}+\underline{u})\right) \\
&\quad - \frac{1}{2}\sin(\tfrac{\varphi}{2}-\tfrac{\psi}{2})\left(-\left\langle\frac{i\underline{\xi}}{|\underline{\xi}|}\hat{f}, \underline{v}+\underline{u}\right\rangle + \frac{i\underline{\xi}}{|\underline{\xi}|}\hat{f}\times(\underline{v}-\underline{u})\right).
\end{aligned}$$

That is indeed a complicated formula. We look therefore at some special cases.

1. Let be $\underline{u} = \underline{v}$ and $\varphi = \psi$, i.e., $p = q$ and $\underline{uv} = \underline{u}^2 = -1$ then
$$\hat{r}_{p,p} = 1 + \sin\varphi\left(\frac{i\underline{\xi}}{|\underline{\xi}|} \times \underline{u}\right).$$
2. Let be $\underline{v} = -\underline{u}$ and $\varphi = \psi$, i.e., $p = -q$ and $\underline{uv} = -\underline{u}^2 = 1$ then
$$\hat{r}_{p,-p} = \cos\varphi + \sin\varphi\left\langle\frac{i\underline{\xi}}{|\underline{\xi}|}, \underline{u}\right\rangle = \cos\varphi + \sin\varphi\frac{i(u_1\xi_1 + u_2\xi_2 + u_3\xi_3)}{|\underline{\xi}|}.$$
That is the symbol of the operator
$$\mathcal{R}_{\underline{u},\varphi,-\underline{u},\varphi} f = \cos\varphi f + \sin\varphi\langle\underline{u}, \mathcal{R}f\rangle,$$
where $\langle\underline{u}, \mathcal{R}f\rangle$ is the generalized Riesz transform by M. Unser and D. Van De Ville [15].

4.1. The isoclinic fractional Riesz transform

More promising are the cases of isoclinic rotations. The property of an rotation to be an isoclinic is nicely explained in [13] and we will cite it. Let P be an arbitrary 4D point, represented as a quaternion $P = w\mathbf{1} + x\mathbf{i} + y\mathbf{j} + z\mathbf{k}$. Let p and q be unit quaternions. Consider the left- and right-multiplication mappings $P \mapsto pP$ and $P \mapsto Pq$. Both mappings have the property of rotating all half-lines originating from $\mathcal{O}$ through the same angle ($\arccos p_0$ and $\arccos q_0$ respectively); such rotations are denoted as **isoclinic**. Because the left- and the right-multiplication are different from each other resulting in different rotations we have to distinguish between left- and right-isoclinic rotations. Conversely, an isoclinic 4D rotation about $\mathcal{O}$ (different from the non-rotation I and from the central reversion $-I$) is represented by either a left-multiplication or a right-multiplication by a unique unit quaternion. This theorem is presumably due to R.S. Ball; in [1] the author does not mention it explicitly as a theorem, but nevertheless gives a proof. However, Ball's proof is slightly incomplete, a complete proof is given by J.E. Mebius in [12]. Hence, the multiplication with an unit quaternion from the right (or left) only describes a (right- or left-) isoclinic rotation in $\mathbb{R}^4$. Moreover, any rotation in $\mathbb{R}^4$ is a combination of a left- and a right-isoclinic rotation (see Theorem 2.2).

We consider the case of right isoclinic rotations, i.e., $p = e^0 = 1 = \mathbf{1}$ and $q = e^{\underline{v}\psi}$. In this case we get
$$\begin{aligned}\hat{r}_{\underline{v},\psi} = \hat{r}_{\mathbf{1},q} &= C_{\mathbf{1},q^{-1}}\chi_-(\underline{\xi}) + C_{\mathbf{1},q}\chi_+(\underline{\xi}) = e^{\underline{u}\cdot 0}\chi_-(\underline{\xi})e^{-\underline{v}\psi} + e^{-\underline{u}\cdot 0}\chi_+(\underline{\xi})e^{\underline{v}\psi}\\ &= \frac{1}{2}\cos\psi(1 + \underline{uv} + 1 - \underline{uv}) + \frac{1}{2}\sin\psi\left(\frac{i\underline{\xi}}{|\underline{\xi}|}\underline{v} - \underline{u}\frac{i\underline{\xi}}{|\underline{\xi}|} + \frac{i\underline{\xi}}{|\underline{\xi}|}\underline{v} + \underline{u}\frac{i\underline{\xi}}{|\underline{\xi}|}\right)\\ &= \cos\psi + \sin\psi\,\frac{i\underline{\xi}}{|\underline{\xi}|}\,\underline{v}\end{aligned}$$
and hence
$$\hat{r}_{\underline{v},\psi}\hat{f} = \cos\psi\hat{f} + \sin\psi\,\frac{i\underline{\xi}}{|\underline{\xi}|}\hat{f}\,\underline{v}.$$

For $\psi = 0$ we get $\hat{r}_{\underline{v},0}\hat{f} = \hat{f}$ and for $\psi = \frac{\pi}{2}$ we get $\hat{r}_{\underline{v},\frac{\pi}{2}}\hat{f} = \frac{i\xi}{|\underline{\xi}|}\hat{f}\,\underline{v}$. That is very similar to the complex case. This leads to the definition of the fractional monogenic signal.

4.2. Properties of the fractional Riesz operator

Theorem 4.1. *Let be $f, f_1, f_2 \in L^p(\mathbb{R}^3), 1 < p < \infty$, then $\mathcal{R}_{\underline{u},\varphi}f$, $\mathcal{R}_{\underline{u},\varphi}f_1$, $\mathcal{R}_{\underline{u},\varphi}f_2 \in L^p(\mathbb{R}^3)$ and $\mathcal{R}_{\underline{u},\varphi}$ fulfills the following properties:*

(P1) *Linearity:*
$(\mathcal{R}_{\underline{u},\varphi}(\alpha_1 f_1 + \alpha_2 f_2))(\underline{x}) = \alpha_1(\mathcal{R}_{\underline{u},\varphi}f_1)(\underline{x}) + \alpha_2(\mathcal{R}_{\underline{u},\varphi}f_2)(\underline{x}), \ \forall\alpha_1, \alpha_2 \in \mathbb{C}.$

(P2) *Shift-invariance: $\mathcal{S}_{\underline{\tau}}(\mathcal{R}_{\underline{u},\varphi}f)(\underline{x}) = \mathcal{R}_{\underline{u},\varphi}(\mathcal{S}_{\underline{\tau}}f)(\underline{x})$, $\underline{\tau} \in \mathbb{R}^m$, where $(\mathcal{S}_{\underline{\tau}}f)(\underline{x}) = f(\underline{x} - \underline{\tau})$,*

(P3) *Scale-invariance: $\mathcal{D}_\sigma(\mathcal{R}_{\underline{u},\varphi}f)(\underline{x}) = \mathcal{R}_{\underline{u},\varphi}(\mathcal{D}_\sigma f)(\underline{x})$, $\sigma \in \mathbb{R}^+$, where $(\mathcal{D}_\sigma f)(\underline{x}) = f(\sigma^{-1}\underline{x})$,*

(P4) *Orthogonality:*
If $f, g \in L^2(\mathbb{R}^3)$ such that $\langle\langle f,\, g\rangle\rangle = 0$, then $\langle\langle \mathcal{R}_{\underline{u},\varphi}f,\, \mathcal{R}_{\underline{u},\varphi}g\rangle\rangle_{\mathbb{CH}} = 0$. Here, $\langle\langle\cdot,\, \cdot\rangle\rangle$ denotes the usual L^2-scalar product and $\langle\langle\cdot,\, \cdot\rangle\rangle_{\mathbb{CH}}$ will be defined in Remark 4.2.

Remark 4.2. Let p and q be complex quaternions which can also be interpreted as vectors in $\mathbb{C}^4$. As vectors in $\mathbb{C}^4$ their scalar product is $p \cdot q = \sum_{j=0}^{3} \overline{p}_j^{\mathbb{C}} q_j$ and can be rewritten in terms of the scalar part of a product of complex quaternions involving complex and quaternionic conjugation:

$$\langle p,\, q\rangle = \mathrm{Sc}\, \left(\overline{p}^{\mathbb{CH}}\, q\right) = \sum_{j=0}^{3} \overline{p}_j^{\mathbb{C}} q_j.$$

Based on that the function space of quaternionic-valued functions, where each component is an L^2-function, can be equipped with the following scalar product

$$\langle\langle f,\, g\rangle\rangle_{\mathbb{CH}} := \mathrm{Sc} \int_{\mathbb{R}^3} \overline{f}^{\mathbb{CH}}(x)\, g(x)\, dx = \mathrm{Sc} \left(\sum_{j=0}^{3} \int_{\mathbb{R}^3} \overline{f}_j^{\mathbb{C}}(x)\, g_j(x)\, dx \right),$$

which is a complex scalar product for the L^2-space of complex quaternionic-valued functions.

Proof: (P1), (P2) and (P3) follow from the linearity, shift-invariance and scale-invariance of the Riesz transform. Property (P4) involves the scalar products of the corresponding spaces. We use the definition of $\langle \tilde{f},\, \tilde{g}\rangle_{\mathbb{CH}}$ and Parseval's identity which allows to replace a complex-valued (scalar-valued) function $\tilde{f}_A(\underline{x})$ by its Fourier transform $\hat{\tilde{f}}_A(\underline{\xi})$. We have

$$\langle\langle \mathcal{R}_{\underline{u},\varphi}f,\, \mathcal{R}_{\underline{u},\varphi}g\rangle\rangle_{\mathbb{CH}} = \mathrm{Sc} \int_{\mathbb{R}^3} \overline{(\mathcal{R}_{\underline{u},\varphi}f)(\underline{x})}^{\mathbb{CH}} (\mathcal{R}_{\underline{u},\varphi}g(\underline{x}))\, d\underline{x}$$
$$= \mathrm{Sc} \int_{\mathbb{R}^3} \overline{(\hat{r}_{\underline{u},\varphi}(\underline{\xi})\hat{f}(\xi)}^{\mathbb{CH}}\, \hat{r}_{\underline{u},\varphi}(\underline{\xi})\hat{g}(\underline{\xi})\, d\underline{\xi}$$

$$
\begin{aligned}
&= \mathrm{Sc} \int_{\mathbb{R}^3} \overline{\left(\cos\varphi\, \hat{f}(\underline{\xi}) - \tfrac{i\underline{\xi}}{|\underline{\xi}|} \hat{f}(\underline{\xi})\underline{u} \sin\varphi\right)}^{\mathbb{CH}} \left(\cos\varphi\, \hat{g}(\underline{\xi}) - \tfrac{i\underline{\xi}}{|\underline{\xi}|} \hat{g}(\underline{\xi})\underline{u} \sin\varphi\right) d\underline{\xi} \\
&= \mathrm{Sc} \int_{\mathbb{R}^3} \left(\cos\varphi\, \overline{\hat{f}(\underline{\xi})}^{\mathbb{C}} + \underline{u}\overline{\hat{f}(\underline{\xi})}^{\mathbb{C}} \tfrac{i\underline{\xi}}{|\underline{\xi}|} \sin\varphi\right) \left(\cos\varphi\, \hat{g}(\underline{\xi}) - \tfrac{i\underline{\xi}}{|\underline{\xi}|} \hat{g}(\underline{\xi})\underline{u} \sin\varphi\right) d\underline{\xi} \\
&= \mathrm{Sc} \int_{\mathbb{R}^3} \left(\cos\varphi + \underline{u}\tfrac{i\underline{\xi}}{|\underline{\xi}|} \sin\varphi\right) \left(\cos\varphi - \tfrac{i\underline{\xi}}{|\underline{\xi}|}\underline{u} \sin\varphi\right) \overline{\hat{f}(\underline{\xi})}^{\mathbb{C}} \hat{g}(\underline{\xi})\, d\underline{\xi} \\
&= \mathrm{Sc} \int_{\mathbb{R}^3} \left(\cos^2\varphi - \underline{u}\left(\tfrac{i\underline{\xi}}{|\underline{\xi}|}\right)^2 \underline{u} \sin^2\varphi\right) \overline{\hat{f}(\underline{\xi})}^{\mathbb{C}} \hat{g}(\underline{\xi})\, d\underline{\xi} \\
&= \int_{\mathbb{R}^3} (\cos^2\varphi + \sin^2\varphi)\, \overline{\hat{f}(\underline{\xi})}^{\mathbb{C}} \hat{g}(\underline{\xi})\, d\underline{\xi} = \int_{\mathbb{R}^m} \overline{f(\underline{x})}^{\mathbb{C}} g(\underline{x})\, d\underline{x} = \langle\langle f, g\rangle\rangle.
\end{aligned}
$$

It has to be mentioned that this equation is only true for the scalar product taken as the scalar part of the inner product. The adjoint operator

$$\mathcal{R}^*_{\underline{u},\varphi} = (\cos\varphi) f - (\sin\varphi)\underline{u}(\mathcal{R}f)$$

is not the inverse operator but we have that

$$
\begin{aligned}
\mathcal{R}^*_{\underline{u},\varphi}\mathcal{R}_{\underline{u},\varphi} f &= (\cos\varphi\,\mathcal{I} - \sin\varphi\,\underline{u}\,\mathcal{R})(\cos\varphi\,\mathcal{I} + \sin\varphi\,\mathcal{R}\,\underline{u}) f \\
&= \cos^2\varphi\, f - \sin^2\varphi\,\underline{u}\,\mathcal{R}^2 f\,\underline{u} - \cos\varphi\sin\varphi(\underline{u}\,(\mathcal{R}f) - (\mathcal{R}f)\underline{u}) \\
&= (\cos^2\varphi + \sin^2\varphi) f - \tfrac{1}{2}\sin(2\varphi)(\underline{u} \times (\mathcal{R}f)) \\
&= f - \tfrac{1}{2}\sin(2\varphi)\,(\underline{u} \times (\mathcal{R}f))
\end{aligned}
$$

and thus $\mathrm{Sc}\,(\mathcal{R}^*_{\underline{u},\varphi}\mathcal{R}_{\underline{u},\varphi} f) = f$.

5. Fractional monogenic signal

Definition 5.1. Let $f \in L^p(\mathbb{R}^3)$, $1 < p < \infty$, the fractional monogenic signal with respect to the right isoclinic rotation $q = e^{\underline{v}\psi} \in \mathbb{H}_1$ is defined in Fourier domain as

$$\widehat{f^{\underline{v},\psi}_{\mathcal{M}}}(\xi) := \hat{f}(\xi) + \hat{r}_{\underline{v},\psi}(\xi)\hat{f}(\xi)e^{\underline{v}(\pi+\psi)}.$$

We demonstrate that the fractional monogenic signal are the boundary values of a right monogenic function in the upper half-space and hence $\widehat{f^{\underline{v},\psi}_{\mathcal{M}}} \in L^{p,+}(\mathbb{R}^3)$. We have

$$
\begin{aligned}
1 + \hat{r}_{\underline{v},\psi}(\xi)e^{\underline{v}(\pi+\psi)} &= 1 + \left(\cos\psi + \sin\psi\,\frac{i\underline{\xi}}{|\underline{\xi}|}\underline{v}\right) e^{\underline{v}(\pi+\psi)} \\
&= 1 - \left(\cos\psi + \sin\psi\,\frac{i\underline{\xi}}{|\underline{\xi}|}\underline{v}\right)(\cos\psi + \underline{v}\sin\psi) \\
&= 1 - \cos^2\psi - \underline{v}\cos\psi\sin\psi - \sin\psi\cos\psi\,\frac{i\underline{\xi}}{|\underline{\xi}|}\underline{v} + \sin^2\psi\,\frac{i\underline{\xi}}{|\underline{\xi}|} \\
&= \sin^2\psi\left(1 + \frac{i\underline{\xi}}{|\underline{\xi}|}\right) - \sin\psi\cos\psi\left(1 + \frac{i\underline{\xi}}{|\underline{\xi}|}\right)\underline{v}
\end{aligned}
$$

$$= \sin\psi \left(1 + \frac{i\underline{\xi}}{|\underline{\xi}|}\right) (\sin\psi - \underline{v}\cos\psi)$$
$$= \sin\psi \left(1 + \frac{i\underline{\xi}}{|\underline{\xi}|}\right) \left(-\cos\left(\tfrac{\pi}{2} + \psi\right) - \underline{v}\left(\sin\left(\tfrac{\pi}{2} + \psi\right)\right)\right)$$
$$= \chi_+(\underline{\xi}) \left(-\sin\psi\, e^{\underline{v}\left(\frac{\pi}{2}+\psi\right)}\right).$$

5.1. Properties of the fractional monogenic signal

The fractional monogenic signal is given by

$$f_{\mathcal{M}}^{\underline{u},\varphi}(\underline{x}) = (\mathcal{M}_{\underline{u},\varphi} f)(\underline{x}), \quad \text{where} \quad M_{\underline{u},\varphi} = \mathcal{I} + M^{e^{\underline{u}(\pi-\varphi)}} \mathcal{R}_{\underline{u},\varphi}.$$

The fractional monogenic signal are the boundary values of an right-monogenic function in the upper half-space and the *amplitude* of the fractional monogenic signal is the amplitude of the monogenic signal modulated by $|\sin\psi|$, i.e.,

$$|f_{\mathcal{M}}^{\underline{u},\varphi}(\underline{x})| = |\sin\varphi||f_{\mathcal{M}}|.$$

The *phase* of the fractional monogenic signal is also different from that of the monogenic signal because the scalar part of the fractional monogenic signal is a combination of the function f and the scalar product of the Riesz transform of f (considered as a vector) and the vector $\underline{u}$.

Theorem 5.2. *Let be $f, f_1, f_2 \in L^p(\mathbb{R}^3), 1 < p < \infty$, then $\mathcal{R}_{\underline{u},\varphi} f$, $\mathcal{R}_{\underline{u},\varphi} f_1$, $\mathcal{R}_{\underline{u},\varphi} f_2 \in L^p(\mathbb{R}^3)$ and $\mathcal{R}_{\underline{u},\varphi}$ fulfills the following properties:*

(M1) *Linearity:*
$(\mathcal{M}_{\underline{u},\varphi}(\alpha_1 f_1 + \alpha_2 f_2)(\underline{x}) = \alpha_1(\mathcal{M}_{\underline{u},\varphi} f_1)(\underline{x}) + \alpha_2(\mathcal{M}_{\underline{u},\varphi} f_2))(\underline{x}), \ \forall \alpha_1, \alpha_2 \in \mathbb{C}.$

(M2) *Shift-invariance:* $\mathcal{S}_{\underline{\tau}}(\mathcal{M}_{\underline{u},\varphi} f)(\underline{x}) = \mathcal{M}_{\underline{u},\varphi}(\mathcal{S}_{\underline{\tau}} f)(\underline{x}), \ \underline{\tau} \in \mathbb{R}^m,$

(M3) *Scale-invariance:* $\mathcal{D}_\sigma(\mathcal{M}_{\underline{u},\varphi} f)(\underline{x}) = \mathcal{M}_{\underline{u},\varphi}(\mathcal{D}_\sigma f)(\underline{x}), \ \sigma \in \mathbb{R}^+,$

6. Concluding remarks

It is also of some interest to consider a monogenic signal of two variables (for images). In this case

$$f_{\mathcal{M}}(x, y) = f(x, y) + \mathbf{i}(\mathcal{R}_1 f)(x, y) + \mathbf{j}(\mathcal{R}_2 f)(x, y).$$

That can be interpreted as something living in $\mathbb{R}^3$ and it should be even easier to consider rotations for this case. A rotation in $\mathbb{R}^3$ can be described by quaternions in the following way [10]. Any rotation in $\mathbb{R}^3$ can be described by the mapping

$$\mathbb{R}^3 \to \mathbb{R}^3, \quad \underline{r} \mapsto a\underline{r}a^{-1},$$

where a is a unit quaternion and $a = \cos(\frac{|\underline{a}|}{2}) + \frac{\underline{a}}{|\underline{a}|}\sin(\frac{|\underline{a}|}{2}) = e^{\underline{u}\varphi}$ with $\underline{u} = \frac{\underline{a}}{|\underline{a}|}$ and $\varphi = \frac{|\underline{a}|}{2}$. The first problem arises from the fact that $(\mathcal{I} + \mathcal{R})f$ is not a vector in $\mathbb{R}^3$ even though it can be identified with a vector. The second problem consists in the fact that the rotations in $\mathbb{R}^3$ are defined by multiplication from the right and the left which will destroy monotonicity. Therefore we suggest to embed the problem into $\mathbb{R}^4$ and use a Riesz operator $\mathcal{R} = \mathbf{i}R_1 + \mathbf{j}R_2$ and a unit vector $\underline{u} = \mathbf{i}u_1 + \mathbf{j}u_2$. Then the definition

$$\hat{r}_{\underline{u},\varphi}(\underline{\xi}) = \chi_-(\underline{\xi})e^{-\underline{v}\frac{\psi}{2}} + \chi_+(\underline{\xi})e^{\underline{v}\frac{\psi}{2}} = \cos\tfrac{\psi}{2} + \sin\tfrac{\psi}{2}\,\frac{i\underline{\xi}}{|\underline{\xi}|}\,\underline{v}$$

$$= \cos\tfrac{\psi}{2} + \sin\tfrac{\psi}{2}\left\langle\frac{i\underline{\xi}}{|\underline{\xi}|}, \underline{v}\right\rangle + \sin\tfrac{\psi}{2}\left(\frac{i\underline{\xi}}{|\underline{\xi}|}\times\underline{v}\right)$$

seems to give a useful definition for a fractional Riesz operator in this case. Also higher dimensions are of interest. Here, the description of a rotation is even more complicated. A way around could be to look for isoclinic and pseudo-isoclinic rotations and their behavior [14]. Another possibility to define a fractional Riesz transform for 2D signals (images) is to embed the signal f as $f\mathbf{k}$ into $\mathbb{H}$. Then the projections $\frac{1}{2}\left(1 \pm \frac{i\underline{\xi}}{|\underline{\xi}|}\mathbf{k}\right)$ multiplied with $\hat{f}\mathbf{k}$ can be identified with vectors in $\mathbb{R}^3$ and the vector can be rotated in $\mathbb{R}^3$ using quaternions.

References

[1] R.S. Ball, ed. H. Gravelius, *Theoretische Mechanik starrer Systeme. Auf Grund der Methoden und Arbeiten mit einem Vorworte.* Berlin: Georg Reimer, 1889.

[2] S. Bernstein, J.-L. Bouchot, M. Reinhardt, B. Heise, *Generalized Analytic Signals in Image Processing:Comparison, Theory and Applications,* in: E. Hitzer and S.J. Sangwine (eds.), Quaternion and Clifford Fourier Transforms and Wavelets, Trends in Mathematics, Birkhäuser, (2013), 221–246.

[3] A. Cusmariu, *Fractional analytic signals.* Signal Processing, **82** (2002), 267–272.

[4] J.A. Davis, D.E. McNamara, D.M. Cottrell, *Analysis of the fractional Hilbert transform.* Appl. Optics, **37** (1998), 6911–6913.

[5] R. Delanghe, *Clifford Analysis: History and Perspective.* Comp. Meth. Func. Theory, **1**(1) (2001), 107–153.

[6] M. Felsberg, G. Sommer, *The monogenic signal.* IEEE Trans. Signal Proc., **49**(12) (2001), 3136–3144.

[7] D. Gabor, *Theory of communication.* J. of the Institution of Electrical Engineers – Part III: Radio and Communication Engineering, **93**(26) (1946), 429–457.

[8] S.L. Hahn, *Multidimensional complex signals with single-orthant spectra.* Proc. IEEE, **80**(8) (1992), 1287–1300.

[9] A.W. Lohmann, D. Mendlovic, Z. Zalevsky, *Fractional Hilbert transform.* Optics Letters, **21** (1996), 281–283.

[10] P. Lounesto, *Clifford Algebras and Spinors*, Cambridge Univ. Press, 1997.

[11] A. McIntosh, *Fourier theory, singular integrals and harmonic functions on Lipschitz domains,* in: J. Ryan (ed.), Clifford Algebras in Analysis and Related Topics, CRC Press, (1996), 33–88.

[12] J.E. Mebius, *Applications of quaternions to dynamical simulation, computer graphics and biomechanics*, Ph.D. Thesis Delft University of Technology, Delft, 1994.

[13] J.E. Mebius, *A Matrix-based Proof of the Quaternion Representation Theorem for Four-Dimensional Rotations*, http://arxiv.org/abs/math/0501249v1.

[14] A. Richard, L. Fuchs, E. Andres, G. Largeteau-Skapin, *Decomposition of nD-rotations: classification, properties and algorithm*, Graphical Models, **73**(6) (2011), 346–353.

[15] M. Unser, D. Van De Ville, *Wavelet Steerability and the Higher-Order Riesz Transform*, IEEE Trans. Image Proc., **19**(3) (2010), 636–652.

[16] A. Venkitaraman, C.S. Seelamantula, *Fractional Hilbert transform extensions and associated analytic signal construction.* Signal Processing, **94** (2014), 359–372.

[17] J.L. Weiner, G.R. Wilkens, *Quaternions and Rotations in* $\mathbb{E}^4$. Amer. Math. Monthly, **112**(1) (2005), 69–76.

[18] A.I. Zayed, *Hilbert transform associated with the fractional Fourier transform.* IEEE Signal Processing Letters, **5** (1998), 206–208.

Swanhild Bernstein
TU Bergakademie Freiberg
Institute of Applied Analysis
D-09599 Freiberg, Germany
e-mail: swanhild.bernstein@math.tu-freiberg.de

Hypercomplex Analysis: New Perspectives and Applications
Trends in Mathematics, 89–110

Weighted Bergman Spaces

Luís Javier Carmona L., Lino Feliciano Reséndis Ocampo
and Luis Manuel Tovar Sánchez

Abstract. In this paper we study weighted Bergman spaces, through Green function and Möbius transformations, and its relationship and remarkable differences with the $F(p,q,s)$ Zhao spaces and so with other classical weighted function spaces.

Mathematics Subject Classification (2010). Primary 30C45; Secondary 30J.

Keywords. Bergman spaces, $F(p,q,s)$ Zhao spaces.

1. Introduction

Let $0 < r < \infty$. Define $\mathbb{D}_r(a) := \{z \in \mathbb{C} : |z-a| < r\}$, $\mathbb{D}_r = \mathbb{D}_r(0)$ and $\mathbb{A}_r = \mathbb{D}\setminus\mathbb{D}_r$. We denote by $\mathbb{D} = \mathbb{D}_1$ the open unit disk in the complex plane $\mathbb{C}$.

Let $\varphi_z : \mathbb{C} \setminus \{\frac{1}{\overline{z}}\} \to \mathbb{C}$ be the Möbius transformation,

$$\varphi_z(w) = \frac{z-w}{1-\overline{z}w},$$

with pole at $w = 1/\overline{z}$ that verifies $\varphi_z^{-1} = \varphi_z$ and

$$1 - |\varphi_z(w)|^2 = \frac{(1-|z|^2)(1-|w|^2)}{|1-\overline{z}w|^2} = (1-|w|^2)|\varphi_z'(w)| \ . \tag{1.1}$$

For $z,\ w \in \mathbb{D}$, we denote the Green function of $\mathbb{D}$, with logarithmic singularity at z, by

$$g(w,z) = \ln\frac{|1-\overline{z}w|}{|z-w|} = \ln\frac{1}{|\varphi_z(w)|} \ . \tag{1.2}$$

Let $\mathcal{H}$ be the space of analytic functions $f : \mathbb{D} \to \mathbb{C}$. For $0 < p < \infty$, $-2 < q < \infty$, $0 \le s < \infty$, let $f \in \mathcal{H}$ be such that,

$$h_{p,q,s}(f)(z) := \int_{\mathbb{D}} |f(w)|^p g^s(w,z)\, dA_q(w) < \infty, \tag{1.3}$$

where $dA_q(w) := (1-|w|^2)^q dA(w)$.

This work was completed with the support of Conacyt.

We define the q,s-weighted p-Bergman space as

$$ {}_s\mathcal{A}_q^p := \{ f \in \mathcal{H} \;:\; \sup_{z\in\mathbb{D}} h_{p,q,s}(f)(z) < \infty \} $$

and for $0 < s < \infty$, the little q,s-weighted p-Bergman space as

$$ {}_{s,0}\mathcal{A}_q^p := \{ f \in \mathcal{H} \;:\; \lim_{|z|\to 1^-} h_{p,q,s}(f)(z) = 0 \} . $$

In the same way, for $0 < p < \infty$, $-2 < q < \infty$ and $0 \leq s < \infty$, let $f \in \mathcal{H}$ be such that,

$$ l_{p,q,s}(f)(z) := \int_{\mathbb{D}} |f(w)|^p (1 - |\varphi_z(w)|^2)^s \, dA_q(w) < \infty \tag{1.4} $$

and define

$$ L(p,q,s) := \{ f \in \mathcal{H} \;:\; \sup_{z\in\mathbb{D}} l_{p,q,s}(f)(z) < \infty \} $$

and for $0 < s < \infty$

$$ L_0(p,q,s) := \{ f \in \mathcal{H} \;:\; \lim_{|z|\to 1^-} l_{p,q,s}(f)(z) = 0 \} . $$

We write $L_p = L(p,0,0)$ and observe that ${}_0\mathcal{A}_q^2 = {}_0L_q^2 = \mathcal{A}_q^2$ is the classical Bergman space of analytic functions. Following Zhu ([8]), we refer also the spaces $\mathcal{A}_q^p$ as the classical Bergman spaces. We define further

$$ \| f \|_g = \| f \|_{g,p,q,s} = \sup_{z\in\mathbb{D}} (h_{p,q,s}(f)(z))^{\frac{1}{p}} $$

and

$$ \| f \|_\varphi = \| f \|_{\varphi,p,q,s} = \sup_{z\in\mathbb{D}} (l_{p,q,s}(f)(z))^{\frac{1}{p}} . $$

Let $0 < \alpha < \infty$. We say that $f \in \mathcal{H}$ belongs to the α-Bloch–Bergman space B^α if

$$ \| f \|_\alpha = \sup_{z\in\mathbb{D}} (1 - |z|^2)^\alpha |f(z)| < \infty $$

and belongs to the little α-Bloch–Bergman space B_0^α if

$$ \| f \|_\alpha = \lim_{|z|\to 1^-} (1 - |z|^2)^\alpha |f(z)| = 0 . $$

It is clear that $B_0^\alpha \subset B^\alpha$.

The aim of this paper is to obtain explicitly properties of the weighted Bergman spaces, in particular we study the nested scale between the classical Bergman spaces and the Bloch–Bergman spaces. We also study the relationships with the $F(p,q,s)$ spaces introduced by Ruhan Zhao, (see [6]). Quaternionic Bergman spaces will be studied in a forthcoming paper [4].

2. The α-Bloch–Bergman space

The sets B^α are vectorial spaces and is immediate consequence of an argument of normality that B^α and B_0^α are complete spaces with the norm $\| \ \|_\alpha$; moreover convergence in this norm implies uniform convergence by compact sets on the unit disk.

Let $0 < R < 1$. The pseudohyperbolic disk is defined by

$$D(z,R) := \varphi_z(\mathbb{D}_R) = \{ \, w \in \mathbb{D} \ : \ |\varphi_z(w)| < R \, \} \, .$$

In fact $D(z,R)$ is a Euclidean disk with center and radius given by

$$c = \frac{1-R^2}{1-R^2|z|^2} z, \qquad R = \frac{1-|z|^2}{1-R^2|z|^2} R \tag{2.1}$$

and we denote by $|D(z,R)|$ its area.

We need the following results:

Lemma 2.1 ([8]). *Let $t > -1$, $c \in \mathbb{R}$. Define $I_{t,c} : \mathbb{D} \to \mathbb{R}$ by*

$$I_{t,c}(z) = \int_{\mathbb{D}} \frac{(1-|w|^2)^t}{|1-z\overline{w}|^{2+t+c}} \, dA(w)$$

and $J_c : \mathbb{D} \to \mathbb{R}$ by

$$J_c(z) = \int_0^{2\pi} \frac{d\theta}{|1-ze^{-i\theta}|^{1+c}}$$

Then

(i) *If $c < 0$, then $I_{t,c}(z) \approx J_c(z) \approx 1$ when $|z| \to 1^-$.*

(ii) *If $c = 0$, then $I_{t,c}(z) \approx J_c(z) \approx \ln \dfrac{1}{1-|z|^2}$, when $|z| \to 1^-$.*

(iii) *If $c > 0$, then $I_{t,c}(z) \approx J_c(z) \approx \dfrac{1}{(1-|z|^2)^c}$.*

As a consequence of the previous lemma we have

Lemma 2.2. *Let $-2 < q < \infty$ and $0 < s < \infty$. Then*

$$\lim_{|z|\to 1^-} \int_{\mathbb{D}} g^s(w,z) \, dA_q(w) = 0$$

$$\lim_{|z|\to 1^-} \int_{\mathbb{D}} (1-|\varphi_z(w)|^2)^s \, dA_q(w) = 0 \, .$$

Proof. By the change of variable $w = \varphi_z(v)$ we have

$$\begin{aligned} \int_{\mathbb{D}} g^s(w,z) \, dA_q(w) &= \int_{\mathbb{D}} \ln^s \frac{1}{|v|} (1-|\varphi_z(v)|^2)^q |\varphi_z'(v)|^2 \, dA(v) \\ &= (1-|z|^2)^{q+2} \int_{\mathbb{D}} \ln^s \frac{1}{|v|} \frac{(1-|v|^2)^{q+2}}{|1-\overline{z}v|^{4+2q}} \, dA(v) \, . \end{aligned}$$

The result follows by taking $t = q+2$ in the previous lemma. The other proof is similar. □

The α-Bloch–Bergman space has the following characterizations:

Theorem 2.3. *Let $0 < p < \infty$, $0 \le \alpha < \infty$, $0 < r < 1 < s < \infty$ and $f \in \mathcal{H}$. The following quantities are equivalent, that is, if one of them is finite then the other ones are finite too:*

(i) $\sup_{z\in\mathbb{D}}(1-|z|^2)^{p\alpha}|f(z)|^p$;

(ii) $\sup_{z\in\mathbb{D}} \frac{1}{|D(z,r)|^{1-\frac{p\alpha}{2}}} \int_{D(z,r)} |f(w)|^p\, dA(w)$;

(iii) $\sup_{z\in\mathbb{D}} \int_{D(z,r)} |f(w)|^p\, dA_{p\alpha-2}(w)$;

(iv) $\sup_{z\in\mathbb{D}} \int_{\mathbb{D}} |f(w)|^p(1-|\varphi_z(w)|^2)^s\, dA_{p\alpha-2}(w)$;

(v) $\sup_{z\in\mathbb{D}} \int_{\mathbb{D}} |f(w)|^p g^s(w,z)\, dA_{p\alpha-2}(w)$;

(vi) $\sup_{z\in\mathbb{D}} \int_{\mathbb{D}} |f(w)|^p \left(\log\frac{1}{|w|}\right)^{p\alpha} |\varphi_z'(w)|^2\, dA(w)$.

Proof. The subharmonicity of $|f(z)|^p$ for $0 < p < \infty$ is the key of the proof. The proof in general is similar to the proof Theorem 1 of R. Zhao in [6]. □

An analogous result is true with the little α-Bloch–Bergman space, (see Theorem 2 of R. Zhao in [6]).

Since we will need some estimations, we include a proof of the equivalence of (i) and (v). Analogous proof can be given for the equivalence of (i) and (iv).

Theorem 2.4. *Let $0 < p < \infty$, $-2 < q < \infty$. If $0 < s < \infty$, then ${}_s\mathcal{A}^p_q \subset B^{\frac{q+2}{p}}$ and ${}_{s,0}\mathcal{A}^p_q \subset B_0^{\frac{q+2}{p}}$. If $1 < s < \infty$, then ${}_s\mathcal{A}^p_q = B^{\frac{q+2}{p}}$ and ${}_{s,0}\mathcal{A}^p_q = B_0^{\frac{q+2}{p}}$.*

Proof. Let $f \in {}_s\mathcal{A}^p_q$ and $0 < R < 1$ be fixed. Then by the change of variable $w = \varphi_z(\zeta)$ and subharmonicity we have

$$\begin{aligned}
\int_{\mathbb{D}} |f(w)|^p g^s(w,z)\, dA_q(w) &\ge \int_{D(z,R)} |f(w)|^p g^s(w,z)\, dA_q(w)\\
&\ge \ln^s\frac{1}{R}\int_{\mathbb{D}_R} |f(\varphi_z(\zeta))|^p(1-|\varphi_z(\zeta)|^2)^q|\varphi_z'(\zeta)|^2\, dA(\zeta)\\
&\ge \frac{1}{2^{2q+4}}\ln^s\frac{1}{R}(1-|z|^2)^{q+2}\int_0^R(1-\rho^2)^q\rho\int_0^{2\pi}|f(\varphi_z(\rho e^{i\theta}))|^p\, d\theta d\rho\\
&\ge \frac{1}{2^{2q+4}}\ln^s\frac{1}{R}(1-|z|^2)^{q+2}|f(z)|^p\int_0^R(1-\rho^2)^q\rho\, d\rho\\
&= C(R)(1-|z|^2)^{q+2}|f(z)|^p
\end{aligned} \tag{2.2}$$

and the first two claims follow from the previous estimation.

Now, if $f \in B^{\frac{q+2}{p}}$, again by using the change of variable $w = \varphi_z(\zeta)$

$$\begin{aligned}\int_{\mathbb{D}} |f(w)|^p g^s(w,z)\,dA_q(w) &\leq \| f \|^p_{\frac{q+2}{p}} \int_{\mathbb{D}} \frac{1}{(1-|w|^2)^2} g^s(w,z)\,dA(w)\\ &= \| f \|^p_{\frac{q+2}{p}} \int_{\mathbb{D}} \frac{1}{(1-|\varphi_z(\zeta)|^2)^2} \ln^s \frac{1}{|\zeta|} |\varphi_z'(\zeta)|^2\,dA(\zeta)\\ &= \| f \|^p_{\frac{q+2}{p}} \int_{\mathbb{D}} \frac{1}{(1-|\zeta|^2)^2} \ln^s \frac{1}{|\zeta|}\,dA(\zeta)\ ;\end{aligned}$$

so as $1 < s < \infty$, the last integral is finite (see [6], Lemma 3.2).

Let $f \in B_0^{\frac{q+2}{p}}$. Given $\varepsilon > 0$, there exists $0 < R < 1$ such that

$$|f(w)|^p \leq \frac{\varepsilon}{(1-|w|^2)^{q+2}} \qquad \text{for all } w \in \mathbb{A}_R.$$

Thus

$$\int_{\mathbb{A}_R} |f(w)|^p g^s(w,z)\,dA_q(w) \leq \varepsilon \int_{\mathbb{A}_R} \frac{g^s(w,z)}{(1-|w|^2)^2}\,dA(w)$$

and

$$\int_{\mathbb{D}_R} |f(w)|^p g^s(w,z)\,dA_q(w) \leq \max_{w \in \overline{\mathbb{D}}_R} |f(w)|^p \int_{\mathbb{D}_R} g^s(w,z)\,dA_q(w)$$

and we conclude the proof applying Lemma 2.2. □

If we write the estimation (2.2), with $s = 0$, we obtain that the classical Bergman spaces are included in some α-Bloch–Bergman space.

Corollary 2.5. *Let $0 < p < \infty$, $-2 < q < \infty$. Then $\mathcal{A}^p_q \subset B_0^{\frac{q+2}{p}}$ and $\| \ \|_{\frac{q+2}{p}} \leq C \| \ \|_g$.*

Proof. By (2.1), $|D(z,R)| \to 0$ when $|z| \to 1^-$ then apply absolute continuity of the integral. □

3. Properties of ${}_s\mathcal{A}^p_q$ and $L(p,q,s)$

In this section we obtain basic properties of the Bergman spaces ${}_s\mathcal{A}^p_q$ and $L(p,q,s)$.

Let $0 < p < \infty$ and $0 < s < s' < \infty$. It is immediate that

$$L(p,q,s) \subset L(p,q,s') \qquad \text{and} \qquad L_0(p,q,s) \subset L_0(p,q,s')\ ;$$

since $1 - x^2 \leq 2 \ln \frac{1}{x}$ for all $x \in (0,1]$ then

$${}_s\mathcal{A}^p_q \subset L(p,q,s) \qquad \text{and} \qquad {}_{s,0}\mathcal{A}^p_q \subset L_0(p,q,s)\ .$$

It is easy verify that ${}_{s,0}\mathcal{A}^p_q$ is a vectorial space.

For $0 < p < 1$, $\| \ \|^p_g$ and $\| \ \|^p_\varphi$ define metrics and for $1 \leq p < \infty$, $\| \ \|_g$ and $\| \ \|_\varphi$ define norms.

Theorem 3.1. *Let $0 < p < \infty$, $-2 < q < \infty$ and $0 < s < \infty$, with $-1 < q + s$. If $f,\ g \in L(p,q,s)$ (or $L_0(p,q,s)$), then $\lambda f + \eta g \in L(p,q,s)$ (or $L_0(p,q,s)$) for $\lambda,\ \eta \in \mathbb{C}$. Moreover, if $0 < p < r < \infty$, then $L(r,q,s) \subset L_0(p,q,s)$.*

Proof. Define for each measurable set $E \subset \mathbb{D}$,

$$\nu_{z,q,s}(E) = \int_E (1 - |\varphi_z(w)|^2)^s \, dA_q(w) .$$

Thus $\nu_{z,q,s}(\mathbb{D}) < \infty$ by Lemma 2.1 and by Lemma 2.2 we have

$$\lim_{|z|\to 1^-} \nu_{z,q,s}(\mathbb{D}) = 0 .$$

Moreover $\nu_{z,q,s}$ is absolutely continuous with respect to the Lebesgue measure on $\mathbb{D}$ and

$$\int_{\mathbb{D}} |f(w)|^p \, d\nu_{z,q,s}(w) = \int_{\mathbb{D}} |f(w)|^p (1 - |\varphi_z(w)|^2)^s \, dA_q(w) .$$

Let $f,\ g \in L(p,q,s)$. The theorem will follow from the next estimation

$$|\lambda f(w) + \eta g(w)|^p \leq 2^p (|\lambda|^p |f(w)|^p + |\eta|^p |g(w)|^p)$$

and the Hölder inequality (see [2], Theorem 13.17)

$$\left(\int_{\mathbb{D}} |f(w)|^p \, d\nu_{z,q,s}(w) \right)^{\frac{1}{p}} \leq \left(\int_{\mathbb{D}} |f(w)|^r \, d\nu_{z,q,s}(w) \right)^{\frac{1}{r}} \nu_{z,q,s}(\mathbb{D})^{\frac{1}{p} - \frac{1}{r}} . \qquad \square$$

Theorem 3.2. *Let $1 \leq p < \infty$, $-2 < q < \infty$ and $0 < s < \infty$. Then $\| \ \|_g$ and $\| \ \|_\varphi$ define norms on ${}_s\mathcal{A}^p_q$ and $L(p,q,s)$, respectively. With these norms ${}_s\mathcal{A}^p_q$ and $L(p,q,s)$ become Banach spaces.*

Proof. It is immediate that $\| \cdot \|_g$ and $\| \cdot \|_\varphi$ are norms.

We prove only that ${}_s\mathcal{A}^p_q$ is complete. Let $\{f_n\}$ be a Cauchy sequence in ${}_s\mathcal{A}^p_q$. By Corollary 2.5, $\{f_n\}$ is also a Cauchy sequence in $B^{\frac{q+2}{p}}$. Then $\{f_n\}$ converges uniformly on compact sets to $f \in B^{\frac{q+2}{p}}$.

Given $\epsilon > 0$, there exists $N > 0$ such that if $n \geq m \geq N$ we have

$$\int_{\mathbb{D}} |f_n(w) - f_m(w)|^p g^s(w,z) \, dA_q(w) \leq \| f_n - f_m \|_g^p < \left(\frac{\epsilon}{2} \right)^p .$$

By Fatou's Lemma

$$\int_{\mathbb{D}} |f(w) - f_m(w)|^p g^s(w,z) \, dA_q(w)$$
$$\leq \liminf_{n\to\infty} \int_{\mathbb{D}} |f_n(w) - f_m(w)|^p g^s(w,z) \, dA_q(w) \leq \left(\frac{\epsilon}{2} \right)^p .$$

Taking the supremum with respect to $z \in \mathbb{D}$ we see that $f - f_m \in {}_s\mathcal{A}^p_q$, therefore $f = (f - f_m) + f_m \in {}_s\mathcal{A}^p_q$. Moreover $\| f - f_m \|_g \leq \frac{\epsilon}{2}$ if $m \geq N$. $\square$

The following result tells us that $L(p,q,s)$ or ${}_s\mathcal{A}^p_q$ can be trivial.

Proposition 3.3. *Let $0 < p < \infty$, $-2 < q < \infty$ and $1 < s < \infty$ with $q + s \leq -1$. Then the space $L(p,q,s) = {}_s\mathcal{A}^p_q$ and consists only of the constant function 0.*

Proof. Let $f \in L(p,q,s)$ be a nonzero function. Let $0 < b < 1$ be fixed. Since $|f(z)|^p$ is a subharmonic function,

$$\begin{aligned} l_{p,q,s}(f)(0) &= \int_{\mathbb{D}} |f(w)|^p (1-|w|^2)^s \, dA_q(w) \\ &\geq \int_b^1 \int_0^{2\pi} |f(re^{i\theta})|^p (1-r^2)^{q+s} r \, d\theta \, dr \\ &\geq \int_0^{2\pi} |f(be^{i\theta})|^p \, d\theta \int_b^1 (1-r^2)^{q+s} r \, dr = \infty \, . \end{aligned}$$

So we get a contradiction. The other proof is similar. □

Define

$$\mathcal{H}^\infty = \{ \, f \in \mathcal{H} \; : \; f \text{ is bounded} \, \} \, .$$

The spaces ${}_{s,0}\mathcal{A}_q^p$ are not empty, since by Lemma 2.2

$$\mathcal{H}^\infty \subset \bigcap \{ \, {}_{s,0}\mathcal{A}_q^p : \; 0 < p < \infty, \; -2 < q < \infty, \, 0 < s < \infty, \; -1 < q + s \} \, .$$

Thus we can improve Corollary 2.5.

Theorem 3.4. *Let* $0 < p < \infty$, $-2 < q < \infty$. *Then*

$$\mathcal{A}_q^p \subset \bigcap_{0<s<\infty} {}_{s,0}\mathcal{A}_q^p \qquad \text{and} \qquad \mathcal{A}_q^p \subset \bigcap_{0<s<\infty} {}_{s,0}L_q^p.$$

Proof. We prove the first inclusion. The second one is similar. Let $f \in \mathcal{A}_q^p$ and $\varepsilon > 0$. By Corollary 2.5, there exists $0 < R < 1$ such that

$$(1-|w|^2)^{q+2} |f(w)|^p < \varepsilon \qquad \text{for all} \qquad w \in \mathbb{A}_R \tag{3.1}$$

and by absolute continuity of the integral

$$\int_{\mathbb{A}_R} |f(w)|^p \, dA_q(w) < \varepsilon \, . \tag{3.2}$$

We split the integral

$$h_{p,q,s}(f)(z) \;\; = \;\; \int_{\mathbb{D}_R} |f(w)|^p g^s(w,z) \, dA_q(w) + \int_{\mathbb{A}_R} |f(w)|^p g^s(w,z) \, dA_q(w) \, .$$

By Lemma 2.2 the first integral goes to 0 when $|z| \to 1^-$. We split again the second integral: By (2.1) we can choose R', such that $0 < R < R' < 1$ and $D(z,R') \subset \mathbb{A}_R$. Now by (3.2) we have

$$\begin{aligned} \int_{\mathbb{A}_R \setminus D(z,R')} |f(w)|^p g^s(w,z) \, dA_q(w) &\leq \ln^s \frac{1}{R} \int_{\mathbb{A}_R \setminus D(z,R')} |f(w)|^p \, dA_q(w) \\ &< \varepsilon \ln^s \frac{1}{R} \, ; \end{aligned}$$

by (3.1) and the change of variable $w = \varphi_z(\zeta)$ we have

$$\int_{D(z,R')} |f(w)|^p g^s(w,z)\, dA_q(w) \leq \int_{D(z,R')} \frac{\varepsilon}{(1-|w|^2)^{q+2}} g^s(w,z)\, dA_q(w)$$
$$= \varepsilon \int_{\mathbb{D}_{R'}} \frac{1}{(1-|\varphi_z(\zeta)|^2)^2} \ln^s \frac{1}{|\zeta|} |\varphi'(\zeta)|^2\, dA(\zeta)$$
$$\leq \varepsilon \int_{\mathbb{D}} \frac{1}{(1-|\zeta|^2)^2} \ln^s \frac{1}{|\zeta|}\, dA(\zeta)$$

and the last integral is finite. This concludes the proof. □

We will need the following elementary estimations.

Lemma 3.5. *Let $q \in \mathbb{R}$ and $z \in \mathbb{D}$. Then for all $w \in \mathbb{D}$*

$$\frac{1}{\rho(z,q)}(1-|w|^2)^q \leq (1-|\varphi_z(w)|^2)^q \leq \rho(z,q)(1-|w|^2)^q, \tag{3.3}$$

and

$$\frac{1}{\rho(z,q)} \leq |\varphi_z'(w)|^q \leq \rho(z,q) \tag{3.4}$$

where

$$\rho(z,q) = \left(\frac{1+|z|}{1-|z|}\right)^{|q|}.$$

The spaces ${}_s\mathcal{A}^p_q$ and $L(p,q,s)$ are Möbius invariant in the following sense:

Proposition 3.6. *Let $z \in \mathbb{D}$ and $0 < p < \infty$, $-2 < q < \infty$ and $0 \leq s < \infty$. If $f \in {}_s\mathcal{A}^p_q$ (or $f \in L(p,q,s)$) then $f \circ \varphi_z \in {}_s\mathcal{A}^p_q$ (or $f \circ \varphi_z \in L(p,q,s)$), that is,*

$$\sup_{b\in\mathbb{D}} \int_{\mathbb{D}} |f(\varphi_z(w))|^p\ g^s(w,b)\, dA_q(w) < \infty\,.$$

Proof. Let $z \in \mathbb{D}$ be fixed. We denote by $w = \varphi_z(v)$, $b = \varphi_z(c)$. Since the Green function is conformally invariant, $g(\varphi_z(v), \varphi_z(c)) = g(v,c)$. Then by (3.3) and (3.4)

$$\int_{\mathbb{D}} |f(\varphi_z(w))|^p\ g^s(w,b)\, dA_q(w)$$
$$= \int_{\mathbb{D}} |f(\varphi_z(\varphi_z(v)))|^p |\varphi_z'(v)|^2 (1-|\varphi_z(v)|^2)^q g^s(\varphi_z(v),\varphi_z(c))\, dA(v)$$
$$\leq \rho(z,|q|+2) \int_{\mathbb{D}} |f(v)|^p (1-|v|^2)^q g^s(v,c)\, dA(v)$$
$$= \rho(z,|q|+2) \int_{\mathbb{D}} |f(v)|^p g^s(v,c)\ dA_q(v).$$

Taking the supremum the result follows. □

Observe that, the previous result is also true for the little spaces.

We will need the following result.

Lemma 3.7. *Let $0 < p < \infty$, $-2 < q < \infty$, $0 \le s < \infty$ and $z \in \mathbb{D}$. Then*

$$\frac{1}{\rho(z, |q| + 2)} l_{p,q,s}(f)(z) \le l_{p,q,s}(f \circ \varphi_z)(0) \le \rho(z, |q| + 2) l_{p,q,s}(f)(z) \ .$$

Proof. Suppose that $l_{p,q,s}(f)(z) < \infty$. Then by the change of variable formula with $v = \varphi_z(w)$, we have

$$\begin{aligned}
&\int_{\mathbb{D}} |f(\varphi_z(v))|^p (1 - |v|^2)^{q+s} \, dA(v) \\
&\quad = \int_{\mathbb{D}} |f(w)|^p |\varphi_z'(w)|^2 (1 - |\varphi_z(w)|^2)^{q+s} \, dA(w) \\
&\quad \le \rho(z, |q| + 2) \int_{\mathbb{D}} |f(w)|^p (1 - |w|^2)^q (1 - |\varphi_z(w)|^2)^s \, dA(w).
\end{aligned}$$

Conversely, if $l_{p,q,s}(f \circ \varphi_z)(0) < \infty$, then with $v = \varphi_z(w)$:

$$\begin{aligned}
&\int_{\mathbb{D}} |f(\varphi_z(v))|^p (1 - |v|^2)^{q+s} \, dA(v) \\
&\quad = \int_{\mathbb{D}} |f(w)|^p |\varphi_z'(w)|^2 (1 - |\varphi_z(w)|^2)^{q+s} \, dA(w) \\
&\quad \ge \frac{1}{\rho(z, |q| + 2)} \int_{\mathbb{D}} |f(w)|^p (1 - |w|^2)^q (1 - |\varphi_z(w)|^2)^s \, dA(w).
\end{aligned}$$ □

The following results clarify the relationship between $L_0(p, q, s)$ and $L(p, q, s)$.

Lemma 3.8. *Let $0 < p < \infty$, $-2 < q < \infty$ and $0 < s < \infty$. Let $0 < R < 1$ and $f \in \mathcal{A}^p_{q+s}$. Given $\varepsilon > 0$, there exists $0 < R' < 1$ such that for each $z \in \mathbb{D}_R$*

$$\int_{\mathbb{A}_{R'}} |f(w)|^p \ln^s \frac{1}{|\varphi_z(w)|} \, dA_q(w) < \varepsilon \ .$$

Proof. As $f \in \mathcal{A}^p_{q+s}$, given $\varepsilon > 0$, there exists $0 < R'' < 1$ such that

$$\int_{\mathbb{A}_{R''}} |f(w)|^p \, dA_{q+s}(w) < \varepsilon \frac{(1 - R)^{2s}}{2^s} \ .$$

Likewise there exists $\alpha \in (0, 1)$ such that

$$1 - t < \ln \frac{1}{t} < 2(1 - t), \qquad \alpha < t < 1,$$

thus

$$(1 - r^2) < \ln \frac{1}{r^2} < 2(1 - r^2), \qquad \alpha < r^2 < 1.$$

There exists $1 > R' > R'' > 0$, such that $\dfrac{2(1-|w|^2)}{(1-R)^2} < 1-\alpha$ for all $w \in \mathbb{A}_{R'}$. Thus $\alpha < |\varphi_z(w)|^2 < 1$, for all $z \in \mathbb{D}_R$ and $w \in \mathbb{A}_{R'}$. Then

$$\int_{\mathbb{A}_{R'}} |f(w)|^p \ln^s \frac{1}{|\varphi_z(w)|}\, dA_q(w) \leq \frac{2^s}{(1-R)^{2s}} \int_{\mathbb{A}_{R''}} |f(w)|^p (1-|w|^2)^s \, dA_q(w) < \varepsilon \, . \qquad \square$$

Proposition 3.9. *Let $0 < p < \infty$, $-2 < q < \infty$ and $0 \leq s < \infty$. Suppose that $f \in \mathcal{H}$ is such that $h_{p,q,s}(f)(z) < \infty$ for all $z \in \mathbb{D}$. Then $h_{p,q,s}(f)$ is a continuous function on $\mathbb{D}$.*

Proof. Let R and R' be as in Lemma 3.8. Let $\{b_n\} \subset \mathbb{D}_R$ be a sequence such that $b_n \to z \in \mathbb{D}_R$ when $n \to \infty$. By (2.1), we observe that

$$\bigcup_{b \in \mathbb{D}_R} \varphi_b(\mathbb{D}_{R'}) = \mathbb{D}_{R''}, \qquad \text{where} \qquad R'' = \frac{R+R'}{1+RR'} \, .$$

Let $I_n : \mathbb{D}_{R''} \to \mathbb{R}$ be defined by

$$I_n(w) = |f(\varphi_{b_n}(w))|^p \ln^s \frac{1}{|w|} |\varphi'_{b_n}(w)|^2 (1-|\varphi_{b_n}(w)|^2)^q \, \chi_{\varphi_{b_n}(\mathbb{D}_{R'})}(w) \, ,$$

where χ denotes the characteristic function. Thus $I_n(w) \to I_z(w)$ if $n \to \infty$. Moreover $|I_n(w)| \leq M \ln^s \frac{1}{|w|}$ for all $w \in \mathbb{D}_{R''}$ and some constant $M > 0$. Taking the change of variable $w = \varphi_{b_n}(v)$ we have

$$\int_{\mathbb{D}_{R'}} |f(w)|^p \ \ln^s \frac{1}{|\varphi_{b_n}(w)|}\, dA_q(w) = \int_{\varphi_{b_n}(\mathbb{D}_{R'})} |f(\varphi_{b_n}(v))|^p \ln^s \frac{1}{|v|} \cdot |\varphi'_{b_n}(v)|^2 (1-|\varphi_{b_n}(v)|^2)^q \, dA(v) \, .$$

Then by Lebesgue's theorem

$$\lim_{n\to\infty} \int_{\mathbb{D}_{R''}} I_n(w)\, dA(w) = \int_{\mathbb{D}_{R''}} I_z(w)\, dA(w) \, .$$

After the previous lemma, this concludes the proof. $\square$

Proposition 3.10. *Let $0 < p < \infty$, $-2 < q < \infty$, $0 \leq s < \infty$. Suppose that $f \in \mathcal{H}$ is such that $l_{p,q,s}(f)(z) < \infty$ for all $z \in \mathbb{D}$. Then $l_{p,q,s}(f)$ is a continuous function on $\mathbb{D}$.*

Proof. If $f = 0$ on $\mathbb{D}$, it is clear that $l_{p,q,s}(f)$ is continuous. Therefore, we suppose that $f \neq 0$, in particular $l_{p,q,s}(f)(0) \neq 0$. Let $z \in \mathbb{D}$ be fixed and let $\delta > 0$ be such that $\overline{\mathbb{D}}_\delta(z) \subset \mathbb{D}$. The function $l : \overline{\mathbb{D}} \times \overline{\mathbb{D}}_\delta(z) \to \mathbb{R}$ defined by

$$(w,\zeta) \to \frac{(1-|\zeta|^2)^s}{|1-\overline{\zeta} w|^{2s}}$$

is uniformly continuous on $\overline{\mathbb{D}} \times \overline{\mathbb{D}}_\delta(z)$. Then given $\epsilon > 0$, there exists $\rho > 0$ such that if $|w' - w| < \rho$ and $|\zeta' - \zeta| < \rho$ then

$$|l(w', \zeta') - l(w, \zeta)| < \frac{\epsilon}{l_{p,q,s}(f)(0)},$$

therefore if $|z - z'| < \rho$ then

$$\begin{aligned} &|l_{p,q,s}(f)(z) - l_{p,q,s}(f)(z')| \\ &\leq \int_{\mathbb{D}} |f(w)|^p \left(1 - |w|^2\right)^s |l(w, z) - l(w, z')| \; dA_q(w) < \epsilon \,. \end{aligned}$$

□

Corollary 3.11. *Let $0 < p < \infty$, $-2 < q < \infty$, $0 \leq s < \infty$. Then $L_0(p, q, s) \subset L(p, q, s)$ and ${}_{s,0}\mathcal{A}^p_q \subset {}_s\mathcal{A}^p_q$.*

Proof. If $f = 0$ it is clear. Suppose $f \neq 0$ and $f \in L_0(p, q, s)$. Then there exists $0 < R < 1$ such that $l_{p,q,s}(f)(z) < l_{p,q,s}(f)(0)$ for all $R < |z| < 1$. By Proposition 3.10, $l_{p,q,s}(f)$ attains its finite maximum on $\mathbb{D}_R$, then $f \in L(p, q, s)$. The other proof is similar. □

Corollary 3.12. *Let $0 < p < \infty$, $-2 < q < \infty$ and $0 \leq s < \infty$. If $f \in L_0(p, q, s)$, then $l_{p,q,s}(f)$ is a uniformly continuous function on $\overline{\mathbb{D}}$. If $f \in {}_{s,0}\mathcal{A}^p_q$, then $h_{p,q,s}(f)$ is a uniformly continuous function on $\overline{\mathbb{D}}$.*

Considering the estimations in the proof of Theorem 3.2, we get the following result.

Corollary 3.13. *Let $1 \leq p < \infty$, $-2 < q < \infty$ and $0 < s < \infty$ with $-1 < q+s$. Then, ${}_{s,0}\mathcal{A}^p_q$ and $L_0(p, q, s)$ are linear Banach subspaces of ${}_s\mathcal{A}^p_q$ and $L(p, q, s)$ respectively.*

Proof. It is enough to prove that ${}_{s,0}\mathcal{A}^p_q$ is a closed subspace of ${}_s\mathcal{A}^p_q$. Let $\{f_n\} \subset {}_{s,0}\mathcal{A}^p_q$ be a sequence converging to f. By Theorem 3.2, $f \in {}_s\mathcal{A}^p_q$. Therefore for any $\epsilon > 0$, there exists $N \in \mathbb{N}$, such that $\| f - f_n \|_g < \frac{\epsilon}{2}$ for all $n \geq N$. By hypothesis there exists a $0 < R < 1$ such that if $z \in \mathbb{A}_R$ then

$$\int_{\mathbb{D}} |f_N(w)|^p g^s(w, z) \, dA_q(w) < \frac{\epsilon^p}{2^p} \, .$$

Thus

$$\begin{aligned} \left(\int_{\mathbb{D}} |f(w)|^p g^s(w, z) \, dA_q(w)\right)^{\frac{1}{p}} &\leq \left(\int_{\mathbb{D}} |f(w) - f_N(w)|^p g^s(w, z) \, dA_q(w)\right)^{\frac{1}{p}} \\ &\quad + \left(\int_{\mathbb{D}} |f_N(w)|^p g^s(w, z) \, dA_q(w)\right)^{\frac{1}{p}} \\ &< \frac{\epsilon}{2} + \frac{\epsilon}{2} = \epsilon \, . \end{aligned}$$

The other proof is similar. □

Remark 3.14. For $0 < p < 1$ we get similar results with the respective metrics.

4. The equality ${}_s\mathcal{A}^p_q = L(p,q,s)$

In this section we prove the equality between the spaces ${}_s\mathcal{A}^p_q$ and $L(p,q,s)$.

Theorem 4.1. *Let $0 < p < \infty$, $-2 < q < \infty$ and $f \in \mathcal{H}$. Suppose that $l_{p,q,s}(f)(0) < \infty$, then:*

a) *For $0 < s < \infty$, we have*

$$\int_{\mathbb{D}} |f(w)|^p \ln^s \frac{1}{|w|} \, dA_q(w) \le t \int_{\mathbb{D}} |f(w)|^p (1-|w|^2)^s \, dA_q(w), \tag{4.1}$$

where $t = t(q,s,R)$ for some $0 < R < 1$ fixed.

b) *For $0 < s < 1$, we have*

$$\int_{\mathbb{D}} |f(w)|^p |w|^{-2s} (1-|w|^2)^s \, dA_q(w) \le \tilde{t} \int_{\mathbb{D}} |f(w)|^p (1-|w|^2)^s \, dA_q(w) \tag{4.2}$$

where $\tilde{t} = \tilde{t}(q,s,R)$ for some $0 < R < 1$ fixed.

Proof. a) We follow the idea of the proof given in Theorem 2.2 of Aulaskari et al. [1] with all its details. Let $c = .0183403$ be the root of $-\ln x = 4(1-x^2)$. Let $c < R < 1$ be fixed. Define

$$\begin{aligned} 0 < \frac{1}{\tau(q,s,R)} &= \int_c^R (1-r^2)^{q+s} r \, dr \\ &= \frac{1}{2(1+q+s)} ((1-c^2)^{1+q+s} - (1-R^2)^{1+q+s}) \\ &= \frac{1}{2(1+q+s)} (.999664^{1+q+s} - (1-R^2)^{1+q+s}) \, . \end{aligned}$$

As $|f|^p$ is subharmonic

$$\begin{aligned} \frac{1}{\tau(q,s,R)} \int_0^{2\pi} \left|f(ce^{i\theta})\right|^p d\theta &= \int_c^R (1-r^2)^{q+s} r \, dr \int_0^{2\pi} \left|f(ce^{i\theta})\right|^p d\theta \\ &\le \int_c^R (1-r^2)^{q+s} r \, dr \int_0^{2\pi} \left|f(re^{i\theta})\right|^p d\theta \\ &= \int_{\mathbb{D}_R \setminus \mathbb{D}_c} |f(w)|^p (1-|w|^2)^s \, dA_q(w) \, . \end{aligned}$$

Therefore

$$\int_0^{2\pi} |f(ce^{i\theta})|^p \, d\theta \le \tau(q,s,R) \int_{\mathbb{D}_R} |f(w)|^p (1-|w|^2)^s \, dA_q(w) \, . \tag{4.3}$$

Define

$$0 < \tilde{\tau}(q,s) = \int_0^c r(1-r^2)^q \ln^s \frac{1}{r} \, dr. \tag{4.4}$$

By subharmonicity and (4.3), we have the estimation

$$\begin{aligned}
&\int_{\mathbb{D}_c} |f(w)|^p (1-|w|^2)^q \ln^s \frac{1}{|w|}\, dA(w) \\
&\quad = \int_0^c \int_0^{2\pi} |f(re^{i\theta})|^p r(1-r^2)^q \ln^s \frac{1}{r}\, d\theta\, dr \\
&\quad \le \int_0^c \int_0^{2\pi} |f(ce^{i\theta})|^p r(1-r^2)^q \ln^s \frac{1}{r}\, d\theta\, dr \\
&\quad = \int_0^c r(1-r^2)^q \ln^s \frac{1}{r}\, dr \int_0^{2\pi} |f(ce^{i\theta})|^p\, d\theta \\
&\quad \le \tau(q,s,R)\tilde{\tau}(q,s) \int_{\mathbb{D}_R} |f(w)|^p (1-|w|^2)^s\, dA_q(w) \\
&\quad \le \tau(q,s,R)\tilde{\tau}(q,s) \int_{\mathbb{D}} |f(w)|^p (1-|w|^2)^s\, dA_q(w).
\end{aligned}$$

From the inequality

$$-\ln x \le 4(1-x^2) \quad \text{for each } x \in (c,1], \tag{4.5}$$

we have

$$\begin{aligned}
&\int_{\mathbb{D}\setminus\mathbb{D}_c} |f(w)|^p (1-|w|^2)^q \ln^s \frac{1}{|w|}\, dA(w) \\
&\quad \le 4^s \int_{\mathbb{D}\setminus\mathbb{D}_c} |f(w)|^p (1-|w|^2)^s\, dA_q(w) \qquad (4.6)\\
&\quad \le 4^s \int_{\mathbb{D}} |f(w)|^p (1-|w|^2)^s\, dA_q(w).
\end{aligned}$$

Let $t(q,s,R) = \tau(q,s,R)\tilde{\tau}(q,s) + 4^s$. Combining (4.5) and (4.6), we have

$$\int_{\mathbb{D}} |f(w)|^p (1-|w|^2)^q \ln^s \frac{1}{|w|}\, dA(w) \le t(q,s,R) \int_{\mathbb{D}} |f(w)|^p (1-|w|^2)^s\, dA_q(w).$$

b) For $0 < s < 1$, we need to consider instead of (4.4), the following equality

$$0 < \int_0^c r^{1-2s}(1-r^2)^{q+s}\, dr = \frac{1}{2} B[c^2, 1-s, 1+q+s],$$

where B denotes the incomplete Beta function. Then we prove the formula (4.2) in a similar way to (4.1). □

Theorem 4.2. *Let $0 < p < \infty$, $-2 < q < \infty$ and $0 \le s < \infty$. Then ${}_s\mathcal{A}^p_q = L(p,q,s)$.*

Proof. We have

$$1 - x^2 \le -2\ln x \quad \text{for each } x \in (0,1]\ .$$

Taking $x = \varphi_z(w)$ we have $1 - |\varphi_z(w)|^2 \le 2g(w,z)$. Therefore

$$l_{p,q,s}(f)(z) \le 2h_{p,q,s}(f)(z) \quad \text{for each } z \in \mathbb{D}\ . \tag{4.7}$$

Thus ${}_s\mathcal{A}^p_q \subset L(p,q,s)$.

We prove now that $L(p,q,s) \subset {}_s\mathcal{A}^p_q$. For this, let $f \in L(p,q,s)$, then $l_{p,q,s}(f)(z) < \infty$. By hypothesis and Lemma 3.7, $l_{p,q,s}(f\circ\varphi)(0) < \infty$. Since $|f\circ\varphi_a|^p$ is subharmonic, the formula (4.1) reads

$$\int_{\mathbb{D}} |f(\varphi_z(w)) \ln^s \frac{1}{|w|}\, dA_q(w) \leq t(q,s,R) \int_{\mathbb{D}} |f(\varphi_z(w))|^p (1-|w|^2)^s\, dA_q(w)\ .$$

Consider the change of variable $w = \varphi_z(v)$ to obtain

$$\begin{aligned}&\int_{\mathbb{D}} |f(v)|^p |\varphi_z'(v)|^2 (1-|\varphi_z(v)|^2)^q \ln^s \frac{1}{|\varphi_z(v)|}\, dA(v)\\ &\quad\leq t(q,s,R) \int_{\mathbb{D}} |f(v)|^p |\varphi_z'(v)|^2 (1-|\varphi_z(v)|^2)^{q+s}\, dA(v)\ ,\end{aligned}$$

or, equivalently,

$$\begin{aligned}0 \leq &\int_{\mathbb{D}} |f(v)|^p |\varphi_z'(v)|^2 (1-|\varphi_z(v)|^2)^q\\ &\cdot \left(t(q,s,R)(1-|\varphi_z(v)|^2)^s - \ln^s \frac{1}{|\varphi_z(v)|}\right) dA(v)\\ \leq\ &\rho(z,|q|+2) \int_{\mathbb{D}} |f(v)|^p (1-|v|^2)^q\\ &\cdot \left(t(q,s,R)(1-|\varphi_z(v)|^2)^s - \ln^s \frac{1}{|\varphi_z(v)|}\right) dA(v)\end{aligned}$$

then we obtain

$$\int_{\mathbb{D}} |f(v)|^p \ln^s \frac{1}{|\varphi_z(v)|}\, dA_q(v) \leq t(q,s,R) \int_{\mathbb{D}} |f(v)|^p (1-|\varphi_z(v)|^2)^s\, dA_q(v), \quad (4.8)$$

and the theorem follows. □

Corollary 4.3. *Let $0 < p < \infty$, $-2 < q < \infty$ and $0 \leq s < \infty$. Then ${}_{s,0}\mathcal{A}^p_q = L_0(p,q,s)$.*

Proof. It is a consequence of the formulas (4.7) and (4.8) since

$$l_{p,q,s}(f)(z) \leq 2h_{p,q,s}(f)(z) \leq 2t(q,s,R) l_{p,q,s}(f)(z)\ .$$ □

Observe that we have used a completely different idea in the proof of the previous theorem to the used by Zhao in Theorem 2.4 of [7].

From now on we will use the notation ${}_s\mathcal{A}^p_q$ instead $L(p,q,s)$.

Theorem 4.4. *Let $0 < p < \infty$ and $0 \leq s < \infty$. Then, for $-1 < q < \infty$,*

$$H^p \subset \bigcap_{-1<q,\ 0\leq s} {}_{s,0}\mathcal{A}^p_q\ .$$

Proof. Let $f \in H^p$ and $\epsilon > 0$. Define

$$M = \sup_{0\leq r<1} \int_0^{2\pi} |f(re^{i\theta})|^p d\theta\ < \infty\ .$$

There exists $0 < r < 1$ such that

$$\int_{\mathbb{D}\setminus\mathbb{D}_r} |f(w)|^p \, dA_q(w) = \int_r^1 (1-\rho^2)^q \rho \int_0^{2\pi} |f(\rho e^{i\theta})|^p d\theta \, dr \le \frac{M}{2} \frac{(1-r^2)^{q+1}}{q+1} < \frac{\epsilon}{2} \, .$$

Since $(1 - |\varphi_z(w)|^2)^s < 1$, we decompose

$$l_{p,q,s}(f)(z) \le \max_{w\in\overline{\mathbb{D}}_r} |f(w)|^p \int_{\mathbb{D}_r} (1 - |\varphi_z(w)|^2)^s \, dA_q(z) + \int_{\mathbb{D}\setminus\mathbb{D}_r} |f(z)|^p \, dA_q(z) \, ,$$

then by Lemma 2.2 we conclude the proof by taking the limit when $|z| \to 1^-$. □

It is possible to replace the weight $(1-|\varphi_z(w)|^2)^s$ by its reflection $(|\varphi_z(w)|^{-2} - 1)^s$, as the following theorem shows.

Theorem 4.5. *Let $0 < p < \infty$, $-2 < q < \infty$ and $0 < s < 1$. Then, $f \in {}_s\mathcal{A}^p_q$ if and only if*

$$\sup_{z\in\mathbb{D}} \int_{\mathbb{D}} |f(w)|^p (|\varphi_z(w)|^{-2} - 1)^s \, dA_q(w) < \infty \, .$$

Proof. As $x + \dfrac{1}{x} \ge 2$ for $x > 0$ and x^s is nondecreasing for $s > 0$, we have in general,

$$\int_{\mathbb{D}} |f(w)|^p (1 - |\varphi_z(w)|^2)^s \, dA_q(w) \le \int_{\mathbb{D}} |f(w)|^p (|\varphi_z(w)|^{-2} - 1)^s \, dA_q(w) \, .$$

We claim that

$$\int_{\mathbb{D}} |f(w)|^p (|\phi_z(w)|^{-2} - 1)^s \, dA_q(w) \le \tilde{t} \int_{\mathbb{D}} |f(w)|^p (1 - |\phi_z(w)|^2)^s \, dA_q(w)$$

where $\tilde{t} = \tilde{t}(q, s, R)$ is as in (4.2). If the result were not true, by change of variable formula with $w = \varphi_z(v)$ and transposing terms,

$$\begin{aligned} 0 &< \int_{\mathbb{D}} |f(\varphi_z(v))|^p (1 - |\varphi_z(v)|^2)^q |\varphi_z'(v)|^2 (|v|^{-2s} - \tilde{t}) \, dA_s(v) \\ &\le \rho(z, 2 + |q|) \int_{\mathbb{D}} |f(\varphi_z(v))|^p (1 - |v|^2)^q (|v|^{-2s} - \tilde{t}) dA_s(v), \end{aligned}$$

which leads to a contradiction with the inequality (4.2), since $|f(\varphi_z(v))|^p$ is a subharmonic function. □

5. Strict inclusions of the spaces $A(p, q, s)$

In this section we will prove the strict inclusions between weighted Bergman spaces for $0 < s \le 1$. For this, we need the following two lemmas. The first is due to A. Zygmund, (see [9]).

Lemma 5.1. *Let $0 < p < \infty$. If $\{n_k\}$ is an increasing sequence of positive integers satisfying $\frac{n_{k+1}}{n_k} \ge \lambda > 1$, for all $k \in \mathbb{N}$, then, there exists a constant $A > 0$,* *depending only on p and λ, such that*

$$\frac{1}{A}\Big(\sum_{k=1}^{\infty}|a_k|^2\Big)^{\frac{1}{2}} \le \Big(\frac{1}{2\pi}\int_0^{2\pi}\Big|\sum_{k=1}^{\infty}a_k e^{in_k\theta}\Big|^p\,d\theta\Big)^{\frac{1}{p}} \le A\Big(\sum_{k=1}^{\infty}|a_k|^2\Big)^{\frac{1}{2}},$$

for any numbers a_k, $k\in\mathbb{N}$.

For $n\in\mathbb{N}$, define

$$I_n = \{\, k\in\mathbb{N} \,:\, 2^n \le k < 2^{n+1}\}\,.$$

The following lemma was proved by Mateljevic and Pavlovic in [3].

Lemma 5.2. *Let $0<\alpha<\infty$ and $0<p<\infty$. Let $f(x)=\sum_{n=1}^{\infty}a_n$, with $0\le x<1$, $a_n\ge 0$ for each $n\in\mathbb{N}$. Then there exists a constant $K>0$ depending only on p and α such that*

$$\frac{1}{K}\sum_{n=0}^{\infty}\frac{t_n^p}{2^{n\alpha}} \le \int_0^1(1-x)^{\alpha-1}f(x)^p\,dx \le K\sum_{n=0}^{\infty}\frac{t_n^p}{2^{n\alpha}},$$

where $t_n=\sum_{k\in I_n}a_k$.

Theorem 5.3. *Let $0<p<\infty$, $-2<q<\infty$ and $0<s<1$ with $-1<q+s$. If $f\in\mathcal{H}$ has a power series expansion given by*

$$f(w)=\sum_{n=0}^{\infty}a_n w^n\,,$$

and

$$\int_0^1\Big(\sum_{n=1}^{\infty}|a_n|r^n\Big)^p(1-r^2)^{q+s}\,dr<\infty\,, \tag{5.1}$$

then $f\in {}_{s,0}\mathcal{A}_q^p$.

Proof. Let $0<R<1$. Consider the following estimation

$$\begin{aligned} l_{p,q,s}(a) \le{} & \sup_{w\in\mathbb{D}_R}|f(w)|^p\int_{\mathbb{D}_R}(1-|\varphi_z(w)|^2)^s\,dA_q(w)\\ &+\int_R^1\Big(\sum_{n=1}^{\infty}|a_n|r^n\Big)^p(1-r^2)^{q+s}r(1-|a|^2)^s\int_0^{2\pi}\frac{d\theta}{|1-\bar{a}re^{i\theta}|^{2s}}\,dr. \end{aligned}$$

We conclude the proof applying Lemmas 2.2, 2.1 and taking the limit when $|z|\to 1^-$. □

Corollary 5.4. *Let $0<p<\infty$, $-2<q<\infty$ and $0<s\le 1$, with $-1<q+s$. Suppose that $f\in\mathcal{H}$ has a power series expansion given by*

$$f(w)=\sum_{n=0}^{\infty}a_n w^n\,.$$

If

$$\sum_{n=0}^{\infty} \frac{1}{2^{n(q+s+1)}} \left(\sum_{k \in I_n} |a_k| \right)^p < \infty$$

then $f \in {}_{s,0}\mathcal{A}^p_q$.

Proof. It is an immediate consequence of Theorem 5.3, condition (5.1) and Lemma 5.2. □

Given $f \in \mathcal{H}$ with power series expansion

$$f(w) = \sum_{n=0}^{\infty} a_n w^n$$

we say that f has Hadamard gaps of length $\lambda > 1$, if there exists an increasing sequence $\{n_k\} \subset \mathbb{N}$ such that

$$a_n = \begin{cases} 0 & \text{si } n \neq n_k \\ a_{n_k} & \text{si } n = n_k, \end{cases}$$

with

$$\frac{n_{k+1}}{n_k} \geq \lambda > 1 , \qquad \text{for } k \in \mathbb{N}.$$

We rewrite simply

$$f(w) = \sum_{k=0}^{\infty} a_k w^{n_k} .$$

Observe that the number of Taylor coefficients a_j is at most $[\log_\lambda 2] + 1$ when $n_j \in I_k$. The following theorem characterizes Lacunary series with Hadamard gaps in ${}_{s,0}\mathcal{A}^p_q$ and ${}_s\mathcal{A}^p_q$

Theorem 5.5. *Let* $0 < p < \infty$, $-2 < q < \infty$ *and* $0 < s \leq 1$ *with* $-1 < q + s$. *Suppose that* $f \in \mathcal{H}$ *has a power series expansion*

$$f(z) = \sum_{k=0}^{\infty} a_k z^{n_k}$$

and Hadamard gaps of length $\lambda > 1$. *Then the following statements are equivalent:*

i) $f \in {}_{s,0}\mathcal{A}^p_q$;
ii) $f \in {}_s\mathcal{A}^p_q$;
iii) $f \in \mathcal{A}^p_{q+s}$;
iv) *the series*

$$\sum_{k=0}^{\infty} \frac{1}{2^{k(1+q+s)}} \left(\sum_{n_j \in I_k} |a_{n_j}|^2 \right)^{\frac{p}{2}} \tag{5.2}$$

is convergent.

Proof. By Corollary 3.11, it follows that i) implies ii). Now ${}_s\mathcal{A}^p_q \subset \mathcal{A}^p_{q+s}$, so ii) implies iii). We see now that iii) implies iv). By Lemmas 5.1 and 5.2, there exist $A > 0$ and $K > 0$, such that

$$\begin{aligned}\int_{\mathbb{D}} |f(w)|^p \, dA_{q+s}(w) &= \int_{\mathbb{D}} \left| \sum_{k=0}^{\infty} a_k z^{n_k} \right|^p dA_{q+s}(w) \\ &= \int_0^1 \int_0^{2\pi} \left| \sum_{k=0}^{\infty} a_k r^{n_k} e^{in_k\theta} \right|^p (1-r^2)^{q+s} \, r \, d\theta \, dr \\ &\geq \frac{2\pi}{A^p} \int_0^1 \left(\sum_{k=0}^{\infty} |a_k|^2 r^{2n_k} \right)^{\frac{p}{2}} r(1-r^2)^{q+s} \, dr \\ &\geq \frac{\pi}{KA^p} \sum_{k=0}^{\infty} \frac{1}{2^{k(q+s+1)}} \left(\sum_{n_j \in I_k} |a_j|^2 \right)^{\frac{p}{2}}\end{aligned}$$

and the series (5.2) is convergent.

We prove that iv) implies i). Since

$$\left(\sum_{n_j \in I_n} |a_j| \right)^p \leq 2^{\frac{p}{2}} ([\log_\lambda 2] + 1)^{\frac{p}{2}} \left(\sum_{n_j \in I_k} |a_j|^2 \right)^{\frac{p}{2}}$$

by Corollary 5.4 we have that $f \in {}_{s,0}\mathcal{A}^p_q$. □

Corollary 5.6. *Let $0 < p < \infty$, $-2 < q < \infty$ and $0 < s < 1$ be with $-1 < q + s$. Then the inclusions*

$${}_s\mathcal{A}^p_q \subset \bigcap_{s<t\leq 1} {}_t\mathcal{A}^p_q \quad \text{and} \quad {}_{s,0}\mathcal{A}^p_q \subset \bigcap_{s<t\leq 1} {}_{t,0}\mathcal{A}^p_q$$

are strict.

Proof. Define $f : \mathbb{D} \to \mathbb{C}$ by

$$f(w) = \sum_{n=0}^{\infty} 2^{\frac{n(q+s+1)}{p}} w^{2^n} .$$

Then, $f \in \mathcal{H}$ and for $s < t \leq 1$,

$$\sum_{n=0}^{\infty} \frac{1}{2^{n(1+q+t)}} (|a_n|^2)^{\frac{p}{2}} = \sum_{n=0}^{\infty} \frac{2^{n(q+s+1)}}{2^{n(1+q+t)}} = \sum_{n=0}^{\infty} \frac{1}{2^{(t-s)n}} < \infty$$

and

$$\sum_{n=0}^{\infty} \frac{1}{2^{n(1+q+s)}} (|a_n|^2)^{\frac{p}{2}} = \sum_{n=0}^{\infty} \frac{2^{n(q+s+1)}}{2^{n(1+q+s)}} = \sum_{n=0}^{\infty} 1 = \infty .$$

Therefore by Theorem 5.5

$$f \in \bigcap_{s<t\leq 1} {}_{t,0}\mathcal{A}^p_q \quad \text{but} \quad f \notin {}_{s,0}\mathcal{A}^p_q .$$

□

Theorem 5.7. *Let $0 < p < \infty$, $-2 < q < \infty$ and $0 < s < 1$ with $-1 < q + s$. If $p = 2k$ with $k \in \mathbb{N}$ then, $f \in {}_s\mathcal{A}^p_q$ if and only if*

$$\sup_{z\in\mathbb{D}} \sum_{n=0}^{\infty} \frac{(1-|z|^2)^s}{(n+1)^{q+s+1}} \left| \sum_{m=0}^{n} \frac{a_n^{\{k\}}\Gamma(n-m+q+s)}{(n-m)!}\overline{z^{n-m}} \right|^2 < \infty \, ,$$

where

$$a_n^{\{k\}} = \frac{1}{2\pi i} \int_0^{2\pi} \frac{f^k(\zeta)}{\zeta^{n+1}}\, d\zeta \, . \tag{5.3}$$

Proof. As $|f(w)|^p = |f^k(w)|^2$ with $a_n^{\{k\}}$ given by (5.3), we have

$$\begin{aligned}
\frac{f^k(w)}{(1-\overline{z}w)^s} &= \sum_{n=0}^{\infty} a_n^{\{k\}} w^n \sum_{n=0}^{\infty} \frac{\Gamma(n+s)}{n!\Gamma(s)}\overline{z^n} w^n \\
&= \sum_{n=0}^{\infty}\sum_{m=0}^{n} \frac{a_m^{\{k\}}\Gamma(n-m+s)}{(n-m)!\Gamma(s)}\overline{z^{n-m}} w^m w^{n-m} = \sum_{n=0}^{\infty}\left(\sum_{m=0}^{n} f_{n,m}\right) w^n ,
\end{aligned}$$

where,

$$f_{m,n} = f_{a,m,n,k,s} = \frac{\Gamma(n-m+s)a_m^{\{k\}}\overline{z^{n-m}}}{(n-m)!\Gamma(s)} \, .$$

With this notation, we have

$$\begin{aligned}
&\int_{\mathbb{D}} \frac{|f(w)|^p}{|1-\overline{z}w|^{2s}}(1-|w|^2)^s\, dA_q(w) = \int_{\mathbb{D}} \frac{|f^k(w)|^2}{|1-\overline{z}w|^{2s}}(1-|w|^2)^s\, dA_q(w) \\
&= \int_{\mathbb{D}} \left| \sum_{n=0}^{\infty} \Big(\sum_{m=0}^{n} f_{n,m}\Big) w^n \right|^2 (1-|w|^2)^{q+s}\, dA(w) \\
&= \int_0^1 \int_0^{2\pi} \left| \sum_{n=0}^{\infty}\left(\sum_{m=0}^{n} f_{n,m}\right) e^{in\theta} r^n \right|^2 r(1-r^2)^{q+s}\, d\theta\, dr \, .
\end{aligned}$$

Next, we define

$$q(\theta) = \sum_{n=0}^{\infty}\left(\sum_{m=0}^{n} f_{n,m}\right) e^{in\theta}\ r^n \, . \tag{5.4}$$

We now calculate the Fourier coefficient of $q(\theta)$, that is,

$$\begin{aligned}
\langle q(\theta), e^{-ik\theta}\rangle &= \sum_{n=0}^{\infty}\left(\sum_{m=0}^{n} f_{n,m}\right) \int_0^{2\pi} r^n\ e^{i(n-k)\theta}\, d\theta = 2\pi\ r^k \sum_{m=0}^{k} f_{k,m} \\
&= 2\pi\ r^n \sum_{m=0}^{n} f_{n,m} \, .
\end{aligned}$$

By Parseval's identity

$$\int_0^{2\pi} \left| \sum_{n=0}^{\infty}\left(\sum_{m=0}^{n} f_{n,m}\right) e^{in\theta}\ r^n \right|^2 d\theta = \frac{1}{2\pi} \sum_{n=0}^{\infty} \left| \sum_{m=0}^{n} f_{n,m} \right|^2 r^{2n} \, .$$

Now, we have

$$\int_0^1 r^{2n+1}(1-r^2)^{q+s}dr = \frac{1}{2}B(n+1,q+s+1) \approx \frac{\Gamma(q+s+1)}{2(n+1)^{q+s+1}}\ .$$

So finally, we get the theorem. □

Taking $z=0$ in the previous theorem we obtain the following result.

Corollary 5.8. *Let $0<p<\infty$, $-2<q<\infty$ and $0<s<1$ with $-1<q+s$. Let $f\in\mathcal{H}$ with power series given by $\sum_{n=0}^{\infty} a_n w^n$. If $p=2k$ with $k\in\mathbb{N}$, then $f\in\mathcal{A}^p_{q+s}$ if and only if*

$$\sum_{n=0}^{\infty}\frac{|a_n|^2}{(n+1)^{q+s}}<\infty\ .$$

6. $F(p,q,s)$ and ${}_s\mathcal{A}^p_q$ spaces

For $S\subset\mathcal{H}$, we define the primitive set of S as

$$P(S)=\{\ h\in\mathcal{H}\ :\ h'\in S\ \}\ .$$

Observe that if $f\in {}_s\mathcal{A}^p_q$ for $0<p<\infty$, $-2<q<\infty$, $0\le s<\infty$, and h is a primitive of f, then $h\in F(p,q,s)$. Since by definition

$$\sup_{z\in\mathbb{D}}\int_{\mathbb{D}}|h'(z)|^p g^s(z,a)\,dA_q(z)<\infty\ ,$$

that is, $P({}_s\mathcal{A}^p_q)=F(p,q,s)$. Then by Theorem 4.2.1 of J. Rättyä we have a characterization of the ${}_s\mathcal{A}^p_q$ spaces in terms of higher derivatives with $1<p<\infty$.

Theorem 6.1 ([5, Theorem 4.2.1]). *Let $f\in\mathcal{H}$ and h a primitive of f. Let $1<p<\infty$, $-2<q<\infty$, $0\le s<\infty$ and $n\in\mathbb{N}$ be with $-1<q+s$. Then, the following conditions are equivalent:*

a) $h\in F(p,q,s)$;
b) $f\in {}_s\mathcal{A}^p_q$;
c) $\sup_{a\in\mathbb{D}}\int_{\mathbb{D}}|f^{(n-1)}(z)|^p(1-|z|^2)^{np-p}(1-|\varphi_a(z)|^2)^s\,dA_q(z)<\infty$;
d) $\sup_{a\in\mathbb{D}}\int_{\mathbb{D}}|f^{(n-1)}(\varphi_a(z))|^p|\varphi_a'(z)|^{np-p+q+2}(1-|z|^2)^{np-p+s}\,dA_q(z)<\infty$;
e) $\sup_{a\in\mathbb{D}}\int_{\mathbb{D}}|f^{(n-1)}(z)|^p(1-|z|^2)^{np-p}g^s(z,a)\,dA_q(z)<\infty$.

By Theorem 4.2.2 of J. Rättyä we have some unexpected relationships between ${}_s\mathcal{A}^p_q$ spaces and $F(p,q,s)$.

Theorem 6.2 **([5, Theorem 4.2.2]).** *Let $h \in \mathcal{H}$. Let $1 < p < \infty$, $-2 < q < \infty$ and $0 \le s < \infty$ with $-1 < q + s - p$. Then the following conditions are equivalent:*

a) $h \in F(p, q, s)$;

b) $\sup_{a\in\mathbb{D}} \int_{\mathbb{D}} |h(z)|^p (1-|z|^2)^{-p}(1-|\varphi_a(z)|^2)^s \, dA_q(z) < \infty$;

c) $\sup_{a\in\mathbb{D}} \int_{\mathbb{D}} |h(z)|^p (1-|z|^2)^{-p} g^s(z,a) \, dA_q(z) < \infty$.

Corollary 6.3. *For $h \in \mathcal{H}$, let $1 < p < \infty$, $-2 < q < \infty$ and $0 \le s < \infty$ with $-1 < q + s - p$ and $-2 < q - p$. Then $F(p, q, s) =_s \mathcal{A}^p_{q-p}$. In particular,*

$$P(F(p,q,s)) = P({}_s\mathcal{A}^p_{q-p}) = F(p, q-p, s) .$$

7. Carleson measures

In this section we use Carleson type measures to characterize ${}_s\mathcal{A}^p_q$.
For $0 < s < \infty$, we say that a positive measure μ defined on $\mathbb{D}$ is a bounded s-Carleson measure provided

$$\sup_{I\subset\partial\mathbb{D}} \frac{\mu(S(I))}{|I|^s} < \infty$$

where $|I|$ denotes the arc length of $I \subset \partial\mathbb{D}$ and $S(I)$ denotes the Carleson box based on I, that is

$$S(I) = \left\{ w \in \mathbb{D} \ : \ \frac{w}{|w|} \in I, \ 1-|w| \le \frac{|I|}{2\pi} \right\} .$$

If the next limit exists

$$\lim_{|I|\to 0} \frac{\mu(S(I))}{|I|^s}$$

for $I \subset \partial\mathbb{D}$, we say that μ is a compact s-Carleson measure. The following lemma characterizes s-Carleson measures (see Lemma 2.2 from [1]).

Lemma 7.1. *Let $0 < s < \infty$ and μ be a positive measure on $\mathbb{D}$. Then*

i) *μ is a bounded s-Carleson measure if and only if*

$$\sup_{z\in\mathbb{D}} \int_{\mathbb{D}} |\varphi_z'(w)|^s \, d\mu(w) < \infty ;$$

ii) *μ is a compact s-Carleson measure if and only if*

$$\lim_{|z|\to 1} \int_{\mathbb{D}} |\varphi_z'(w)|^s \, d\mu(w) = 0 .$$

The following result is immediate.

Theorem 7.2. *Let $0 < s < \infty$, $0 < p < \infty$ and $-2 < q < \infty$. Consider $f \in \mathcal{H}$ and $d\mu_f(w) = |f(w)|^p (1-|w|^2)^{q+s} \, dA(w)$. Then,*

i) *$f \in {}_s\mathcal{A}^p_q$ if and only if $d\mu_f(w)$ is a bounded s-Carleson measure.*

ii) *$f \in {}_{s,0}\mathcal{A}^p_q$ if and only if $d\mu_f(w)$ is a compact s-Carleson measure.*

Acknowledgment

Many thanks to the Referee.

References

[1] R. Aulaskari, D. Stegenga and J. Xiao, *Some subclasses of $BMOA$ and their characterization in terms of Carleson measures*, Rocky Mountain J. Math. **26**, (1996), 485–506.

[2] E. Hewitt, K. Stromberg *Real and Abstract Analysis*, Springer-Verlag, 1975.

[3] M. Mateljevic, M. Pavlovic, *L^p behaviour of power series with positive coefficients and Hardy spaces*, Proc. Amer. Math. Soc. **87**, (1983), 309–316.

[4] J. Pérez H., L.F. Reséndis O. and L.M. Tovar S., *Quaternionic Bergman Spaces*, Preprint (2014), 1–12.

[5] J. Rättyä, *On some complex Function Spaces and Classes*, Ann. Acad. Scie. Fenn. Math. Diss. **124**, (2001), 1–73.

[6] R. Zhao, *On α-Bloch functions and $VMOA$*, Acta Mathematica Scientia, **16** (3), (1996), 349–360.

[7] R. Zhao, *On a general family of function spaces*, Ann. Acad. Scie. Fenn. Math. Diss. **105**, (1996), 1–56.

[8] K. Zhu, *Operator Theory in Function Spaces*, Marcel Dekker, New York, 1990.

[9] A. Zygmund, *Trigonometric Series. Cambridge*, Univ. Press, London and New York, 1959.

Luís Javier Carmona L.
Universidad Autónoma Metropolitana
Unidad Iztapalapa, Departamento de Matemáticas
Av. San Rafael Atlixco Num.186
C.P. 09340, México D.F.
e-mail: carmona_406@hotmail.com

Lino Feliciano Reséndis Ocampo
Universidad Autónoma Metropolitana
Unidad Azcapotzalco, Departamento de Ciencia Básicas
Av. San Pablo 180. Col. Reynosa Tamaulipas
C.P. 02200, México D.F.
e-mail: lfro@correo.azc.uam.mx

Luis Manuel Tovar Sánchez
Escuela Superior de Física y Matemáticas del IPN
Edif. 9, Unidad ALM,
Zacatenco del IPN.
C.P. 07300, México D.F.
e-mail: tovar@esfm.ipn.mx

Hypercomplex Analysis: New Perspectives and Applications
Trends in Mathematics, 111–118

On Appell Sets and Verma Modules for $\mathfrak{sl}(2)$

David Eelbode and Nikolaas Verhulst

Abstract. The aim of this paper is to introduce a general framework which can be used to generalize both Appell sets in multivariate analysis and special polynomials in a complex variable $z \in \mathbb{C}$, inspired by certain special functions appearing in harmonic and Clifford analysis on $\mathbb{R}^m$. As an illustration, we have a closer look at Hermite polynomials.

Mathematics Subject Classification (2010). 30G35; 17B10; 33C45.

Keywords. Appell sets, Verma modules, Hermite polynomials.

1. Introduction

Mathematical analysis and classical representation theory have always gained from a mutual infusion of ideas and techniques. The former often provides concrete examples which may illustrate and initiate more abstract notions which are studied and generalized in the latter. The present paper is written with this philosophy in mind, hereby drawing inspiration from two particular problems arising in harmonic analysis on $\mathbb{R}^m$. First of all, there exist quite a few function theories which are centered around a set of operators which realize a copy of the simple Lie algebra $\mathfrak{sl}(2)$ under the commutator bracket. Classical harmonic analysis itself, for example, is centered around the Laplace operator Δ_m, acting as an endomorphism on $\mathbb{C}[x_1, \ldots, x_m]$. The associated Lie algebra $\mathfrak{sl}(2)$ is then given by

$$\mathfrak{sl}(2) \cong \mathrm{Alg}(X, Y, H) \cong \mathrm{Alg}\left(\frac{1}{2}\Delta_m, -\frac{1}{2}|\underline{x}|^2, -\mathbb{E}_x - \frac{m}{2}\right), \tag{1.1}$$

where $|\underline{x}|^2$ is the squared norm of $\underline{x} \in \mathbb{R}^m$ and $\mathbb{E}_x = \sum x_j \partial_{x_j}$ stands for the Euler operator on $\mathbb{R}^m$. A celebrated result in harmonic analysis, due to R. Howe [10], describes this Lie algebra as the so-called dual partner of the orthogonal group $\mathrm{SO}(m)$, acting on polynomials in $\mathbb{C}[x_1, \ldots, x_m]$ through the regular representation, leading to a multiplicity-free decomposition of this space in terms of Verma modules for $\mathfrak{sl}(2)$. Also in super analysis (a function theory in which both commuting and anti-commuting variables are taken into account, see, e.g., [5]) and the theory of Dunkl operators (in which the rotational symmetry is reduced to a finite subgroup, see, e.g., [6, 7]), a key ingredient is the existence of a subalgebra $\mathcal{A} \cong \mathfrak{sl}(2)$ inside the full endomorphism algebra acting on the polynomials (or smooth functions in general). This algebra is then used to define generalizations of the classical Hermite polynomials, which lead to an expression for the Fourier integral as an exponential operator $\exp(A)$, with $A \in \mathcal{A}$.

A second motivation comes from the theory on Appell sequences. Classically, these are defined in terms of a complex variable z, as sets of holomorphic polynomials $\{P_k(z) : k \in \mathbb{Z}^+\}$ satisfying the relation $P_k'(z) := kP_{k-1}(z)$, where $\deg P_k(z) = k$ and $P_0 \neq 0$. Denoting the derivation operator by means of $P = \partial_z$, one can also interpret these Appell sequences as representations for the Heisenberg–Weyl algebra $\mathfrak{h}_1$, provided there exists an associated raising operator M satisfying $MP_k(z) = P_{k+1}(z)$ for all $k \in \mathbb{Z}^+$. Formally, this operator thus performs the integration within the Appell sequence. One can indeed verify that these defining relations imply that P and M satisfy the canonical commutation relation $[P, M] = 1 \in \mathbb{C} = \mathfrak{z}(\mathfrak{h}_1)$. Notable examples are the Hermite polynomials, but also other orthogonal polynomials such as Bernoulli and Euler polynomials. More generally, we say:

Definition 1.1. A representation for the Heisenberg algebra (denoted by means of $\mathfrak{h}_1$) is an algebra morphism $\rho_V : \mathfrak{h}_1 \to \mathrm{End}(\mathbb{V})$. If there exists a vector v_0 in $\ker \rho_V(P)$, we will from now on refer to the set $\mathbb{A} := \{\rho_V^k(M)[1] : k \in \mathbb{Z}^+\}$ as an Appell sequence related to the operators $(\rho_V(P), \rho_V(M))$.

In the classical situation, the vector space $\mathbb{V} = \mathrm{span}(P_k(z))_k$ is a polynomial family, with $\rho_V(P)$ a differential operator and $\rho_V(M)$ an conjugation thereof (a formal integration operator). In this paper, we will consider generalizations of Appell sequences, for which $\mathbb{V}$ may be a more abstract representation space. Note that the representation $\mathbb{V}$ must be infinite dimensional, which follows from the fact that $\mathrm{tr}([\rho_V(P), \rho_V(M)]) = 0$, as it is the trace of a commutator. This means that finite-dimensional representations must necessarily satisfy $\dim(\mathbb{V}) = 1$. This suggests looking for generalizations of Appell sequences in canonical infinite-dimensional modules for $\mathfrak{sl}(2)$, i.e., Verma modules.

Recently, the problem of constructing analogues of complex Appell sequences in multivariate analysis has gained new interest, see a.o. [3, 9, 11, 8, 2]. These sequences are defined as polynomial sets $\mathbb{V} = \mathrm{span}(P_k(\underline{x}))_k$ containing scalar-valued (resp. Clifford algebra-valued) null solutions for the Laplace or Dirac operator in the variable $\underline{x} \in \mathbb{R}^m$, for which the lowering operator P is a differential operator belonging to the Clifford–Weyl algebra $\mathcal{W}_m^C$, defined by $\mathcal{W}_m^C := \mathrm{Alg}(x_i; \partial_{x_j}) \otimes \mathbb{C}_m$ (with $\mathbb{C}_m$ the universal Clifford algebra in m dimensions). It turns out that some of these Appell sequences can be related to the branching problem for certain irreducible representations for the spin group, the (inductive) construction of orthonormal bases for these spaces and generalizations of the classical Fueter theorem, see, e.g., [1, 8]. This has given rise to Gegenbauer and Jacobi polynomials in harmonic (resp. Clifford) analysis, for which the formal integration operator was obtained in terms of an operator containing fractions.

2. Appell sets in Verma modules for $\mathfrak{sl}(2)$

Adopting the classical definition (which can be given for any Lie algebra $\mathfrak{g}$) to our case of interest, we have the following:

Definition 2.1. Let $\lambda \in \mathbb{C}$ be a complex number, and consider the vector space $V_\lambda = \mathbb{C}v_\lambda$ on which $\mathfrak{sl}(2)$ acts as $H[v_\lambda] = \lambda v_\lambda$ and $X[v_\lambda] = 0, Y[v_\lambda] = 0$. One may then define the highest-weight Verma module $M_\lambda := \mathcal{U}(\mathfrak{n}^-) \otimes_{\mathcal{U}(\mathfrak{b}^+)} V_\lambda$, where $\mathfrak{n}^- = \mathbb{C}Y$ and $\mathfrak{b}^+ = \mathbb{C}H \oplus \mathbb{C}X$, with highest weight $\lambda \in \mathbb{C}$.

In view of the PBW-theorem for the universal enveloping algebra $\mathcal{U}(\mathfrak{sl}(2))$, the weight space decomposition is given by $M_\lambda = \text{span}_\mathbb{C}\big(Y^k \otimes v_\lambda : k \in \mathbb{Z}^+\big)$. The following fact about Verma modules is well known:

Proposition 2.2. *The Verma modules M_λ are irreducible for $\lambda \notin \mathbb{Z}^+$.*

Remark 2.3. In this paper, we have chosen to work with highest-weight Verma modules M_λ, but the construction for lowest-weight modules is similar.

The conditions on λ ensuring that M_λ is irreducible are very convenient, in the sense that we will only be able to introduce generalizations of Appell sequences as algebra morphisms σ_λ from $\mathfrak{h}_1$ into $\text{End}(M_\lambda)$ under the same conditions on λ as in the proposition above (which means that the action of $\sigma_\lambda(P)$ and $\sigma_\lambda(M)$ will only be well defined for $\lambda \in \mathbb{C} \setminus \mathbb{Z}^+$). In what follows, we consider the subring $R := \mathcal{U}(\mathfrak{b}^-) \subset \mathcal{U}(\mathfrak{sl}(2))$, where we use the notation $\mathfrak{b}^- := \mathbb{C}H \oplus \mathbb{C}Y$ for the (other) Borel subalgebra (compare with $\mathfrak{b}^+$).

Remark 2.4. Fractions will play a crucial role throughout this paper, and we therefore define them as $A/B := AB^{-1}$, where the inversion always appears at the right-hand side. This is important, because in general A and B will belong to a non-commutative ring.

Definition 2.5. For all $\alpha \in \mathbb{C}$, we define $S_\alpha^- := \big\langle (H + \alpha + 2j) : j \in \mathbb{Z}^- \big\rangle \subset R$, as the set which is multiplicatively generated by the elements between brackets.

It is then easily verified that $S_\alpha^- \subset R$ satisfies the right Øre condition for arbitrary $\alpha \in \mathbb{C}$, see, e.g., [4]. This condition is needed whenever one wants to consider the right ring of fractions $R(S^{-1})$ with respect to a multiplicatively closed subset $S \subset R$. Indeed, for arbitrary $\xi \in R$ and $\sigma \in S_\alpha^-$ one has that $\xi S_\alpha^- \cap \sigma R \neq \emptyset$. Indeed, in view of the PBW-theorem for $\mathcal{U}(\mathfrak{sl}(2))$, it suffices to consider an element ξ of the form $\xi = Y^a H^b$ (hereby omitting the tensor product symbols and with $a, b \in \mathbb{Z}^+$). For $\sigma = H + \alpha + 2j$ (with $j \in \mathbb{Z}^-$), one then clearly has that $(H + \alpha + 2j)Y^a H^b = Y^a H^b\big(H + \alpha + 2(j - a)\big)$, since $2(j - a) \in \mathbb{Z}^-$. We can thus define the localization w.r.t. the set S_α^-, which will be denoted by means of $R_\alpha^- := R\big((S_\alpha^-)^{-1}\big)$.

Remark 2.6. Since $(H + \alpha + 2j)Y^a H^b X^c = Y^a H^b X^c\big(H + \alpha + 2(j + c - a)\big)$, it is clear that one can also consider the localization of the full enveloping algebra $\mathcal{U}(\mathfrak{sl}(2))$ with respect to the (enlarged) subset $S_\alpha := \big\langle (H + \alpha + 2j) : j \in \mathbb{Z} \big\rangle$, where $j \in \mathbb{Z}$ is now an arbitrary integer. We therefore also introduce the notation $\mathcal{U}_\alpha(\mathfrak{sl}(2)) := \mathcal{U}(\mathfrak{sl}(2))(S_\alpha^{-1})$. Obviously, for all $\alpha \in \mathbb{C}$ one has that $R_\alpha^- \subset \mathcal{U}_\alpha(\mathfrak{sl}(2))$. Note that these are all inside the skew field over $\mathcal{U}(\mathfrak{sl}(2))$.

The motivation for considering this particular localization R_α^-, instead of $\mathcal{U}_\alpha(\mathfrak{sl}(2))$, comes from the fact that the action of the latter localization will not always be well defined on the Verma modules we would like to consider for practical purposes. In full generality, one has the following:

Proposition 2.7. *Whenever $(\lambda + \alpha) \notin 2\mathbb{Z}^+$, the action of R_α^- is well defined on irreducible Verma modules M_λ.*

Proof. To prove that the action on the localization w.r.t. S_α^- is well defined, we first of all note that the elements in S_α^- act as a constant on the weight spaces. It then suffices to verify that for all integers $k \geq 0$ and $j \leq 0$ one has that $(H + \alpha + 2j)[Y^k \otimes v_\lambda] = (\alpha + \lambda - 2k + 2j)Y^k \otimes v_\lambda \neq 0$. This is indeed guaranteed whenever $\alpha + \lambda \neq 2(k + |j|) \in 2\mathbb{Z}^+$. □

Corollary 2.8. *The action of R_λ^- is well defined on irreducible Verma modules M_λ (i.e., for arbitrary $\lambda \in \mathbb{C}$ such that λ is not a positive integer).*

Corollary 2.9. *For arbitrary $\alpha \in \mathbb{C}$, the action of $\mathcal{U}_\alpha(\mathfrak{sl}(2))$ is well defined on irreducible Verma modules M_λ whenever $\alpha + \lambda \notin 2\mathbb{Z}$.*

Remark 2.10. Note that the action of $\mathcal{U}_\lambda(\mathfrak{sl}(2))$ on M_λ is not always well defined, in view of the fact that, e.g., for $\lambda = l \in \mathbb{Z}^-$ and $j = k - l$ we get that $(H + \lambda + 2j)[Y^k \otimes v_\lambda] = 0$.

Let us then prove the main result of this section:

Theorem 2.11. *Suppose M_λ is an irreducible Verma module for the algebra $\mathfrak{sl}(2)$, which means that $\lambda \notin \mathbb{Z}^+$. One can then define an action of $\mathfrak{h}_1$ on M_λ, by means of the algebra morphism*

$$\sigma_\lambda : \mathfrak{h}_1 \to End(M_\lambda) : (P, M) \mapsto \left(X, \frac{2Y}{H + \lambda}\right) .$$

Note that the operator $\sigma_\lambda(M)$ actually belongs to the localization R_λ^-, which means that its (repeated) action on M_λ is well defined (see Corollary 2.8).

Remark 2.12. Recalling the notation from the introduction, we thus have that the basis for M_λ defined by $\mathbb{A} := \{\sigma_\lambda^k(M)[1 \otimes v_\lambda] : k \in \mathbb{Z}^+\}$ defines an Appell sequence.

Proof. It suffices to verify that the action of $\sigma_\lambda(P)$ and $\sigma_\lambda(M)$ on the module M_λ satisfies the Heisenberg relation $[\sigma_\lambda(P), \sigma_\lambda(M)] = 1$. For that purpose, we note that for all $k > 0$, we have:

$$\begin{aligned}[\sigma_\lambda(P), \sigma_\lambda(M)](Y^k \otimes v_\lambda) &= \left(XY\frac{2}{H + \lambda} - Y\frac{2}{H + \lambda}X\right)(Y^k \otimes v_\lambda) \\ &= Y^k \otimes v_\lambda .\end{aligned}$$

For $k = 0$, the statement is trivial, which proves the theorem. □

Invoking the definition $(\lambda)_k = \lambda(\lambda-1)\cdots(\lambda-k+1)$, we then have:

Definition 2.13. Suppose M_λ is a highest-weight Verma module for $\mathfrak{sl}(2)$, with $\lambda \in \mathbb{C}\setminus\mathbb{Z}^+$. The monomial basis for M_λ is given by the weight vectors

$$v_\lambda(k) := \frac{1}{(\lambda)_k} Y^k \otimes v_\lambda \qquad (k \in \mathbb{Z}^+) \ .$$

Note that the embedding of $\mathfrak{h}_1$ into $\mathcal{U}_\lambda(\mathfrak{sl}(2))$ also gives rise to another Appell sequence. To see this, we will calculate the commutator $[\sigma_\lambda(P), \sigma_\lambda(M)]$ inside the localization and then investigate its action on Verma modules M_μ (with $\mu \in \mathbb{C}$ arbitrary). In view of the fact that $(H+\lambda)X = X(H+\lambda+2)$, we get:

$$[\sigma_\lambda(P), \sigma_\lambda(M)] = \frac{2H}{H+\lambda} + 2Y\left[X, \frac{1}{H+\lambda}\right] = 2\frac{H(H+\lambda+2)+2YX}{(H+\lambda)(H+\lambda+2)} \ .$$

When acting on an arbitrary weight space in the module M_μ, we thus get:

$$[\sigma_\lambda(P), \sigma_\lambda(M)]Y^k \otimes v_\mu = 2\frac{(\mu-2k)(\mu+\lambda-2k+2)+2k(\mu+1-k)}{(\mu+\lambda-2k)(\mu+\lambda-2k+2)} Y^k \otimes v_\mu \ .$$

It is then easily verified that the Appell condition is verified (i.e., the constant in front of the weight vector is equal to 1) for $\mu \in \{\lambda, 2-\lambda\}$. This means that we have now obtained the following (somewhat stronger) result:

Corollary 2.14. *Consider a complex number $\lambda \in \mathbb{C}\setminus\mathbb{Z}$. One can then define*

$$\sigma_\lambda : \mathfrak{h}_1 \to End(M_\mu) : (P, M) \mapsto \left(X, \frac{2Y}{H+\lambda}\right) ,$$

for $\mu \in \{\lambda, 2-\lambda\}$. Both Verma modules M_λ and $M_{2-\lambda}$ then become Appell sequences for the operators $\sigma_\lambda(P)$ and $\sigma_\lambda(M)$. Note that the latter operator belongs to R_λ^-, which means that its action is always well defined.

Note that we imposed the condition $\lambda \notin \mathbb{Z}$ in the corollary above, to ensure that both $\lambda, \mu = 2-\lambda \in \mathbb{C}\setminus\mathbb{Z}^+$. Let us then consider a few examples:

(i) Consider the classical realization for $\mathfrak{sl}(2)$ in harmonic analysis on $\mathbb{R}^m$, see (1.1). It is then clear that we can start from an arbitrary harmonic function $f_\alpha(\underline{x})$ on an open subset $\Omega \subset \mathbb{R}^m$ which is homogeneous of degree $\alpha \in \mathbb{C}$. This function then plays the role of a highest weight vector for which $\lambda = -\alpha - \frac{m}{2} \notin \mathbb{Z}^-$, leading to the Appell sequence

$$\mathbb{A}_\lambda := \left\{ \frac{|\underline{x}|^{2k} f_\alpha(\underline{x})}{2^k\left(\alpha+\frac{m}{2}\right)\left(\alpha+\frac{m}{2}+1\right)\cdots\left(\alpha+\frac{m}{2}+k-1\right)} \right\} ,$$

for the lowering operator $P = \frac{1}{2}\Delta_x$. In case $\alpha + \frac{m}{2} \notin \mathbb{Z}$, we can also consider an Appell sequence starting from M_μ, with $\mu = 2-\lambda$. In the context of harmonic analysis, there is a well-known realization for the highest weight vector for M_λ in terms of the Kelvin inversion:

$$\mathcal{J}_0 : f_\alpha(\underline{x}) \mapsto \frac{1}{|\underline{x}|^{m-2}} f\left(\frac{\underline{x}}{|\underline{x}|^2}\right) = |\underline{x}|^{2-m-2\alpha} f_\alpha(\underline{x}) \ .$$

This gives rise to the Appell sequence

$$\mathbb{A}_\mu := \left\{ \frac{(-1)^k |\underline{x}|^{2-m-2\alpha+2k} f_\alpha(\underline{x})}{2^k \left(\alpha + \frac{m}{2} - 2\right)\left(\alpha + \frac{m}{2} - 3\right) \cdots \left(\alpha + \frac{m}{2} - k - 1\right)} \right\} .$$

(ii) In [8], we have obtained the harmonic (resp. monogenic) Gegenbauer polynomials through the knowledge of a particular subalgebra of the Weyl algebra $\mathcal{W}_m$ (resp. $\mathcal{W}_m^C$), for all $m \geq 3$ given by

$$\mathfrak{sl}(2) \cong \mathrm{Alg}\big(-\partial_{x_m}, x_m(2\mathbb{E}_x + m - 2) - r^2\partial_{x_m}, -2\mathbb{E}_x - (m-2)\big) .$$

It is then clear that the polynomial set

$$\mathcal{G}_{2-m} := \big\{ (x_m(2\mathbb{E}_x + m - 2) - r^2\partial_{x_m})^k[1] : k \in \mathbb{N} \big\}$$

can be considered as a highest-weight Verma module M_λ with highest weight vector $1 \in \mathbb{C}$, for $\lambda = -(m-2)$.

3. Hermite bases in Verma modules for $\mathfrak{sl}(2)$

One can now develop a general framework to define special polynomials (e.g., Hermite polynomials). Traditionally, such polynomials can be defined through an explicit formula of the form $S_k(z) = \sum_{j=0}^k c_{j,k}(S)z^j$, with $z \in \mathbb{C}$ and $c_{j,k}(S)$ a certain coefficient that determines the special function under consideration. We will generalize this picture, hereby using the following idea: instead of using a complex variable z, we will use the operator $\sigma_\lambda(Y)$ which creates the monomial basis for an arbitrary (fixed) Verma module M_λ, hereby fixing the realization $\mathfrak{sl}(2) = \mathrm{Alg}(X, Y, H)$. For example, the Hermite basis for the Verma module M_λ is then defined through the repeated action of the following operators in $\mathcal{U}_\lambda(\mathfrak{sl}(2))$:

$$\sigma_\lambda^{(h)}(P) := X \quad \text{and} \quad \sigma_\lambda^{(h)}(M) := \sigma_\lambda(M - P) = \frac{2Y}{H+\lambda} - X .$$

Note that we have added a superscript (h) to indicate that these generate the Hermite basis, corresponding to the *probabilists' Hermite polynomials*, as opposed to the *physicists' Hermite polynomials* which would require adding a factor 2 to the term $\sigma_\lambda(M)$. The raising operator can also be defined as

$$\sigma_\lambda^{(h)}(M) \quad = \quad -\exp\left(\frac{1}{2}\sigma_\lambda^2(M)\right)\sigma_\lambda(P)\exp\left(-\frac{1}{2}\sigma_\lambda^2(M)\right) , \tag{3.1}$$

where the exponential is defined through its formal Taylor expansion.

Definition 3.1. Suppose M_λ is a Verma module, with $\lambda \in \mathbb{C}\backslash\mathbb{Z}^+$ and weight spaces $Y^k \otimes v_\lambda$ $(k \in \mathbb{Z}^+)$. The Hermite basis for M_λ is then given by the following set of vectors:

$$v_\lambda^{(h)}(k) := \left(\frac{2Y}{H+\lambda} - X\right)^k [1 \otimes v_\lambda] \qquad (k \in \mathbb{Z}^+) .$$

As a result of expression (3.1), this can also be written as follows:

$$v_\lambda^{(h)}(k) = (-1)^k \exp\left(\frac{1}{2}\sigma_\lambda^2(M)\right)\sigma_\lambda^k(P)\exp\left(-\frac{1}{2}\sigma_\lambda^2(M)\right)[1\otimes v_\lambda]\,.$$

The fact that this defines a *basis* follows from the following proposition, the essence of which is encoded in a technical lemma:

Lemma 3.2. *For all $k \in \mathbb{Z}^+$, we have the following expansion of binomial type when acting on the highest weight vector $1 \otimes v_\lambda$:*

$$\big(\sigma_\lambda(M) - \sigma_\lambda(P)\big)^{2k} = \sum_{j=0}^{k} \frac{(-1)^j k!}{2^j j!(k-2j)!}\sigma_\lambda^{2k-2j}(M)$$

$$\big(\sigma_\lambda(M) - \sigma_\lambda(P)\big)^{2k+1} = \sum_{j=0}^{k} \frac{(-1)^j k!}{2^j j!(k-2j)!}\sigma_\lambda^{1+2k-2j}(M).$$

Proof. The theorem can be easily proved by induction, taking into account that factors $\sigma_\lambda(P)$ may safely be ignored once they are at the right-hand side (in view of the fact that the expression is meant to act on the highest weight vector $1 \otimes v_\lambda \in M_\lambda$). □

Proposition 3.3. *The explicit expression for the Hermite basis vectors for a Verma module M_λ in terms of the monomial basis, is given by:*

$$v_\lambda^{(h)}(k) = \sum_{j=0}^{\kappa} \frac{(-1)^j k!}{2^j j!(k-2j)!(\lambda)_{k-2j}} Y^{k-2j} \otimes v_\lambda\ ,$$

hereby introducing the integer $\kappa := \lfloor \frac{k}{2} \rfloor \in \mathbb{Z}^+$.

Proof. This immediately follows from the previous lemma, hereby making use of the fact that $\sigma_\lambda(M)$ generates the monomial basis for M_λ. □

The classical Hermite polynomial $H_k(x)$ in a real variable $x \in \mathbb{R}$, as in the probabilistic normalization, corresponds to the case where monomial basis vectors $v_\lambda(k) \in M_\lambda$ are identified with monomials x^k. Note that the Hermite basis vectors satisfy the following recurrence relations:

$$\begin{aligned} v_\lambda^{(h)}(k+1) &= \frac{1}{(\lambda-k)} Y v_\lambda^{(h)}(k) - X v_\lambda^{(h)}(k) \\ &= \frac{1}{(\lambda-k)} Y v_\lambda^{(h)}(k) - k v_\lambda^{(h)}(k-1)\ , \end{aligned}$$

where we explicitly made use of the fact that the Hermite basis vectors define an Appell sequence. This gives then rise to the following eigenvalue problem for the linear operator $\mathcal{L}_\lambda \in \mathcal{U}_\lambda(\mathfrak{sl}(2))$, which is the equivalent of the Hermite equation in the classical context:

$$\mathcal{L}_\lambda v_\lambda^{(h)}(k) := \left(X^2 - \frac{2Y}{H+\lambda}\right) v_\lambda^{(h)}(k) = -k v_\lambda^{(h)}(k)\ .$$

References

[1] S. Bock, K. Gürlebeck, R. Lávička, V. Souček, V., *The Gelfand–Tsetlin bases for spherical monogenics in dimension* 3, Rev. Mat. Iberoam. **28**, Issue 4 (2012), 1165–1192.

[2] I. Cação, D. Eelbode, *Jacobi polynomials and generalized Clifford algebra-valued Appell sequences*, to appear in Math. Meth. Appl. Sc., doi: 10.1002/mma.2914.

[3] I. Cação, I. Falcão, H. Malonek, *Laguerre derivative and monogenic Laguerre polynomials: an operational approach*, Math. Comput. Model. **53** (2011), 1084–1094.

[4] P.M. Cohn, *Skew fields, Theory of general division rings*, Cambridge University Press, 1995.

[5] H. De Bie, *Fourier transform and related integral transforms in superspace*, J. Math. Anal. Appl. **345** (2008), 147–164.

[6] H. De Bie, *An alternative definition of the Hermite polynomials related to the Dunkl Laplacian*, SIGMA 4 (2008), 093, 11 pages.

[7] C.F. Dunkl, *Differential-difference operators associated to reflection groups*, Trans. Am. Math. Soc. **311** (1989), 167–183.

[8] D. Eelbode, *Monogenic Appell sequences as representations of the Heisenberg algebra*, Adv. Appl. Cliff. Alg. **22**, Issue 4 (2012), 1009–1023.

[9] I. Falcão, H. Malonek, *Generalized exponentials through Appell sets in* $\mathbb{R}^{n+1}$ *and Bessel functions* in: Simos, T.E., Psihoyios, G., Tsitouras, C. (eds.), AIP Conference Proceedings, Vol. 936 (2007), 738–741.

[10] R. Howe, *Remarks on classical invariant theory*, Trans. Amer. Math. Soc. **313** No. 2 (1989), 539–570.

[11] Malonek, H.R., Tomaz, G., *Bernoulli polynomials and Pascal matrices in the context of Clifford analysis*, Discr. Appl. Math. **157** No. 4 (2009), pp. 838–847.

David Eelbode and Nikolaas Verhulst
University of Antwerp Campus Middelheim (Building G)
Middelheimlaan 1
B-2020 Antwerp, Belgium
e-mail: `david.eelbode@uantwerpen.be`
`nikolaas.verhulst@uantwerpen.be`

Hypercomplex Analysis: New Perspectives and Applications
Trends in Mathematics, 119–132

Integral Formulas for *k*-hypermonogenic Functions in $\mathbb{R}^3$

Sirkka-Liisa Eriksson, Heikki Orelma and Nelson Vieira

Abstract. We consider harmonic functions with respect to the Laplace–Beltrami operator of the Riemannian metric $ds^2 = x_2^{-2k}\left(\sum_{i=0}^{2} dx_i^2\right)$ and their quaternion function theory in $\mathbb{R}^3$. Leutwiler noticed around 1990 that if the usual Euclidean metric is changed to the hyperbolic one, that is $k = 1$, then the power function $(x_0 + x_1e_1 + x_2e_2)^n$, calculated using quaternions, is the conjugate gradient of a hyperbolic harmonic function. We study generalized holomorphic functions, called k-hypermonogenic functions satisfying the modified Dirac equation. Note that 0-hypermonogenic are monogenic and 1-hypermonogenic functions are hypermonogenic defined by H. Leutwiler and the first author.

We prove the Cauchy type integral formulas for k-hypermonogenic where the kernels are calculated using the hyperbolic distance of the Poincaré upper half-space model. Earlier these results have been proved for hypermonogenic functions.

Mathematics Subject Classification (2010). Primary 30A05; Secondary 30A45.

Keywords. Hypermonogenic, hyperbolic, Laplace–Beltrami, monogenic, quaternions.

1. Introduction

We study generalized function theory connected to the hyperbolic metric. In this frame work the generalized holomorphic functions are called k-hypermonogenic functions introduced in [1]. The theory of k-hypermonogenic functions is connecting the theory of monogenic functions with the value $k = 0$ and theory of hypermonogenic functions with value $k = n - 1$. Moreover, it also connected to the eigenfunctions of Laplace–Beltrami operator with respect to the hyperbolic metric of the Poincaré upper half-space model.

Hypermonogenic functions were introduced in $\mathbb{R}^{n+1}$ by H. Leutwiler and the first author in [6]. An introduction to the theory is presented in [8] or in [10].

The first Cauchy type integral formulas for them were proved in [7] and the total formula with two kernels in [2]. They were improved to contain just one single kernel in [9] and [4]. Later in [5] it was noticed a surprising result that the kernel is the Cauchy kernel of monogenic function shifted to the Euclidean center of the hyperbolic ball.

The general Cauchy type theorem for k-hypermonegenic functions was proved in [3] but the kernels have a complicated formula. In the main result of this paper we present a different way to compute the kernels and an explicit integral formula in $\mathbb{R}^3$. In consecutive papers the integral operators will be studied. The results may be also generalized to higher dimensions.

2. Preliminaries

We review the main notations and concepts. Let $\mathbb{H}$ be the real associative division algebra of quaternions generated by e_1, e_2, e_3 satisfying the properties $e_1e_2 = e_3$ and $e_ie_j + e_je_i = -2\delta_{ij}$ where δ_{ij} is the usual Kronecker delta. Elements $x = x_0 + x_1e_1 + x_2e_2$ for $x_0, x_1, x_2 \in \mathbb{R}$ are called *paravectors*. The vector space $\mathbb{R}^3$ is identified with the real vector space of paravectors and therefore elements $x_0 + x_1e_1 + x_2e_2$ and (x_0, x_1, x_2) are identified. The field of complex numbers is identified with the set $\{x_0 + x_1e_1 | \; x_0, x_1 \in \mathbb{R}\}$.

We use several involutions. Let $q = q_0+q_1e_1+q_2e_2+q_3e_3$. The *main involution* is the mapping $q \to q'$ defined by

$$q' = q_0 - q_1e_1 - q_2e_2 + q_3e_3.$$

Similarly the *reversion* is the anti-involution $q \to q^*$ defined by

$$q' = q_0 + q_1e_1 + q_2e_2 - q_3e_3.$$

The *conjugation* is the anti-involution $q \to \overline{q}$ that is the composition of the preceding involutions, that is $\overline{q} = (q')^* = (q^*)'$ and

$$\bar{q} = q_0 - q_1e_1 - q_2e_2 - q_3e_3.$$

The following product rules hold

$$(ab)' = a'b' \qquad (ab)^* = b^*a^* \qquad \overline{ab} = \overline{b}\,\overline{a}$$

for all quaternions a and b. Using the unique decomposition $q = u+ve_2$ for $u, v \in \mathbb{C}$ we define the mappings $P : \mathbb{H} \to \mathbb{C}$ and $Q : \mathbb{H} \to \mathbb{C}$ by $Pq = u$ and $Qq = v$ (see [6]). In order to compute the P- and Q-parts, we define the involution $q \to \widehat{q}$ by

$$\widehat{q} = q_0 + q_1e_1 - q_2e_2 - q_3e_3.$$

Then we obtain the formulas

$$Pq = \frac{1}{2}\left(q + \widehat{q}\right), \tag{2.1}$$

$$Qq = -\frac{1}{2}\left(q - \widehat{q}\right)e_2. \tag{2.2}$$

Note that the product rule $\widehat{ab} = \widehat{a}\widehat{b}$ holds for all quaternions a and b. The following product rules ([6]) hold

$$P(ab) = (Pa)\,Pb + (Qa)\,Q(b'), \tag{2.3}$$

$$\begin{aligned} Q(ab) &= (Pa)\,Qb + (Qa)\,P'(b) \\ &= aQb + (Qa)\,b'. \end{aligned} \tag{2.4}$$

Note that if $a \in \mathbb{H}$ then

$$a'e_2 = e_2\widehat{a}.$$

Moreover, if $a \in \mathbb{C}$ then

$$ae_2 = e_2a'. \tag{2.5}$$

Our general assumption is that a function $f(x_0, x_1, x_2) : \Omega \to \mathbb{H}$ is defined on an open subset Ω of $\mathbb{R}^3$ and its components are continuously differentiable. The left Dirac operator and the right Dirac operator in $\mathbb{H}$ are defined by

$$D_l f = \frac{\partial f}{\partial x_0} + e_1\frac{\partial f}{\partial x_1} + e_2\frac{\partial f}{\partial x_2}, \qquad D_r f = \frac{\partial f}{\partial x_0} + \frac{\partial f}{\partial x_1}e_1 + \frac{\partial f}{\partial x_2}e_2.$$

Their conjugate operators $\overline{D}_l$ and $\overline{D}_r$ are introduced by

$$\overline{D}_l f = \frac{\partial f}{\partial x_0} - e_1\frac{\partial f}{\partial x_1} - e_2\frac{\partial f}{\partial x_2}, \qquad \overline{D}_r f = \frac{\partial f}{\partial x_0} - \frac{\partial f}{\partial x_1}e_1 - \frac{\partial f}{\partial x_2}e_2.$$

These operators are also called generalized Cauchy–Riemann operators.

The modified Dirac operators M_k^l, $\overline{M}_k^l$, M_k^r and $\overline{M}_k^r$ are introduced in ([1]) by

$$M_k^l f(x) = D_l f(x) + k\frac{\overline{Qf(x)}}{x_2}, \qquad M_k^r f(x) = D_r f(x) + k\frac{Qf(x)}{x_2},$$

$$\overline{M}_k^l f(x) = \overline{D}_l f(x) - k\frac{\overline{Qf(x)}}{x_2}, \qquad \overline{M}_k^r f(x) = \overline{D}_r f(x) + k\frac{Qf(x)}{x_2}$$

for $x \in \{x \in \Omega \mid x_2 \neq 0\}$. Since the operator M_1^l is directly connected to the hyperbolic Poincaré upper half-space model it is specially important and is also denoted briefly by M (introduced in ([6]).

Definition 2.1. Let $\Omega \subset \mathbb{R}^3$ be an open set. Let $k \in \mathbb{R}$. A mapping $f : \Omega \to \mathbb{H}$ is called **left k-hypermonogenic**, if $f \in C^1(\Omega)$ and $M_k^l f(x) = 0$ for any $x \in \{x \in \Omega \mid x_2 \neq 0\}$. The 0-hypermonogenic functions are called **monogenic**. The 1-hypermonogenic functions are called briefly **hypermonogenic**. The **right k-hypermonogenic functions** are defined similarly. A twice continuously differentiable function $f : \Omega \to \mathbb{H}$ is called **k-hyperbolic harmonic** if $\overline{M}_k^l M_k^l f = 0$ for any $x \in \{x \in \Omega \mid x_2 \neq 0\}$.

3. Integral formulas for k-hypermonogenic functions

The main idea is that we use the hyperbolic metric

$$ds^2 = \frac{dx_0^2 + dx_1^2 + dx_2^2}{x_2^2} \tag{3.1}$$

of the Poincaré upper half-space model instead of the Euclidean one. Note that the operator

$$\Delta_h f = x_2^2 \Delta f - x_2 \frac{\partial f}{\partial x_2}$$

is the hyperbolic Laplace–Beltrami operator with respect to this metric. The hyperbolic distance may be computed as follows (see [14]).

Lemma 3.1. *The hyperbolic distance $d_h(x,a)$ with respect to the metric* (3.1) *between the points x and a in $\mathbb{R}^3_+$ is*

$$d_h(x,a) = \operatorname{arcosh} \lambda(x,a),$$

where

$$\lambda(x,a) = \frac{|x-a|^2 + |x-\hat{a}|^2}{4x_2 a_2} = \frac{|x-a|^2}{2x_2 a_2} + 1$$

and $|x-a|$ is the usual Euclidean distance between the points a and x.

We note also the relation between the Euclidean and hyperbolic balls.

Proposition 3.2. *The hyperbolic ball $B_h(a, r_h)$ in $\mathbb{R}^3_+$ with the hyperbolic center $a = a_0 + a_1 e_1 + a_2 e_2$ and the radius r_h is the same as the Euclidean ball with the Euclidean center*

$$c_a(r_h) = a_0 + a_1 e_1 + a_2 \cosh r_h e_2$$

and the Euclidean radius $r_e = a_2 \sinh r_h$.

There are two main relations between k-hyperbolic harmonic functions and k-hypermonogenic functions recalled as follows.

Theorem 3.3 ([6]). *Let $\Omega \subset \mathbb{R}^3$ be an open set and $f : \Omega \to \mathbb{H}$ be twice continuously differentiable. Then f is k-hypermonogenic if and only if f and $xf(x)$ are k-hyperbolic harmonic functions.*

Proposition 3.4 ([6]). *Let $\Omega \subset \mathbb{R}^3$ be an open set. Let $k \in \mathbb{R}$. If a mapping $h : \Omega \to \mathbb{H}$ is k-hyperbolic harmonic then $\overline{M}_k^l h$ is k-hypermonogenic. Conversely, if a mapping $f : \Omega \to \mathbb{H}$ is k-hypermonogenic there exists locally a complex k-hyperbolic harmonic function h satisfying $f = \overline{D}h$.*

We use the following calculation rules, proved in [11]

Lemma 3.5. *If $c(x,a) = Pa + a_2 \cosh d_h(x,a) e_2$ then*

$$x_2 \overline{D}^x \lambda(x,a) = \frac{\overline{x - c(x,a)}}{a_2},$$

$$x_2 \overline{D}^x d_h(x,a) = \frac{\overline{x - c(x,a)}}{a_2 \sinh d_h(x,a)} = \frac{\overline{x - c(x,a)}}{|x - c(x,a)|}.$$

Our key tool is the relation between k-hyperbolic harmonic functions and the eigenfunctions of the hyperbolic Laplace–Beltrami operator, stated slightly more general form next.

Proposition 3.6 ([12]). *If u is a solution of the equation*

$$x_2^2 \triangle u(x) - kx_2 \frac{\partial u}{\partial x_2}(x) + lu(x) = 0 \tag{3.2}$$

in an open subset $\Omega \subset \mathbb{R}^3_+$, then $f(x) = x_2^{\frac{1-k}{2}} u(x)$ is an eigenfunction of the hyperbolic Laplace operator corresponding to the eigenvalue $\frac{1}{4}\left(k^2 + 2k - 3 - 4l\right)$. Conversely, if f is an eigenfunction of the hyperbolic Laplace operator corresponding to the eigenvalue γ in an open subset $\Omega \subset \mathbb{R}^3_+$ then $u(x) = x_2^{\frac{k-1}{2}} f(x)$ is the solution of the equation (3.2) *in Ω with $\gamma = \frac{1}{4}\left(k^2 + 2k - 3 - 4l\right)$.*

We are looking for eigenfunctions of the hyperbolic Laplace depending just on the hyperbolic distance $d_h(x, e_2)$. We first recall a formula of the hyperbolic Laplace for functions depending only on $d_h(x, e_2)$.

Lemma 3.7 ([13]). *If f is twice continuously differentiable depending only on $r_h = d_h(x, e_2)$ then the hyperbolic Laplace in $\mathbb{R}^3_+$ is given by*

$$\triangle_h f(r_h) = \frac{\partial^2 f}{\partial r_h^2} + 2\coth r_h \frac{\partial f}{\partial r_h}.$$

Eigenfunctions of the hyperbolic Laplace depending on r_h are related to an easier differential equation.

Lemma 3.8. *If f is a solution of the equation*

$$\frac{\partial^2 f}{\partial r_h^2} + \gamma f = 0$$

then the function $g(r_h) = \frac{1}{\sinh r_h}\frac{\partial f}{\partial r_h}$, depending only on r_h, is an eigenfunction of the hyperbolic Laplace operator corresponding to the eigenvalue $-(\gamma + 1)$.

Proof. We just compute

$$\frac{\partial^2 g}{\partial r_h^2} = \left(-\frac{1}{\sinh r_h} + 2\frac{\cosh^2 r_h}{\sinh^3 r_h}\right)\frac{\partial f}{\partial r_h} - 2\frac{\cosh r_h}{\sinh^2 r_h}\frac{\partial^2 f}{\partial r_h^2} + \frac{1}{\sinh r_h}\frac{\partial^3 f}{\partial r_h^3},$$

$$2\frac{\partial g}{\partial r_h}\frac{\cosh r_h}{\sinh r_h} = -\frac{2\cosh^2 r_h}{\sinh^3 r_h}\frac{\partial f}{\partial r_h} + \frac{2\cosh r_h}{\sinh^2 r_h}\frac{\partial^2 f}{\partial^2 r_h}.$$

Since $\frac{\partial^2 f}{\partial r_h^2} = -\gamma f$ we conclude that $\frac{1}{\sinh r_h}\frac{\partial^3 f}{\partial r_h^3} = g$ and

$$\frac{\partial^2 g}{\partial r_h^2} + 2\frac{\partial g}{\partial r_h}\frac{\cosh r_h}{\sinh r_h} = \frac{1}{\sinh r_h}\frac{\partial^3 f}{\partial r_h^3} - g = -(\gamma + 1) g. \qquad \square$$

Applying the previous lemma we obtain the general solution depending on the hyperbolic distance.

Theorem 3.9. *The general solution of the equation*

$$\frac{\partial^2 f}{\partial r_h^2} + 2\coth r_h \frac{\partial f}{\partial r_h} + \gamma f = 0 \tag{3.3}$$

is

$$f(r_h) = \begin{cases} \frac{C\cosh\left(\sqrt{1-\gamma}r_h\right)}{\sinh r_h} + \frac{C_0\sinh\left(\sqrt{1-\gamma}r_h\right)}{\sinh r_h}, & \text{if } \gamma < 1, \\ \frac{C}{\sinh r_h} + \frac{C_0 r_h}{\sinh r_h}, & \text{if } \gamma = 1, \\ \frac{C\cos\left(\sqrt{\gamma-1}r_h\right)}{\sinh r_h} + \frac{C_0\sin\left(\sqrt{\gamma-1}r_h\right)}{\sinh r_h}, & \text{if } \gamma > 1, \end{cases}$$

for some real constants C and C_0.

Corollary 3.10. *The bounded solutions of* (3.3) *are*

$$f(r_h) = \begin{cases} \frac{C\sinh\left(\sqrt{1-\gamma}r_h\right)}{\sinh r_h}, & \text{if } \gamma < 1, \\ \frac{C r_h}{\sinh r_h}, & \text{if } \gamma = 1, \\ \frac{C\sin\left(\sqrt{\gamma-1}r_h\right)}{\sinh r_h}, & \text{if } \gamma > 1. \end{cases}$$

Corollary 3.11. *The particular solution of* (3.3) *with a singularity at e_2 is*

$$f(r_h) = \begin{cases} \frac{\cosh\left(\sqrt{1-\gamma}r_h\right)}{\sinh r_h}, & \text{if } \gamma < 1, \\ \frac{1}{\sinh r_h}, & \text{if } \gamma = 1, \\ \frac{\cos\left(\sqrt{\gamma-1}r_h\right)}{\sinh r_h}, & \text{if } \gamma > 1. \end{cases}$$

Corollary 3.12. *The particular solution of* (3.3) *with $\gamma = \frac{1}{4}(4-(k+1)^2)$ outside the point e_2 is*

$$F(x) = \frac{\cosh\left(\frac{d_h(x,e_2)(k+1)}{2}\right)}{\sinh d_h(x,e_2)} = \frac{\cosh\left(\frac{d_h(x,e_2)(k+1)}{2}\right)}{|x - \cosh d_h(x,e_2)|}$$

and $x_2^{\frac{k-1}{2}} F(x)$ is k-hyperbolic harmonic.

If we transform the preceding function with the hyperbolic translation $\tau(x) = a_2 x + Pa$, we note that

$$F(x) = \frac{\cosh\left(\frac{d_h(\tau(x),a)(k+1)}{2}\right)}{\sinh d_h(\tau(x),a)}$$

and

$$F\left(\tau^{-1}(u)\right) = \frac{\cosh\left(\frac{d_h(u,a)(k+1)}{2}\right)}{\sinh d_h(u,a)}$$

is also an eigenfunction of the hyperbolic Laplace corresponding to the value $\frac{1}{4}((k+1)^2-4)$.

Corollary 3.13. *The function*

$$F_h(x,a) = \frac{\cosh\left(\frac{d_h(x,a)(k+1)}{2}\right)}{|x - Pa - a_2 \cosh d_h(x,a)\, e_2|} = \frac{\cosh\left(\frac{d_h(x,a)(k+1)}{2}\right)}{a_2 \sinh d_h(x,a)}$$

satisfies the equation

$$\frac{\partial^2 f}{\partial r_h^2} + 2\coth r_h \frac{\partial f}{\partial r_h} + \gamma f = 0$$

with $\gamma = \frac{1}{4}(4-(k+1)^2)$ *outside* $x = a$ *and* $x_2^{\frac{k-1}{2}} F_h(x,a)$ *is* k*-hyperbolic outside* $x = a$.

Using (3.4), we may directly compute the corresponding k-hypermonogenic function.

Theorem 3.14. *Denote* $r_h = d_h(x,a)$ *and set* $s = \frac{k+1}{2}$. *The function*

$$h_k(x,a) = a_2^{s+1} x_2^{s-1} w_k(x,a)\, p(x,a)$$

is paravector valued k*-hypermonogenic with respect to* x *outside* $x = a$ *when*

$$w_k(x,a) = (s-1)\cosh(sr_h)\, e_2 \frac{x - Pa}{a_2} + s\cosh((s-1)\, r_h)$$

and the function

$$p(x,a) = \frac{(x - c(x,a))^{-1}}{x_2\, |x - c(x,a)|}$$

is hypermonogenic with respect to x.

Proof. Denote $s = \frac{k+1}{2}$. Applying the previous corollary and Proposition 3.4, we note that the function

$$g(x_2, r_h) = -\frac{a_2^{s-1} x_2^{s-1} \cosh(sr_h)}{\sinh r_h}$$

is k-hyperbolic harmonic and $h_k = \overline{D}^x g(x_{2,} r_h)$ is k-hypermonogenic. We just make simple calculations

$$\frac{h_k(x,a)}{a_2^{s-1} x_2^{s-1}} = (s-1)\frac{e_2 \cosh(sr_h)}{x_2 \sinh r_h} - \left(\frac{s\sinh(sr_h) - \cosh(sr_h)\coth r_h}{\sinh r_h}\right)\overline{D}^x r_h.$$

Applying Lemma 3.5, we obtain

$$\frac{x_2 \overline{D}^x r_h}{a_2^2 \sinh^2 r_h} = \frac{\overline{x - c(x,a)}}{|x - c(x,a)|^3} = \frac{(x - c(x,a))^{-1}}{|x - c(x,a)|}$$

and

$$\frac{x - c(x,a)}{a_2} \frac{(x - c(x,a))^{-1}}{|x - c(x,a)|} = \frac{1}{a_2\, |x - c(x,a)|}$$
$$= \frac{1}{a_2^2 \sinh r_h}.$$

Hence we obtain

$$\frac{h_k(x,a)}{a_2^{s+1}x_2^{s-1}} = w_k(x,a)\frac{(x-c(x,a))^{-1}}{x_2|x-c(x,a)|}$$

and

$$\begin{aligned} w_k(x,a) = {} & (s-1)\cosh(sr_h)\, e_2 \frac{x-c(x,a)}{a_2} \\ & - s\sinh r_h \sinh(sr_h) + \cosh r_h \cosh(sr_h) \\ = {} & (s-1)\cosh(sr_h)\, e_2 \frac{x-Pa}{a_2} \\ & + s(\cosh r_h \cosh(sr_h) - \sinh r_h \sinh(sr_h)). \end{aligned}$$

Using the rule of hyperbolic cosines, we conclude

$$w_k(x,a) = (s-1)\cosh(sr_h)\, e_2 \frac{x-Pa}{a_2} + s\cosh((s-1)r_h)$$

and

$$h_k(x,a) = a_2^{s+1}x_2^{s-1} w_k(x,a)\, p(x,a).$$

Applying [11], we infer that the function $p(x,a)$ is hypermonogenic, completing the proof. □

Denote the real surface measure by dS and a real weighted volume measure by

$$dm_k = \frac{1}{x_2^k} dm.$$

The generalized Stokes theorem for M_k operators is the following result.

Theorem 3.15 ([2]). *Let Ω be an open subset of $\mathbb{R}^3_+$ (or $\mathbb{R}^3_-$) and $\overline{K} \subset \Omega$ be a smoothly bounded compact set with the outer unit normal field ν. If $f, g \in \mathcal{C}^1(\Omega, \mathbb{H})$, then*

$$\int_{\partial K} g\nu f \frac{dS}{x_2^k} = \int_K \left((M_k^r g) f + g M_k^l f - \frac{k}{x_2} P(gf')\, e_n \right) \frac{dm}{x_2^k}.$$

By taking the P-part from both sides of the equation of the previous theorem we obtain.

Theorem 3.16 ([2]). *Let Ω be an open subset of $\mathbb{R}^3 \setminus \{x_2 = 0\}$ and $\overline{K} \subset \Omega$ be a smoothly bounded compact set with the outer unit normal field ν. If $f, g \in \mathcal{C}^1(\Omega, \mathbb{H})$, then*

$$\int_{\partial K} P(g\nu f) \frac{dS}{x_2^k} = \int_K P((M_k^r g) f + g M_k^l f) \frac{dm}{x_2^k}.$$

Theorem 3.17. *Let Ω be an open subset of $\mathbb{R}^3_+$ (or $\mathbb{R}^3_-$) and $\overline{K} \subset \Omega$ be a smoothly bounded compact set with the outer unit normal field ν. Let $s = \frac{k+1}{2}$ and*

$$h_k(x,a) = a_2^{s+1}x_2^{s-1} w_k(x,a)\, p(x,a)$$

be the same function as in Theorem 3.14. *If f is k-hypermonogenic in Ω and $a \in K$, then*

$$Pf(a) = \frac{1}{4\pi}\int_{\partial K} P\left(h_k(x,a)\,\nu f(x)\right)\frac{dS}{x_2^k}.$$

Proof. Using Theorem 3.16 we obtain

$$\begin{aligned}
&\int_{\partial K} P\left(h_k(x,a)\,\nu f\right)\frac{dS}{x_2^k}\\
&= \int_{\partial\left(K\setminus B_r^h(a)\right)} P\left(h_k(x,a)\,\nu f\right)\frac{dS}{x_2^k} - \int_{\partial B_{r_h}(a)} P\left(h_k(x,a)\,\nu f\right)\frac{dS}{x_2^k}\\
&= -\int_{\partial B_{r_h}(a)} P\left(h_k(x,a)\,\nu f\right)\frac{dS}{x_2^k} = -P\left(\int_{\partial B_{r_h}(a)} h_k(x,a)\,\nu f\frac{dS}{x_2^k}\right).
\end{aligned}$$

Since the hyperbolic ball $\partial B_{r_h}(a)$ is the Euclidean ball with the Euclidean center $c_a(r_h) = Pa + a_2\cosh r_h e_2$ and the radius $r = a_n \sinh r_h$ we have

$$\nu(x) = \frac{x - c_a(r_h)}{a_2\sinh r_h}$$

and we deduce

$$\begin{aligned}
&P\left(\int_{\partial B_{r_h}(a)} h_k(x,a)\,\nu f\frac{dS}{x_2^k}\right)\\
&= \frac{1}{a_2^2\sinh^2 r_h}P\left(\int_{\partial B_{r_h}(a)} a_2^{\frac{k+3}{2}}\left((s-1)\cosh(sr_h)\,e_2\frac{x-Pa}{a_2}\right.\right.\\
&\qquad\qquad\left.\left. + s\cosh((s-1)\,r_h)\right)\frac{fdS}{x_2^{\frac{k+3}{2}}}\right).
\end{aligned}$$

If $r_h \to 0$ then

$$\lim_{r_h\to 0} P\left(\int_{\partial B_{r_h}(a)} h_k(x,a)\,\nu f\frac{dS}{x_2^k}\right) = 4\pi Pf(a). \qquad \square$$

The Q-part satisfies the following generalized Stokes theorem.

Theorem 3.18 ([2]). *Let Ω be an open subset of $\mathbb{R}_+^3$ (or $\mathbb{R}_-^3$) and $\overline{K} \subset \Omega$ be a smoothly bounded compact set with the outer unit normal field ν. If $f, g \in \mathcal{C}^1(\Omega, \mathbb{H})$, then*

$$\int_{\partial K} g\nu f dS = \int_K \left(\left(M_{-k}^r g\right) f + gM_k^l f + \frac{k}{x_n}Q\left(gf'\right)\right)dm.$$

Applying the operator Q to the previous result, we directly conclude the following result:

Theorem 3.19 ([2]). *Let Ω be an open subset of $\mathbb{R}^3_+$ (or $\mathbb{R}^3_-$) and $\overline{K} \subset \Omega$ be a smoothly bounded compact set with the outer unit normal field ν. If $f, g \in \mathcal{C}^1(\Omega, \mathbb{H})$, then*

$$\int_{\partial K} Q(g\nu f)\, dS = \int_K Q\left(\left(M^r_{-k} g\right) f + g M^l_k f\right) dm.$$

Theorem 3.20. *Let Ω be an open subset of $\mathbb{R}^3_+$ (or $\mathbb{R}^3_-$) and $\overline{K} \subset \Omega$ be a smoothly bounded compact set with the outer unit normal field ν. Let $s = \frac{k+1}{2}$. The function*

$$v_{-k}(x,a) = a_2^{s+1} x_2^{-s} w_{-k}(x,a)\, p(x,a)$$

is paravector valued $-k$-hypermonogenic outside $x = a$ with respect to x, when

$$w_{-k}(x,a) = -s\cosh\left((s-1) r_h\right) e_2 \frac{x - Pa}{a_2} - (s-1)\cosh\left(s r_h\right).$$

If f is k-hypermonogenic in Ω and $a \in K$, then

$$Qf(a) = \frac{1}{4\pi}\int_{\partial K} Q\left(v_{-k}(x,a)\, \nu f(x)\right) dS.$$

The shifted Euclidean kernel may also be computed using the hat involution.

Lemma 3.21 ([5]). *If x and a belong to $\mathbb{R}^3_+$ then*

$$p(x,a) = 4x_2 \frac{(x-a)^{-1}}{|x-a|} e_2 \frac{(x-\widehat{a})^{-1}}{|x-\widehat{a}|} = 4x_2 \frac{(x-\widehat{a})^{-1}}{|x-\widehat{a}|} e_2 \frac{(x-a)^{-1}}{|x-a|}.$$

Theorem 3.22. *Let Ω be an open subset of $\mathbb{R}^3_+ \setminus \{x_2 = 0\}$ and $\overline{K} \subset \Omega$ be a smoothly bounded compact set with the outer unit normal field ν. If f is k-hypermonogenic in Ω and $y \in K$, then*

$$f(a) = \frac{a_2}{2\pi}\int_{\partial K} \frac{g_k(x,a)\left((x-a)^{-1}\nu(x) f(x) - (\widehat{x}-a)^{-1}\widehat{\nu}(x)\widehat{f}(x)\right) dS}{|x-a|\,|x-\widehat{a}|}$$

where $s = \frac{k+1}{2}$ and

$$g_k(x,a) = a_2^{s-1} x_2^{1-s}\left((s-1)\cosh(s r_h) - s\cosh\left((s-1) r_h\right)\right).$$

Proof. Applying Theorems 3.17 and 3.22 we deduce

$$\begin{aligned}
8\pi f(a) &= \int_{\partial K} 2P\left(h_k(x,a)\,\nu f(x)\right)\frac{dS}{x_2^k} + \int_{\partial K} 2Q\left(v_{-k}(x,a)\,\nu f(x)\right) dS e_2 \\
&= \left(\int_{\partial K} h_k(x,a)\,\nu f(x)\frac{dS}{x_2^k} + \int_{\partial K} \widehat{h_k(x,a)}\widehat{\nu}(x)\widehat{f}(x)\frac{dS}{x_2^k}\right) \\
&\quad + \left(\int_{\partial K} v_{-k}(x,a)\,\nu f(x)\, dS - \int_{\partial K} \widehat{v_{-k}(x,a)}\widehat{\nu}(x)\widehat{f}(x)\, dS\right) \\
&= \int_{\partial K} a_2^{\frac{k+3}{2}}\left(w_k(x,a) + w_{-k}(x,a)\right) p(x,a)\,\nu f(x)\frac{dS}{x_2^{\frac{k+1}{2}}} \\
&\quad + \int_{\partial K} a_2^{\frac{k+3}{2}}\left(\widehat{w_k(x,a)} - \widehat{w_{-k}(x,a)}\right)\widehat{p(x,a)}\widehat{\nu}(x)\widehat{f}(x)\frac{dS}{x_2^{\frac{k+1}{2}}}
\end{aligned}$$

where

$$w_k(x,a) = (s-1)\cosh(sr_h)\, e_2 \frac{x-Pa}{a_2} + s\cosh((s-1)\,r_h),$$

$$w_{-k}(x,a) = -s\cosh((s-1)\,r_h)\, e_2 \frac{x-Pa}{a_2} - (s-1)\cosh(sr_h).$$

Since

$$e_2 \frac{x-Pa}{a_2} - 1 = e_2 \frac{x-Pa+a_2e_2}{a_2} = e_2 \frac{x-\hat{a}}{a_2}$$

we have

$$w_k(x,a) + w_{-k}(x,a) = ((s-1)\cosh(sr_h) - s\cosh((s-1)\,r_h))\, e_2 \frac{x-\hat{a}}{a_2}.$$

Applying

$$e_2 \frac{x-Pa}{a_2} + 1 = e_2 \frac{x-Pa-a_2e_2}{a_2} = e_2 \frac{x-a}{a_2}$$

we conclude that

$$\widehat{w_k(x,a)} - \widehat{w_{-k}(x,a)} = (s-1)\cosh(sr_h) - s\cosh((s-1)\,r_h)\, \frac{e_2(\widehat{x}-\widehat{a})}{a_2},$$

completing the proof. □

Theorem 3.23. *Let Ω be an open subset of $\mathbb{R}^3_+$ and $\overline{K} \subset \Omega$ a smoothly bounded compact set with outer unit normal field ν. If f is k-hypermonogenic in Ω and $a \in intK$ then*

$$f(a) = \frac{1}{4\pi} \int_{\partial K} g_k(x,a)\, p(a,x) \left(Q(x\overline{\nu}(x) f'(x)) + aQ(\overline{\nu}(x) f'(x))\right) dS_x$$

where the function

$$p(a,x) = \frac{(a - c(a,x))^{-1}}{a_2 \left|a - c(a,x)\right|}$$

is hypermonogenic with respect to a and

$$g_k(x,a) = a_2^{s-1} x_2^{1-s} \left((s-1)\cosh(sr_h) - s\cosh((s-1)\,r_h)\right)$$

for $s = \frac{k+1}{2}$.

Proof. Decomposing

$$\nu(x) f(x) = P(\nu(x) f(x)) + Q(\nu(x) f(x))\, e_2$$

and

$$\widehat{\nu(x)}\widehat{f}(x) = \widehat{\nu(x) f(x)} = P(\nu(x) f(x)) - Q(\nu(x) f(x))\, e_2$$

we obtain

$$2\pi f(a) = a_2 \int_{\partial K} g_k(x,a) \frac{(x-a)^{-1} - (\widehat{x}-a)^{-1}}{|a-x|\,|a-\widehat{x}|} P(\nu(x) f(x))\, d\sigma_x$$

$$+ a_2 \int_{\partial K} g_k(x,a) \frac{(x-a)^{-1} + (\widehat{x}-a)^{-1}}{|a-x|\,|a-\widehat{x}|} e_2 Q'(\nu(x) f(x))\, d\sigma_x.$$

If we factor $(x-a)^{-1}$ to the left side of the difference $(x-a)^{-1}-(\widehat{x}-a)^{-1}$ we obtain

$$(x-a)^{-1}-(\widehat{x}-a)^{-1}=(x-a)^{-1}\left(1-(x-a)(\widehat{x}-a)^{-1}\right).$$

Then factoring $(\widehat{x}-a)^{-1}$ to the right side of the difference $\left(1-(x-a)(\widehat{x}-a)^{-1}\right)$ we infer further

$$\begin{aligned}(x-a)^{-1}-(\widehat{x}-a)^{-1}&=(x-a)^{-1}\left(1-(x-a)(\widehat{x}-a)^{-1}\right)\\&=(x-a)^{-1}(\widehat{x}-a-x+a)(\widehat{x}-a)^{-1}\\&=(x-a)^{-1}(\widehat{x}-x)(\widehat{x}-a)^{-1}\\&=2x_2(x-a)^{-1}e_2(\widehat{x}-a)^{-1}.\end{aligned}$$

Applying Lemma 3.21, we obtain

$$a_2\frac{(x-a)^{-1}-(\widehat{x}-a)^{-1}}{|a-x|\,|a-\widehat{x}|}=\frac{2x_2a_2(a-x)^{-1}e_2(a-\widehat{x})^{-1}}{|a-x|\,|a-\widehat{x}|}=\frac{x_2(a-c(a,x))^{-1}}{2a_2|a-c(a,x)|}.$$

Decomposing $x=Px+x_2e_2$ and $\widehat{x}=Px-x_2e_2$ we obtain

$$(x-a)^{-1}(Px+x_2e_2-a)=1=(\widehat{x}-a)^{-1}(Px-x_2e_2-a).$$

Hence

$$(x-a)^{-1}x_2e_2-(x-a)^{-1}(a-Px)=-(\widehat{x}-a)^{-1}a_2e_2-(\widehat{x}-a)^{-1}(a-Px),$$

which implies

$$\left((x-a)^{-1}+(\widehat{x}-a)^{-1}\right)x_2e_2=\left((x-a)^{-1}-(\widehat{x}-a)^{-1}\right)(a-Px)$$

and further

$$\begin{aligned}\frac{\left((x-a)^{-1}+(\widehat{x}-a)^{-1}\right)x_2a_2e_2}{|a-x|\,|a-\widehat{x}|}&=a_2\frac{(x-a)^{-1}-(\widehat{x}-a)^{-1}}{|a-x|\,|a-\widehat{x}|}(a-Px)\\&=\frac{x_2(a-c(a,y))^{-1}}{2a_2|a-c(a,y)|}(a-Px).\end{aligned}$$

Combining the previous equalities we have

$$\begin{aligned}2\pi f(a)=&\int_{\partial K}\frac{g_k(x,a)(a-c(a,x))^{-1}}{2a_2|a-c(a,x)|}x_2P(\nu(x)f(x))\,dS\\&+\int_{\partial K}g_k(x,a)\frac{(a-c(a,x))^{-1}}{2a_2|a-c(a,x)|}(a-Px)Q'(\nu(xy)f(x))\,dS\\=&\int_{\partial K}g_k(x,a)\frac{(a-c(a,x))^{-1}}{2a_2|a-c(a,x)|}u(x)\,dS\\&+\int_{\partial K}g_k(x,a)\frac{(a-c(a,x))^{-1}}{2a_2|a-c(a,x)|}aQ'(\nu(x)f(x))\,dS,\end{aligned}$$

where

$$u(x) = x_2 P(\nu(x) f(x)) - PxQ'(\nu(x) f(x)).$$

Applying (2.3) and (2.4) we deduce

$$\begin{aligned}-PxQ'(\nu(x) f(x)) + x_2 P(\nu(x) f(x)) &= PxQ(\nu'(x) f'(x)) + QxP'(\nu'(x) f'(x)) \\ &= Q(x\nu'(x) f'(x)),\end{aligned}$$

completing the proof. □

Conclusion. We have proved integral formulas for k-hypermonogenic functions in $\mathbb{R}^3$. In consecutive papers we are going to study properties of corresponding integral operators and generalize results to higher dimensions.

Acknowledgment

N. Vieira was supported by Portuguese funds through the CIDMA – Center for Research and Development in Mathematics and Applications, and the Portuguese Foundation for Science and Technology ("FCT – Fundação para a Ciência e a Tecnologia"), within project PEst-OE/MAT/UI4106/2014.

References

[1] Eriksson-Bique, S.-L., *k-hypermonogenic functions*, In: Progress in Analysis, Vol. I, World Scientific (2003), 337–348.

[2] Eriksson, S.-L., *Integral formulas for hypermonogenic functions*, Bull. Bel. Math. Soc., **11**, No. 5, (2004), 705–717.

[3] Eriksson, S.-L., *Cauchy-type integral formulas for k-hypermonogenic functions*, in: More progress in Analysis, Vol. I, Proceedings of the 5th international ISAAC congress, H.G.W. Bergehr (ed.), World Scientific, Hackensack, NJ, (2009), 337–348.

[4] Eriksson, S.-L., *Hyperbolic Extensions of Integral Formulas*, Adv. Appl. Clifford Alg., **20**, No. 3-4, (2010), 575–586.

[5] Eriksson, S.-L., *A Hyperbolic Dirac Operator and its Kernels*, Complex Var. Elliptic Equ., **58**, No. 6, (2013), 767–781.

[6] Eriksson-Bique, S.-L. and Leutwiler, H., *Hypermonogenic functions*, In: Clifford Algebras and their Applications in Mathematical Physics, Vol. 2, Ryan, John et al. (eds.), Birkhäuser, Boston, MA, (2000), 287–302.

[7] Eriksson, S.-L. and Leutwiler, H., *Hypermonogenic functions and their Cauchy-type theorems*, In: Advances in Analysis and Geometry. New developments using Clifford algebras, Tao Qian et al. (eds.), Birkhäuser, Basel, (2004), 97–112.

[8] Eriksson, S.-L. and Leutwiler, H., *Hyperbolic Function Theory*, Adv. Appl. Clifford Alg., **17**, No. 3, (2007), 437–450.

[9] Eriksson, S.-L. and Leutwiler, H., *An Improved Cauchy Formula for Hypermonogenic Functions*, Adv. Appl. Clifford Alg., **19**, No. 2, (2009), 269–282.

[10] Eriksson, S.-L. and Leutwiler, H., *Introduction to Hyperbolic Function Theory*, in: Proceedings of Clifford Algebras and Inverse Problems, Research Report, **90**, (2009), 1–28.

[11] Eriksson, S.-L., H. Orelma, *On Hypermonogenic Functions*, Complex Var. Elliptic Equ., **58**, No. 7, (2013), 975–990.

[12] Eriksson, S.-L. and H. Orelma, *Mean value properties for the Weinstein equation using the hyperbolic metric.* Complex Anal. Oper. Theory, **7**, No. 5, (2013), 1609–1621.

[13] Eriksson, S.-L. and H. Orelma, *Hyperbolic Laplace operator and the Weinstein equation in* $\mathbb{R}^3$, Adv. in Appl. Clifford Alg., **24** (2014), 109–124.

[14] Leutwiler, H., Appendix: Lecture notes of the course *Hyperbolic harmonic functions and their function theory*, Clifford algebras and potential theory, 85–109, Univ. Joensuu Dept. Math. Rep. Ser. **7**, Univ. Joensuu, Joensuu, 2004.

Sirkka-Liisa Eriksson and Heikki Orelma
Department of Mathematics
Tampere University of Technology
P.O.Box 553
FI-33101 Tampere, Finland
e-mail: `Sirkka-Liisa.Eriksson@tut.fi`
`Heikki.Orelma@tut.fi`

Nelson Vieira
Department of Mathematics
University of Aveiro Universitário de Santiago
3810-193 Aveiro, Portugal
e-mail: `nloureirovieira@gmail.com`

Hypercomplex Analysis: New Perspectives and Applications
Trends in Mathematics, 133–143

Spectral Properties of Compact Normal Quaternionic Operators

Riccardo Ghiloni, Valter Moretti and Alessandro Perotti

Abstract. General, especially spectral, features of compact normal operators in quaternionic Hilbert spaces are studied and some results are established which generalize well-known properties of compact normal operators in complex Hilbert spaces. More precisely, it is proved that the norm of such an operator always coincides with the maximum of the set of absolute values of the eigenvalues (exploiting the notion of spherical eigenvalue). Moreover the structure of the spectral decomposition of a generic compact normal operator T is discussed also proving a spectral characterization theorem for compact normal operators.

Mathematics Subject Classification (2010). 46S10, 47C15, 47B07, 30G35, 81R15.

Keywords. Compact operators, quaternionic Hilbert spaces.

1. Introduction

Theory of linear operators in quaternionic Hilbert spaces is a well-established topic of functional analysis with many applications in physics, especially quantum mechanics (see the introduction of [6] for a wide discussion). As in complex functional analysis, compact operators play a relevant role as they share features both with generic operators in infinite-dimensional spaces and with matrices in finite-dimensional spaces. This intermediate role is particularly evident regarding spectral analysis of normal compact operators. In fact, these operators in infinite-dimensional (complex or quaternionic) Hilbert spaces, on the one hand admit a pure point spectrum (except, perhaps, for 0), on the other hand their spectral expansion needs a proper infinite Hilbertian basis. This paper is devoted to focus on these peculiar properties exploiting the general framework established in [6].

Work partially supported by GNSAGA and GNFM of INdAM, MIUR-PRIN project "Varietà reali e complesse: geometria, topologia e analisi armonica" and MIUR-FIRB project "Geometria Differenziale e Teoria Geometrica delle Funzioni".

The notion of spectrum of an operator on quaternionic Hilbert spaces has been introduced only few years ago [2] in the more general context of quaternionic Banach modules. It has been a starting point for developing functional calculus for the classes of slice and slice regular functions on a quaternionic space (see [2, 6]).

Let H be a (right) quaternionic Hilbert space (we refer to Section 2 for basic definitions and to [6] for more details), let $\mathfrak{B}(\mathsf{H})$ be the set of right linear operators on H and let $\mathfrak{B}_0(\mathsf{H})$ be the set of right linear compact operators on H. In [5] some properties of compact operators on quaternionic Hilbert spaces were studied. In particular, it was shown that the *spherical spectrum* (cf. Section 2.3 for complete definitions) of a compact operator T contains only the eigenvalues of T and possibly 0, and the set of eigenvectors relative to a non-zero eigenvalue q is finite dimensional. Another result, similar to what occurs for compact operators in complex Hilbert spaces, is conjectured in [5]:

Conjecture. *If $T \in \mathfrak{B}_0(\mathsf{H})$ is self-adjoint, then either $\|T\|$ or $-\|T\|$ is an eigenvalue of T.*

In the following we will prove the conjecture for the more general class of normal compact operators on a quaternionic Hilbert space. We also prove the spectral decomposition theorem for normal compact operators and its converse. The complex Hilbert space versions of these results can be found for example in [8, §3.3].

1.1. Main theorems

As recalled in Section 2.3, the set of eigenvalues of a linear operator T coincides with the *spherical point spectrum*, denoted by $\sigma_{pS}(T)$. We can then rephrase the conjecture in the following way.

Theorem 1.1. *Given any normal operator $T \in \mathfrak{B}_0(\mathsf{H})$ with spherical point spectrum $\sigma_{pS}(T)$, there exists $\lambda \in \sigma_{pS}(T)$ such that:*

$$|\lambda| = \max\{|\mu| \mid \mu \in \sigma_S(T)\} = \|T\| \,. \tag{1.1}$$

The next result is the spectral decomposition for normal compact operators. In the finite-dimensional case, where the compactness requirement is empty, the result is well known [7] (see [4] for an ample exposition and further references).

Theorem 1.2. *Given a normal operator $T \in \mathfrak{B}_0(\mathsf{H})$ with spherical point spectrum $\sigma_{pS}(T)$, there exists a Hilbert basis $\mathcal{N} \subset \mathsf{H}$ made of eigenvectors of T such that:*

$$Tx = \sum_{z\in\mathcal{N}} z\lambda_z\langle z|x\rangle \quad \textit{for each } x \in \mathsf{H}, \tag{1.2}$$

where $\lambda_z \in \mathbb{H}$ is an eigenvalue relative to the eigenvector z and, if $\lambda_z \neq 0$ only a finite number of distinct other elements $z' \in \mathcal{N}$ verify $\lambda_z = \lambda_{z'}$, moreover the values λ_z are at most countably many.

The set Λ of eigenvalues λ_z with $z \in \mathcal{N}$ has the property that for every $\epsilon > 0$ there is a finite set $\Lambda_\epsilon \subset \Lambda$ with $|\lambda| < \epsilon$ if $\lambda \notin \Lambda_\epsilon$ (following [8] we say that the eigenvalues "vanish at infinity"). Thus 0 is the only possible accumulation point of Λ. If H is infinite dimensional, then 0 belongs to $\sigma_S(T)$.

Remark 1.3. Let $\mathbb{S}$ denote the two-dimensional sphere of imaginary units in $\mathbb{H}$:

$$\mathbb{S} := \{q \in \mathbb{H} \mid q^2 = -1\}.$$

For $\imath \in \mathbb{S}$, let $\mathbb{C}_\imath$ be the real subalgebra of $\mathbb{H}$ generated by $\imath$. We will see in Section 2 that to every normal operator T can be associated an anti self-adjoint operator J. Along with a chosen imaginary unit $\imath \in \mathbb{S}$, J defines (cf. Definition 2.5) two $\mathbb{C}_\imath$-Hilbert subspaces of H, denoted by $\mathsf{H}_+^{J\imath}$ and $\mathsf{H}_-^{J\imath}$.

In the preceding theorem, for every imaginary unit $\imath$, it is possible to choose $\mathcal{N}$ such that:

$$\{\lambda_z \mid z \in \mathcal{N}\} \setminus \{0\} = (\sigma_{pS}(T) \setminus \{0\}) \cap \mathbb{C}_\imath \tag{1.3}$$

as well as $\mathcal{N} \subset \mathsf{H}_+^{J\imath}$. With a conjugation, in general losing the condition $\mathcal{N} \subset \mathsf{H}_+^{J\imath}$, one can always have $\lambda_z \in \mathbb{C}_\imath^+$ but $\mathcal{N} \subset \mathsf{H}_+^{J\imath} \cup \mathsf{H}_-^{J\imath}$.

The spectral theorem for compact operators has the following converse. To state it, we need to recall a definition. Given a subset $\mathcal{K}$ of $\mathbb{C}$, we define the *circularization* $\Omega_{\mathcal{K}}$ *of* $\mathcal{K}$ *(in* $\mathbb{H}$*)* by setting

$$\Omega_{\mathcal{K}} := \{\alpha + \jmath\beta \in \mathbb{H} \mid \alpha, \beta \in \mathbb{R}, \alpha + i\beta \in \mathcal{K}, \jmath \in \mathbb{S}\}. \tag{1.4}$$

Theorem 1.4. *Let $T \in \mathfrak{B}(\mathsf{H})$. Assume that there exist a Hilbert basis $\mathcal{N}$ of H and a map $\mathcal{N} \ni z \mapsto \lambda_z \in \mathbb{H}$ satisfying the following requirements:*

(i) *$Tx = \sum_{z\in\mathcal{N}} z\lambda_z\langle z|x\rangle$ for every $x \in \mathsf{H}$.*
(ii) *For every $z \in \mathcal{N}$ such that $\lambda_z \neq 0$, only a finite number of distinct other elements $z' \in \mathcal{N}$ verify $\lambda_z = \lambda_{z'}$;*
(iii) *The set Λ is countable at most;*
(iv) *For every $\epsilon > 0$, there is a finite set $\Lambda_\epsilon \subset \Lambda$ with $|\lambda| < \epsilon$ if $\lambda \notin \Lambda_\epsilon$.*

Under these conditions T is normal and compact and

$$\sigma_S(T) \setminus \{0\} = \Omega_\Lambda \setminus \{0\}.$$

Remark 1.5. The structure of the whole spherical spectrum (see Definition 2.7) of a compact operator $T \in \mathfrak{B}_0(\mathsf{H})$ has been studied in [5, Corollary 2]:

$$\sigma_S(T) \setminus \{0\} = \sigma_{pS}(T) \setminus \{0\}.$$

If T is normal, then its *spherical residual spectrum* (cf. Section 2.3 for definitions) is empty. Therefore in this case if $0 \in \sigma_S(T) \setminus \sigma_{pS}(T)$ then 0 belongs to the *spherical continuous spectrum* $\sigma_{cS}(T)$.

2. Quaternionic Hilbert spaces

We recall some basic notions about quaternionic Hilbert spaces (see, e.g., [1]). Let $\mathbb{H}$ denote the skew field of quaternions. Let H be a *right* $\mathbb{H}$-module. H is called a *quaternionic pre-Hilbert space* if there exists a Hermitian quaternionic scalar product $\mathsf{H} \times \mathsf{H} \ni (u, v) \mapsto \langle u|v\rangle \in \mathbb{H}$ satisfying the following three properties:

- *Right linearity*: $\langle u|vp + wq\rangle = \langle u|v\rangle p + \langle u|w\rangle q$ if $p, q \in \mathbb{H}$ and $u, v, w \in \mathsf{H}$.
- *Quaternionic Hermiticity*: $\langle u|v\rangle = \overline{\langle v|u\rangle}$ if $u, v \in \mathsf{H}$.
- *Positivity*: If $u \in \mathsf{H}$, then $\langle u|u\rangle \in \mathbb{R}^+$ and $u = 0$ if $\langle u|u\rangle = 0$.

We can define the *quaternionic norm* by setting

$$\|u\| := \sqrt{\langle u|u\rangle} \in \mathbb{R}^+ \quad \text{if } u \in \mathsf{H}.$$

Definition 2.1. A quaternionic pre-Hilbert space H is said to be a *quaternionic Hilbert space* if it is complete with respect to its natural distance $d(u,v) := \|u-v\|$.

Example 2.2. The space $\mathbb{H}^n$ with scalar product $\langle u, v\rangle = \sum_{i=1}^n \bar{u}_i v_i$ is a finite-dimensional quaternionic Hilbert space.

Let H be a quaternionic Hilbert space.

Definition 2.3. A *right* $\mathbb{H}$*-linear operator* is a map $T : D(T) \longrightarrow \mathsf{H}$ such that:

$$T(ua + vb) = (Tu)a + (Tv)b \quad \text{if } u, v \in D(T) \text{ and } a, b \in \mathbb{H},$$

where the *domain* $D(T)$ of T is a (not necessarily closed) right $\mathbb{H}$-linear subspace of H.

It can be shown that an operator $T : D(T) \longrightarrow \mathsf{H}$ is continuous if and only if it is bounded, i.e., there exists $K \geq 0$ such that

$$\|Tu\| \leq K\|u\| \quad \text{for each } u \in D(T).$$

Let $\|T\| := \sup_{u\in D(T)\setminus\{0\}} \frac{\|Tu\|}{\|u\|} = \inf\{K \in \mathbb{R} \mid \|Tu\| \leq K\|u\| \ \forall u \in D(T)\}$. The set $\mathfrak{B}(\mathsf{H})$ of all bounded operators $T : \mathsf{H} \longrightarrow \mathsf{H}$ is a complete metric space w.r.t. the metric $D(T,S) := \|T - S\|$,

Many assertions that are valid in the complex Hilbert spaces case, continue to hold for quaternionic operators. We mention the uniform boundedness principle, the open map theorem, the closed graph theorem, the Riesz representation theorem and the polar decomposition of operators.

As in the complex case, a linear operator $T : \mathsf{H} \to \mathsf{H}$ is called *compact* if it maps bounded sequences to sequences that admit convergent subsequences. We refer to [5] for some properties of compact operators on quaternionic Hilbert spaces. In particular, $\mathfrak{B}_0(\mathsf{H})$ is a closed bilateral ideal of $\mathfrak{B}(\mathsf{H})$ and is closed under adjunction ([5, Theorem 2]).

2.1. Left scalar multiplications

It is possible to equip a (right) quaternionic Hilbert space H with a *left* multiplication by quaternions. It is a non-canonical operation relying upon a choice of a preferred Hilbert basis. So, pick out a Hilbert basis $\mathcal{N}$ of H and define the *left scalar multiplication of* H *induced by* $\mathcal{N}$ as the map $\mathbb{H} \times \mathsf{H} \ni (q, u) \mapsto qu \in \mathsf{H}$ given by

$$qu := \sum_{z\in\mathcal{N}} zq\langle z|u\rangle \quad \text{if } u \in \mathsf{H} \text{ and } q \in \mathbb{H}.$$

For every $q \in \mathbb{H}$, the map $L_q : u \mapsto qu$ belongs to $\mathfrak{B}(\mathsf{H})$. Moreover, the map $\mathcal{L}_\mathcal{N} : \mathbb{H} \longrightarrow \mathfrak{B}(\mathsf{H})$, defined by setting $\mathcal{L}_\mathcal{N}(q) := L_q$ is a norm-preserving real algebra homomorphism.

The set $\mathfrak{B}(\mathsf{H})$ is always a *real Banach C^*-algebra with unity.* It suffices to consider the right scalar multiplication $(Tr)(u) = T(u)r$ for real r and the adjunction $T \mapsto T^*$ as *-involution. By means of a left scalar multiplication, it can be given the richer structure of *quaternionic Banach C^*-algebra.*

Theorem 2.4 ([6]§3.2). *Let H be a quaternionic Hilbert space equipped with a left scalar multiplication. Then the set $\mathfrak{B}(\mathsf{H})$, equipped with the pointwise sum, with the scalar multiplications defined by*

$$(qT)u := q(Tu) \quad \textit{and} \quad (Tq)(u) := T(qu),$$

with the composition as product and with $T \mapsto T^*$ as *-involution, is a quaternionic two-sided Banach C^*-algebra with unity.

Observe that the map $\mathcal{L}_{\mathcal{N}}$ gives a *-representation of $\mathbb{H}$ in $\mathfrak{B}(\mathsf{H})$.

2.2. Imaginary units and complex subspaces

Consider a quaternionic Hilbert space H equipped with a left scalar multiplication $\mathbb{H} \ni q \mapsto L_q$. For short, we write $L_q u = qu$. For every imaginary unit $\imath \in \mathbb{S}$, the operator $J := L_\imath$ is anti self-adjoint and unitary; that is, it holds:

$$J^* = -J \quad \text{and} \quad J^*J = \mathrm{I}.$$

It holds also the converse statement: if an operator $J \in \mathfrak{B}(\mathsf{H})$ is anti self-adjoint and unitary, then $J = L'_\imath$ for some left scalar multiplication of H (see [6, Proposition 3.8]).

In the following, we also need a definition known from the literature [3].

Definition 2.5. Let $J \in \mathfrak{B}(\mathsf{H})$ be an anti self-adjoint and unitary operator and let $\imath \in \mathbb{S}$. Let $\mathbb{C}_\imath$ denote the real subalgebra of $\mathbb{H}$ generated by $\imath$; that is, $\mathbb{C}_\imath := \{\alpha+\imath\beta \in \mathbb{H} \mid \alpha, \beta \in \mathbb{R}\}$. Define the *$\mathbb{C}_\imath$-complex subspaces $\mathsf{H}_+^{J\imath}$ and $\mathsf{H}_-^{J\imath}$ of H associated with J and $\imath$* by setting

$$\mathsf{H}_\pm^{J\imath} := \{u \in \mathsf{H} \mid Ju = \pm u\imath\}.$$

Remark 2.6. $\mathsf{H}_\pm^{J\imath}$ are closed subsets of H, because $u \mapsto Ju$ and $u \mapsto \pm u\imath$ are continuous. However, they are not (right $\mathbb{H}$-linear) subspaces of H. Note also that the space H admits the direct sum decomposition

$$\mathsf{H} = \mathsf{H}_+^{J\imath} \oplus \mathsf{H}_-^{J\imath},$$

with projections $\mathsf{H} \ni x \mapsto P_\pm(x) := \frac{1}{2}(x \mp Jx\imath) \in \mathsf{H}_\pm^{J\imath}$.

2.3. Resolvent and spectrum

It is not clear how to extend the definitions of spectrum and resolvent in quaternionic Hilbert spaces. Let us focus on the simpler case of eigenvalues of a bounded right $\mathbb{H}$-linear operator T. Without fixing any left scalar multiplication of H, the equation determining the eigenvalues reads as follows:

$$Tu = uq.$$

Here a drawback arises: if $q \in \mathbb{H} \setminus \mathbb{R}$ is fixed, the map $u \mapsto uq$ is not right $\mathbb{H}$-linear. Consequently, the eigenspace of q cannot be a right $\mathbb{H}$-linear subspace. Indeed, if

$\lambda \neq 0$, $u\lambda$ is an eigenvector of $\lambda^{-1}q\lambda$ instead of q itself. As a second guess, one could decide to deal with quaternionic Hilbert spaces equipped with a left scalar multiplication and require that

$$Tu = qu.$$

Now both sides are right $\mathbb{H}$-linear. However, this approach is not suitable for physical applications, where self-adjoint operators should have real spectrum. We come back to the former approach and accept that each eigenvalue q brings a whole conjugation class of the quaternions, the *eigensphere*

$$\mathbb{S}_q := \{\lambda^{-1}q\lambda \in \mathbb{H} \,|\, \lambda \in \mathbb{H} \setminus \{0\}\}.$$

We adopt the viewpoint introduced in [2] for quaternionic two-sided Banach modules. Given an operator $T : D(T) \longrightarrow \mathsf{H}$ and $q \in \mathbb{H}$, let

$$\Delta_q(T) := T^2 - T(q+\overline{q}) + \mathrm{I}|q|^2.$$

Definition 2.7. The *spherical resolvent set* of T is the set $\rho_S(T)$ of $q \in \mathbb{H}$ such that:

(a) $\mathrm{Ker}(\Delta_q(T)) = \{0\}$.
(b) $\mathrm{Range}(\Delta_q(T))$ is dense in H.
(c) $\Delta_q(T)^{-1} : \mathrm{Range}(\Delta_q(T)) \longrightarrow D(T^2)$ is bounded.

The *spherical spectrum* $\sigma_S(T)$ of T is defined by $\sigma_S(T) := \mathbb{H} \setminus \rho_S(T)$. It decomposes into three disjoint *circular* (i.e., invariant by conjugation) subsets:

(i) the *spherical point spectrum* of T (the set of *eigenvalues*):

$$\sigma_{pS}(T) := \{q \in \mathbb{H} \,|\, \mathrm{Ker}(\Delta_q(T)) \neq \{0\}\}.$$

(ii) the *spherical residual spectrum* of T:

$$\sigma_{rS}(T) := \left\{q \in \mathbb{H} \,\middle|\, \mathrm{Ker}(\Delta_q(T)) = \{0\}, \overline{\mathrm{Range}(\Delta_q(T))} \neq \mathsf{H}\right\}.$$

(iii) the *spherical continuous spectrum* of T:

$$\sigma_{cS}(T) := \left\{q \in \mathbb{H} \,\middle|\, \Delta_q(T)^{-1} \text{ is densely defined but not bounded}\right\}.$$

The *spherical spectral radius* of T is defined as

$$r_S(T) := \sup\left\{|q| \,\middle|\, q \in \sigma_S(T)\right\} \in \mathbb{R}^+ \cup \{+\infty\}.$$

2.4. Spectral properties

The spherical resolvent and the spherical spectrum can be defined for bounded right $\mathbb{H}$-linear operators on quaternionic two-sided Banach modules in a form similar to that introduced above (see [2]). Several spectral properties of bounded operators on complex Banach or Hilbert spaces remain valid in that general context. Here we recall some of these properties in the quaternionic Hilbert setting (cf. Theorem 4.3 in [6]).

Theorem 2.8 ([6]§4.1). *Let* H *be a quaternionic Hilbert space and let* $T \in \mathfrak{B}(\mathsf{H})$. *Then*

(a) $r_S(T) \leq \|T\|$.

(b) $\sigma_S(T)$ *is a non-empty compact subset of* $\mathbb{H}$.

(c) *Let* $P \in \mathbb{R}[X]$. *Then, if* T *is self-adjoint, the following* spectral map property *holds:*

$$\sigma_S(P(T)) = P(\sigma_S(T)).$$

(d) Gelfand's spectral radius formula *holds:*

$$r_S(T) = \lim_{n \to +\infty} \|T^n\|^{1/n}.$$

In particular, if T *is* normal *(i.e.,* $TT^* = T^*T$*), then* $r_S(T) = \|T\|$.

Regardless different definitions with respect to the complex Hilbert space case, the notions of spherical spectrum and resolvent set enjoy some properties which are quite similar to those for complex Hilbert spaces. Other features, conversely, are proper to the quaternionic Hilbert space case. First of all, it turns out that the spherical point spectrum coincides with the set of eigenvalues of T.

Proposition 2.9. *Let* H *be a quaternionic Hilbert space and let* $T : D(T) \longrightarrow \mathsf{H}$ *be an operator. Then* $\sigma_{pS}(T)$ *coincides with the set of all eigenvalues of* T.

The subspace $\mathrm{Ker}(\Delta_q(T))$ has the role of an eigenspace of T. In particular, $\mathrm{Ker}(\Delta_q(T)) \neq \{0\}$ if and only if $\mathbb{S}_q$ is an eigensphere of T.

Theorem 2.10. *Let* T *be an operator with dense domain on a quaternionic Hilbert space* H.

(a) $\sigma_S(T) = \sigma_S(T^*)$.

(b) *If* $T \in \mathfrak{B}(\mathsf{H})$ *is normal, then*
   (i) $\sigma_{pS}(T) = \sigma_{pS}(T^*)$.
   (ii) $\sigma_{rS}(T) = \sigma_{rS}(T^*) = \emptyset$.
   (iii) $\sigma_{cS}(T) = \sigma_{cS}(T^*)$.

(c) *If* T *is self-adjoint, then* $\sigma_S(T) \subset \mathbb{R}$ *and* $\sigma_{rS}(T)$ *is empty.*

(d) *If* T *is anti self-adjoint, then* $\sigma_S(T) \subset \mathrm{Im}(\mathbb{H})$ *and* $\sigma_{rS}(T)$ *is empty.*

(e) *If* $T \in \mathfrak{B}(\mathsf{H})$ *is unitary, then* $\sigma_S(T) \subset \{q \in \mathbb{H} \,|\, |q| = 1\}$.

(f) *If* $T \in \mathfrak{B}(\mathsf{H})$ *is anti self-adjoint and unitary, then* $\sigma_S(T) = \sigma_{pS}(T) = \mathbb{S}$.

It can be shown that, differently from operators on complex Hilbert spaces, a normal operator T on a quaternionic space is unitarily equivalent to T^*.

2.5. Compact operators

Compact operators have some peculiar spectral properties. Some of them were investigated in [5]. In particular, if $T \in \mathfrak{B}_0(\mathsf{H})$ and $q \in \sigma_{pS}(T)\backslash\{0\}$ is an eigenvalue, then $Ker(\Delta_q(T))$ has finite dimension [5, Theorem 3]. Moreover, the spherical spectrum of $T \in \mathfrak{B}_0(\mathsf{H})$ consists only of the eigenvalues of T and (possibly) 0 (cf. [5, Corollary 2]):

$$\sigma_S(T) \setminus \{0\} = \sigma_{pS}(T) \setminus \{0\}.$$

2.6. Slice nature of normal operators

We recall the "slice" character of $\mathbb{H}$:

- $\mathbb{H} = \bigcup_{\jmath \in \mathbb{S}} \mathbb{C}_\jmath$ where $\mathbb{C}_\jmath$ is the real subalgebra $\langle \jmath \rangle \simeq \mathbb{C}$.
- $\mathbb{C}_\jmath \cap \mathbb{C}_\kappa = \mathbb{R}$ for every $\jmath, \kappa \in \mathbb{S}$ with $\jmath \neq \pm\kappa$.

This decomposition of $\mathbb{H}$ has an "operatorial" counterpart on a quaternionic Hilbert space. It was established in Theorem 5.9 of [6].

Theorem 2.11 ([6]§5.4). *Given any normal operator $T \in \mathfrak{B}(\mathsf{H})$, there exist three operators $A, B, J \in \mathfrak{B}(\mathsf{H})$ such that:*

(i) $T = A + JB$.
(ii) *A is self-adjoint and B is positive.*
(iii) *J is anti self-adjoint and unitary.*
(iv) *A, B and J commute mutually.*

Furthermore, it holds:

- *A and B are uniquely determined by T: $A = (T + T^*)\frac{1}{2}$ and $B = |T - T^*|\frac{1}{2}$.*
- *J is uniquely determined by T on* $\mathrm{Ker}(T - T^*)^\perp$.

(where for $S \in \mathfrak{B}(\mathsf{H})$, $|S|$ denotes the operator defined as the square root of the positive operator S^*S).

In the following, we denote by $\sigma(B)$ and $\rho(B)$ the standard spectrum and resolvent set of a bounded operator B of a complex Hilbert space, respectively.

Proposition 2.12 ([6] §5.4). *Let H be a quaternionic Hilbert space, let $T \in \mathfrak{B}(\mathsf{H})$ be a normal operator, let $J \in \mathfrak{B}(\mathsf{H})$ be an anti self-adjoint and unitary operator satisfying $TJ = JT$, $T^*J = JT^*$, let $\imath \in \mathbb{S}$ and let $\mathsf{H}_\pm^{J\imath}$ be the complex subspaces of H associated with J and $\imath$ (see Definition 2.5). Then we have that*

(a) $T(\mathsf{H}_+^{J\imath}) \subset \mathsf{H}_+^{J\imath}$ *and* $T^*(\mathsf{H}_+^{J\imath}) \subset \mathsf{H}_+^{J\imath}$.

Moreover, if $T|_{\mathsf{H}_+^{J\imath}}$ and $T^*|_{\mathsf{H}_+^{J\imath}}$ denote the $\mathbb{C}_\imath$-complex operators in $\mathfrak{B}(\mathsf{H}_+^{J\imath})$ obtained restricting respectively T and T^* to $\mathsf{H}_+^{J\imath}$, then it holds:

(b) $(T|_{\mathsf{H}_+^{J\imath}})^* = T^*|_{\mathsf{H}_+^{J\imath}}$.
(c) $\sigma(T|_{\mathsf{H}_+^{J\imath}}) \cup \overline{\sigma(T|_{\mathsf{H}_+^{J\imath}})} = \sigma_S(T) \cap \mathbb{C}_\imath$. *Here $\sigma(T|_{\mathsf{H}_+^{J\imath}})$ is considered as a subset of $\mathbb{C}_\imath$ via the natural identification of $\mathbb{C}$ with $\mathbb{C}_\imath$ induced by the real vector isomorphism $\mathbb{C} \ni \alpha + i\beta \mapsto \alpha + \imath\beta \in \mathbb{C}_\imath$.*
(d) $\sigma_S(T) = \Omega_{\mathcal{K}}$, *where* $\mathcal{K} := \sigma(T|_{\mathsf{H}_+^{J\imath}})$.

An analogous statement holds for $\mathsf{H}_-^{J\imath}$.

3. Proofs of the main results

3.1. Proof of Theorem 1.1

If $T = 0$ there is nothing to prove, since $0 \in \sigma_{pS}(T)$ and $\|T\| = 0$ in that case. So, we henceforth assume that $T \neq 0$. Since $T \in \mathfrak{B}(\mathsf{H})$ is normal, Theorem 2.11 assures the existence of an anti self-adjoint unitary right $\mathbb{H}$-linear operator $J : \mathsf{H} \to \mathsf{H}$, commuting with T and T^* and fulfilling $T = (T+T^*)\frac{1}{2} + J|T-T^*|\frac{1}{2}$. Next, if $\mathsf{H}_+^{J\imath}$ is the complex subspace associated with an imaginary unit $\imath \in \mathbb{S}$ as in Definition 2.5, it turns out that T is the unique right $\mathbb{H}$-linear operator, defined on H, whose restriction to $\mathsf{H}_+^{J\imath}$ coincides with the complex-linear operator $S := T|_{\mathsf{H}_+^{J\imath}} : \mathsf{H}_+^{J\imath} \to \mathsf{H}_+^{J\imath}$ (it immediately arises from Propositions 3.11 and 5.11 in [6]). We also know that $\|S\| = \|T\|$, in view of Proposition 3.11 in [6].

By hypotheses T is compact and thus S is compact as well as we go to prove. If $\{u_n\}_{n\in\mathbb{N}} \subset \mathsf{H}_+^{J\imath}$ is a bounded sequence of vectors, it is a bounded sequence of vectors of H too, and thus the sequence $\{Tu_n\}_{n\in\mathbb{N}}$ admits a subsequence $\{Tu_{n_k}\}_{k\in\mathbb{N}}$ converging to some $v \in \mathsf{H}$, because T is compact. However, since $\mathsf{H}_+^{J\imath}$ is closed (because of its definition and the fact that J is continuous), we also have that $v \in \mathsf{H}_+^{J\imath}$ and that $\{Su_{n_k}\}_{k\in\mathbb{N}}$ converges to $v \in \mathsf{H}_+^{J\imath}$, because $Tu_n = Su_n$. We have found that, for every bounded sequence $\{u_n\}_{n\in\mathbb{N}} \subset \mathsf{H}_+^{J\imath}$, there is a subsequence of $\{Su_n\}_{n\in\mathbb{N}}$ converging to some $v \in \mathsf{H}_+^{J\imath}$. Thus S is compact.

To go on, Lemma 3.3.7 in [8] entails that there exists $\lambda \in \sigma_p(S)$ with $|\lambda| = \|S\|$. Notice that $\lambda \neq 0$ otherwise $S = 0$ and thus $T = 0$ by uniqueness of the extension of S. Finally, point (d) of Proposition 2.12 implies that $\lambda \in \sigma_S(T)$. Since T is compact, by Corollary 2 of [5], we have that $\lambda \in \sigma_{pS}(T)$. Summing up, we have obtained that there is $\lambda \in \sigma_{pS}(T)$ with $|\lambda| = \|S\| = \|T\|$, where the absolute value is, indifferently, that in $\mathbb{C}$ or that in $\mathbb{H}$. The remaining identity in (1.1) is now equivalent to: $\sup\{|\mu| \mid \mu \in \sigma_S(T)\} = \|T\|$, i.e., $r_S(T) = \|T\|$. In this form, it was proved in point (d) of Theorem 2.8.

3.2. Proof of Theorem 1.2

Fix an imaginary unit $\imath \in \mathbb{S}$ and consider the normal compact operator $S : \mathsf{H}_+^{J\imath} \to \mathsf{H}_+^{J\imath}$ as in the proof of Theorem 1.1. As a consequence of Theorem 3.3.8 in [8], there exists a Hilbert basis $\mathcal{N} \subset \mathsf{H}_+^{J\imath}$ made of eigenvectors of S and a map $\mathcal{N} \ni z \mapsto \lambda_z \in \mathbb{C}_\imath$ such that each λ_z is an eigenvalue of S in $\mathbb{C}_\imath$ relative to $z \in \mathcal{N}$ and, if $\lambda_z \neq 0$, only a finite number of distinct other elements $z' \in \mathcal{N}$ verify $\lambda_z = \lambda_{z'}$. Moreover, the values λ_z are at most countably many. We know by Lemma 3.10(b) in [6] that $\mathcal{N}$ is also a Hilbert basis of H, so that, if $x \in \mathsf{H}$, then

$$x = \sum_{z\in\mathcal{N}} z\langle z|x\rangle \,.$$

Since T is continuous and $Tz = Sz = z\lambda_z$, we have:

$$Tx = \sum_{z\in\mathcal{N}} z\lambda_z\langle z|x\rangle \,.$$

From Theorem 3.3.8 in [8], we also get that the set $\{\lambda_z\}_{z\in\mathcal{N}}$ of the eigenvalues of S, and therefore those of T, vanish at infinity. Thus 0 is the only possible accumulation point of Λ. If H is not finite dimensional and 0 is not an eigenvalue of T, then 0 must be an accumulation point of Λ, since every set of eigenvectors $Ker(\Delta_{\lambda_z}(T))$ has finite dimension if $\lambda_z \neq 0$. Since $\sigma_S(T)$ is closed, in any case $0 \in \sigma_S(T)$.

Remark 3.1. Since $\lambda_z \in \mathbb{C}_\imath$ for each $z \in \mathcal{N}$, equation (1.3) follows from Corollary 2 in [5].

3.3. Proof of Theorem 1.4

Fix an imaginary unit $\imath \in \mathbb{S}$. For each eigenvector $z \in \mathcal{N}$, with non real eigenvalue $\lambda_z \in \mathbb{H}$, we can choose a unit quaternion μ_z such that $\mu_z^{-1}\lambda_z\mu_z$ belongs to the intersection of the eigensphere of λ_z with the complex plane $\mathbb{C}_\imath$. This means that $z' := z\mu_z$ is still an eigenvector of T, with eigenvalue $\lambda_{z'} := \mu_z^{-1}\lambda_z\mu_z \in \mathbb{C}_\imath$. If $\lambda_z \in \mathbb{R}$, we set $\mu_z = 1$. The set $\mathcal{N}' := \{z\mu_z \mid z \in \mathcal{N}\}$ is still a Hilbert basis of H, such that $Tx = \sum_{z\in\mathcal{N}'} z\lambda_z\langle z|x\rangle$ for every $x \in \mathsf{H}$.

The linear operator J defined by setting

$$Jx := \sum_{z\in\mathcal{N}'} z\imath\langle z|x\rangle$$

is an anti self-adjoint and unitary operator on H (cf. Proposition 3.1 in [6]). Since

$$TJx = \sum_{z\in\mathcal{N}'} z\lambda_z\langle z|Jx\rangle = \sum_{z\in\mathcal{N}'} z\lambda_z\imath\langle z|x\rangle$$

and

$$JTx = \sum_{z\in\mathcal{N}'} z\imath\langle z|Tx\rangle = \sum_{z\in\mathcal{N}'} z\imath\lambda_z\langle z|x\rangle,$$

we have that J and T commute. Moreover, since $T^*x = \sum_{z\in\mathcal{N}'} z\overline{\lambda}_z\langle z|x\rangle$, the same holds for J and T^*. Let $\mathsf{H}_\pm^{J\imath}$ be the complex subspaces of H associated with J and the imaginary unit $\imath \in \mathbb{S}$ as in Definition 2.5. Observe that $\mathcal{N}' \subset \mathsf{H}_+^{J\imath}$, since $Jz = z\imath$ for each $z \in \mathcal{N}'$. Let $\jmath \in \mathbb{S}$ be an imaginary unit orthogonal to $\imath$. Then $\mathcal{N}'\jmath := \{z\jmath \mid z \in \mathcal{N}'\}$ is a Hilbert basis for $\mathsf{H}_-^{J\imath}$ (cf. Lemma 3.10 in [6]). The $\mathbb{C}_\imath$-complex subspaces $\mathsf{H}_\pm^{J\imath}$ of H are invariant for T, since

$$JTu = TJu = TJ(\mp Ju\imath) = \pm(Tu)\imath \quad \text{for each } u \in \mathsf{H}_\pm^{J\imath}.$$

Let $S_\pm := T|_{\mathsf{H}_\pm^{J\imath}}$ be the restrictions of T. Then $S_\pm$ are diagonalizable, since

$$S_+x_+ = \sum_{z\in\mathcal{N}'} z\lambda_z\langle z|x_+\rangle \quad \text{and} \quad S_-x_- = \sum_{z\in\mathcal{N}'} z\jmath\overline{\lambda}_z\langle z\jmath|x_-\rangle$$

for every $x_\pm \in \mathsf{H}_\pm^{J\imath}$. We can then apply Theorem 3.3.8 of [8] and obtain that $S_\pm$ are normal and compact. Using Proposition 3.11 of [6], we get the normality of T.

It remains to prove that T is compact. Recall from Remark 2.6 that $\mathsf{H} = \mathsf{H}_+^{J\imath} \oplus \mathsf{H}_-^{J\imath}$, with projections $P_\pm$ defined by $P_\pm(x) = \frac{1}{2}(x \mp Jx\imath) \in \mathsf{H}_\pm^{J\imath}$. If $\{x_n\}_{n\in\mathbb{N}} \subset \mathsf{H}$ is a bounded sequence of vectors, then also $\{P_+x_n\}_{n\in\mathbb{N}} \subset \mathsf{H}_+^{J\imath}$ and $\{P_-x_n\}_{n\in\mathbb{N}} \subset \mathsf{H}_-^{J\imath}$ are bounded sequences. Since S_+ is compact, the sequence $\{S_+P_+x_n\}_{n\in\mathbb{N}}$ admits a subsequence $\{S_+P_+x_{n_k}\}_{k\in\mathbb{N}}$ converging to some $v_+ \in \mathsf{H}_+^{J\imath}$. Similarly, since S_- is compact, we can extract from the sequence $\{S_-P_-x_{n_k}\}_{k\in\mathbb{N}}$ a subsequence $\{S_-P_-x_{n_{k_l}}\}_{l\in\mathbb{N}}$ converging to some $v_- \in \mathsf{H}_-^{J\imath}$. Then the sequence $\{Tx_{n_{k_l}}\}_{l\in\mathbb{N}}$ converges to $v := v_+ + v_-$, since $Tx = S_+P_+x + S_-P_-x$ for each $x \in \mathsf{H}$.

We have shown that for every bounded sequence $\{x_n\}_{n\in\mathbb{N}} \subset \mathsf{H}$ there is a subsequence of $\{Tx_n\}_{n\in\mathbb{N}}$ converging to some $v \in \mathsf{H}$. Thus T is compact.

The last statement concerning the spectrum of T follows from Proposition 2.12.

References

[1] Brackx, F., Delanghe, R., Sommen, F.: Clifford analysis, *Research Notes in Mathematics*, vol. 76. Pitman (Advanced Publishing Program), Boston, MA (1982).

[2] Colombo, F., Sabadini, I., Struppa, D.C.: Noncommutative functional calculus, *Progress in Mathematics*, vol. 289. Birkhäuser/Springer Basel AG, Basel (2011). DOI 10.1007/978-3-0348-0110-2. URL http://dx.doi.org/10.1007/978-3-0348-0110-2. Theory and applications of slice hyperholomorphic functions.

[3] Emch, G.: Mécanique quantique quaternionienne et relativité restreinte. I. Helv. Phys. Acta **36**, 739–769 (1963).

[4] Farenick, D.R., Pidkowich, B.A.F.: The spectral theorem in quaternions. Linear Algebra Appl. **371**, 75–102 (2003). DOI 10.1016/S0024-3795(03)00420-8. URL http://dx.doi.org/10.1016/S0024-3795(03)00420-8.

[5] Fashandi, M.: Compact operators on quaternionic Hilbert spaces. Facta Univ. Ser. Math. Inform. **28**(3), 249–256 (2013).

[6] Ghiloni, R., Moretti, V., Perotti, A.: Continuous slice functional calculus in quaternionic Hilbert spaces. Rev. Math. Phys. **25**(4), 1350,006, 83 (2013). DOI 10.1142/S0129055X13500062. URL http://dx.doi.org/10.1142/S0129055X13500062.

[7] Jacobson, N.: Normal Semi-Linear Transformations. Amer. J. Math. **61**(1), 45–58 (1939). DOI 10.2307/2371384. URL http://dx.doi.org/10.2307/2371384.

[8] Pedersen, G.K.: Analysis now, *Graduate Texts in Mathematics*, vol. 118. Springer-Verlag, New York (1989). DOI 10.1007/978-1-4612-1007-8. URL http://dx.doi.org/10.1007/978-1-4612-1007-8.

Riccardo Ghiloni, Valter Moretti and Alessandro Perotti
Department of Mathematics
University of Trento
I-38123, Povo-Trento, Italy
e-mail: ghiloni@science.unitn.it
moretti@science.unitn.it
perotti@science.unitn.it

Hypercomplex Analysis: New Perspectives and Applications
Trends in Mathematics, 145–166

Three-dimensional Quaternionic Analogue of the Kolosov–Muskhelishvili Formulae

Yuri Grigor'ev

Abstract. The aim of this work is to construct a three-dimensional quaternionic generalization of the Kolosov–Muskhelishvili formulae. The theory of Moisil–Theodoresco system in terms of regular quaternionic functions of reduced quaternion variable is used. As applications the main problems of an elastic sphere equilibrium are solved.

Mathematics Subject Classification (2010). Primary 30G35; Secondary 74B05.

Keywords. Classical linear elasticity, functions of hypercomplex variables and generalized variables, regular quaternionic functions, Moisil–Theodoresco system, representation formulae.

1. Introduction

In two-dimensional problems of the theory of elasticity the methods of complex variable theory are effectively used. In plane problems the basis of this is the representation of the general solution of the equilibrium equations in terms of two arbitrary analytic functions called the Kolosov–Muskhelishvili formulae [29]. In axially symmetric problems different classes of generalized analytic functions of complex variable [1,39], p- and (p, q)-analytic functions [34] are used. As a generalization of the method of complex functions in multidimensional problems the methods of hypercomplex functions are developed [9, 11, 22–25]. For three-dimensional problems of mathematical physics such an apparatus is the Moisil–Theodoresco system theory, which is developed as the theory of regular quaternion functions of incomplete quaternion variable [15,19,22]. We note that this theory is covered by Clifford analysis. In [28] the first quaternion solution of the equilibrium equations of the theory of elasticity is obtained, but without proof of representation generality. In [4, 5] a spatial quaternionic analog of the complex Kolosov–Muskhelishvili formulae in

The author is partially sponsored by RFBR, grant N 12-01-00507-a.

base on the Papkovich–Neuber representation and polynomial solutions as applications are obtained. In [6, 7] a very special way for a hypercomplex function theory and its application are presented. In [14] three-dimensional representation formulae were derived for every term of the Fourier series expansion of displacement components in cylindrical coordinates. Some other attempts were made in [32, 33]. In [16–20, 31] some preliminary results of using a quaternion function method in the theory of elasticity with solutions of some problems for an elastic sphere are presented. In [26] the method of a regularization of integral equations of elasticity using a Cauchy-type integral for the Moisil–Theodoresco system was proposed. In [27] some connections between equations of continua and hypercomplex functions are noticed. The connection between numerical collocation and quaternionic representations of solutions in three-dimensional elasticity was found out in [36]. Other representations of the 3D elasticity and thermo-elasticity equations through regular quaternion functions were given in [37, 38].

In this paper a variant of three-dimensional quaternion generalization of the Kolosov–Muskhelishvili formulae is proposed, which is effectively applied to solve the basic problems of the theory of elasticity for the ball. It is shown that in particular cases of plane and axially symmetric deformations this representation goes into the Kolosov–Muskhelishvili and Solov'ev formulae.

2. Preliminaries and notations

Let i, j, k be the basic quaternions obeying the following rules of multiplication:

$$i^2 = j^2 = k^2 = -1, \quad ij = -ji = k, \quad jk = -kj = i, \quad ki = -ik = j.$$

An element q of the quaternion algebra $\mathbb{H}$ we write in the form $q = q_0 + iq_x + jq_y + kq_z = q_0 + \mathbf{q}$, where q_0, q_x, q_y, q_z are the real numbers, q_0 is called the scalar part of the quaternion, $\mathbf{q} = iq_x + jq_y + kq_z$ is called the vector part of the quaternion q. The quaternion conjugation is denoted as $\tilde{q} = q_0 - \mathbf{q}$.

Let x, y, z be the Cartesian coordinates in the Euclidean space $\mathbb{R}^3$. Let Ω be a domain of $\mathbb{R}^3$ with a piecewise smooth boundary. A quaternion-valued function or, briefly, $\mathbb{H}$-valued function f of a reduced quaternion variable $\mathbf{r} = ix + jy + kz \in \mathbb{R}^3$ is a mapping

$$f : \Omega \longrightarrow \mathbb{H},$$

such that

$$f(\mathbf{r}) = f_0(\mathbf{r}) + \mathbf{f}(\mathbf{r}) = f_0(x, y, z) + if_x(x, y, z) + jf_y(x, y, z) + kf_z(x, y, z).$$

The functions f_0, f_x, f_y, f_z are real valued defined in Ω. Continuity, differentiability or integrability of f are defined coordinate-wisely. For continuously real-differentiable functions $f : \Omega \subset \mathbb{R}^3 \longrightarrow \mathbb{H}$, which we will denote for simplicity by $f \in C^1(\Omega, \mathbb{H})$, the operator $\nabla = i\partial_x + j\partial_y + k\partial_z$ is called the generalized Cauchy–Riemann operator.

According to R. Fueter [22] a function f is called left-regular in Ω if

$$\nabla f = 0, \quad \mathbf{r} \in \Omega. \tag{2.1}$$

A similar definition can be given for right-regular functions. From now on we use only left-regular functions that, for simplicity, we call regular. With the vectorial notations the regularity condition is given as follows:

$$\nabla f(\mathbf{r}) = -\nabla \cdot \mathbf{f}(\mathbf{r}) + \nabla f_0(\mathbf{r}) + \nabla \times \mathbf{f}(\mathbf{r}) = 0, \tag{2.2}$$

where ∇f_0, $\nabla \cdot \mathbf{f}$, $\nabla \times \mathbf{f}$ are the usual gradient, divergence and curl, respectively. Thus, the coordinate-wise representations of the regularity condition are given as follows:

$$\begin{cases} f_{x,x} + f_{y,y} + f_{z,z} = 0, \\ f_{0,x} + f_{z,y} - f_{y,z} = 0, \\ f_{0,y} + f_{x,z} - f_{z,x} = 0, \\ f_{0,z} + f_{y,x} - f_{x,y} = 0. \end{cases} \tag{2.3}$$

The system (2.3) is called the Moisil–Theodoresco system (MTS) [15, 30] and is a spatial generalization of the Cauchy–Riemann system (CRS). If we assume that f depends only on two variables, for example, x and y, then the MTS (2.3) splits into two CRS and the complex functions $f(\zeta) = f_x(x, y) - i f_y(x, y)$, $g(\zeta) = f_0(x, y) - i f_z(x, y)$ will be the analytic functions of complex variable $\zeta = x + iy$. If the MTS is written in the cylindrical coordinates ρ, φ, z, then in the case of axial symmetry the MTS splits into two generalized by Vekua [39] CRS and $f(\zeta) = f_0(z, \rho) - i f_\varphi(z, \rho)$, $g(\zeta) = f_z(z, \rho) - i f_\rho(z, \rho)$ will be the generalized analytic by Vekua functions of the complex variable $\zeta = z + i\rho$. Exactly these functions are used in axially symmetric problems [1].

The equations of elastic equilibrium are called the Lamé equations:

$$L\mathbf{u} \equiv (\lambda + 2\mu)\nabla(\nabla \cdot \mathbf{u}) - \mu\nabla \times (\nabla \times \mathbf{u}) = 0. \tag{2.4}$$

If we introduce the next notations

$$(\lambda + 2\mu)\nabla \cdot \mathbf{u} = f_0, \quad -\mu\nabla \times \mathbf{u} = \mathbf{f}, \tag{2.5}$$

then the Lamé equation (2.4) is transformed into the MTS:

$$\nabla \cdot \mathbf{f} = 0, \quad \nabla f_0 + \nabla \times \mathbf{f} = 0, \tag{2.6}$$

thus the quaternion function $f = f_0 + \mathbf{f}$ is regular. In the paper [28] it was indicated that such the connection between the Lamé equation and quaternion functions was first pointed by G. Moisil.

3. The radial integration operator

We will introduce the operator of radial integration I^α, acting according to the rule:

$$I^\alpha f = I^\alpha f(\mathbf{r}) = \int\limits_0^1 t^\alpha f(\mathbf{r}t)dt = \int\limits_0^1 t^\alpha f(rt, \theta, \varphi)dt,$$

here r, θ, φ are the spherical coordinates. With the help of the operator I^α it is convenient to solve problems in star, relative to the origin of coordinates, regions, which we will denote as Ω^*. We will give some properties of the operator I^α, which are used in this paper.

1. $(\mathbf{r} \cdot \nabla)\, I^\alpha = I^\alpha(\mathbf{r} \cdot \nabla) = r\partial_r I^\alpha = I^\alpha r \partial_r$.
2. $(\mathbf{r} \cdot \nabla)\, I^\alpha f = r\partial_r I^\alpha f = f - (\alpha + 1)I^\alpha f,\ f \in C^0_\alpha$;
3. $$I^\alpha I^\beta f = I^\beta I^\alpha f = \begin{cases} (\alpha - \beta)^{-1}(I^\beta - I^\alpha)f, & \alpha \neq \beta; \\ -\int_0^1 f(\mathbf{r}t)t^\alpha \ln t dt, & \alpha = \beta. \end{cases}$$

Theorem 3.1 (General solution of biharmonic equation). *The general solution of the biharmonic equation $\triangle\triangle u(\mathbf{r}) = 0$ in Ω^* has the form*

$$u(\mathbf{r}) = v(\mathbf{r}) + \frac{1}{4} r^2 I^{1/2} w(\mathbf{r}), \quad \mathbf{r} \in \Omega^*, \tag{3.1}$$

where v, w are arbitrary harmonic functions in Ω^*, and $\triangle u = w$.

Proof. Let v, w be harmonic functions in Ω^* and u is expressed by the formula (3.1).

Then with the help of the properties of the operator I^α we will find

$$\triangle u = \frac{1}{4}\triangle r^2 I^{1/2} w = \frac{3}{2} I^{1/2} w + \mathbf{r} \cdot \nabla I^{1/2} w = \frac{3}{2} I^{1/2} w + w - \frac{3}{2} I^{1/2} w = w, \tag{3.2}$$

i.e., u is biharmonic. Conversely, let u be the biharmonic function. Let $w = \triangle u$, and let

$$v = u - \frac{1}{4} r^2 I^{1/2} \triangle u.$$

Then the function u is represented in the form (3.1) in terms of two harmonic functions, the function w is harmonic by virtue of (3.2), and $\triangle u = w$ is also by virtue of (3.2). □

With the help of Theorem 3.1 and the properties of the operator I^α the following two statements are proved:

Theorem 3.2. *The general solution of the system*

$$\begin{cases} \nabla \times \mathbf{u}(\mathbf{r}) = \mathbf{v}(\mathbf{r}), \\ \nabla \cdot \mathbf{u}(\mathbf{r}) = v_0(\mathbf{r}), \end{cases}$$

where v_0, $\mathbf{v}$ *are the given harmonic respectively scalar and vector functions in* Ω^*, *and* $\nabla \cdot \mathbf{v} = 0$, *has the form*

$$\mathbf{u}(\mathbf{r}) = -\mathbf{r} \times I^1\mathbf{v} + \nabla\Big[\frac{r^2}{4} I^{1/2}\Big(v_0 - \mathbf{r} \cdot I^2\nabla \times \mathbf{v}\Big)\Big] + \nabla g_0, \quad \mathbf{r} \in \Omega^*,$$

where g_0 *is an arbitrary harmonic function in* Ω^*.

Theorem 3.3 (General solution of MTS). *The general solution of MTS in* Ω^* *has the form*

$$f(\mathbf{r}) = \varphi_0 + \mathbf{r} \times I^1\nabla\varphi_0 + \nabla\psi_0, \quad \mathbf{r} \in \Omega^*,$$

where $\varphi_0(\mathbf{r})$, $\psi_0(\mathbf{r})$ *are arbitrary harmonic functions in* Ω^*.

We will consider in Ω^* the Volterra integral equation of the second kind

$$(1 + aI^\alpha + bI^\beta)u(\mathbf{r}) = f(\mathbf{r}). \tag{3.3}$$

Parameters $a, b, \alpha, \beta \in \mathbb{C}$ of this equation are used to make a quadratic equation

$$p^2 - p(a + b + \alpha + \beta) + a\beta + b\alpha + \alpha\beta = 0, \quad p \in \mathbb{C},$$

whose roots are

$$p_{1,2} = \frac{1}{2}(a + b + \alpha + \beta) \pm \frac{1}{2}[(a - b + \alpha - \beta)^2 + 4ab]^{1/2}.$$

Then with the help of the operator I^α properties it can be proved the next

Theorem 3.4. *Let* $p_1 \neq p_2$ *then the equation* (3.3) *has the unique solution that is expressed in the form:*

$$u = f + (p_2 - p_1)^{-1}[(\alpha - p_1)(\beta - p_1)I^{p_1} - (\alpha - p_2)(\beta - p_2)I^{p_2}]f. \tag{3.4}$$

Radial integration operators are used in mathematical physics, for example, for proving of S.L. Sobolev embedding theorems, in the theory of special functions, etc. In order to solve the biharmonic equation in the theory of elasticity, they were used in [2], [3], [12]. In the theory of quaternionic functions the operator I^α has been used in the work [35] to construct a regular function with a given scalar part. The systematic use of radial integration operators for the theory of Moisil–Theodoresco system was introduced in [16]– [20], [31].

4. A primitive function

According to [28] let us call a $\mathbb{H}$-valued function F as a primitive of the regular function f if $\nabla F = f$, i.e.,

$$\nabla \cdot \mathbf{F} = -f_0, \quad \nabla F_0 + \nabla \times \mathbf{F} = \mathbf{f}. \tag{4.1}$$

Obviously, $\nabla(\nabla\mathbf{F}) = -\Delta\mathbf{F} = \nabla f = 0$ and F is a harmonic function. Such primitive function is not regular.

Theorem 4.1 (A general representation for the primitive function). *A general representation for the primitive F of a regular function f in Ω^* has the form*

$$F(\mathbf{r}) = \chi_0 + \mathbf{r} \times I^1(\nabla\chi_0 - \mathbf{f}) - \nabla\left(\frac{1}{2}r^2 I^{1/2} I^1 f_0\right) + \nabla g_0, \quad \forall \mathbf{r} \in \Omega^*, \tag{4.2}$$

where $\chi_0(\mathbf{r}), g_0(\mathbf{r})$ are arbitrary real-valued harmonic functions in Ω^*.

Proof. Let f be a regular function in Ω^* and F be its primitive. We will show that there will be found harmonic functions χ_0, g_0 such that F is represented in the form (4.2). Really, we will put $\chi_0 = F_0$. Then from (4.1) we will get that $\mathbf{F}$ satisfies the system

$$\nabla \cdot \mathbf{F} = -f_0, \quad \nabla \times \mathbf{F} = \mathbf{f} - \nabla\chi_0, \quad \forall \mathbf{r} \in \Omega^*,$$

where $\nabla \cdot (\mathbf{f} - \nabla\chi_0) = 0$. Therefore, according to Theorem 3.2, $\mathbf{F}$ is represented in the form

$$\mathbf{F} = \mathbf{r} \times I^1(\nabla\chi_0 - \mathbf{f}) - \frac{1}{4}\nabla\left\{r^2 I^{1/2}\left[f_0 + \mathbf{r} \cdot I^2 \nabla \times \mathbf{f}\right]\right\} + \nabla g_0,$$

where g_0 is an arbitrary real-valued harmonic function in Ω^*. Transforming this expression with taking into account the regularity f and with the help of an operator I^α properties, we will get

$$\begin{aligned}\mathbf{F} &= \mathbf{r} \times I^1(\nabla\chi_0 - \mathbf{f}) - \tfrac{1}{4}\nabla\left\{r^2 I^{1/2}\left[f_0 - (\mathbf{r}\cdot\nabla)I^1 f_0\right]\right\} + \nabla g_0 \\ &= \mathbf{r} \times I^1(\nabla\chi_0 - \mathbf{f}) - \tfrac{1}{2}\nabla\left(r^2 I^{1/2} I^1 f_0\right) + \nabla g_0.\end{aligned} \tag{4.3}$$

Thereby, F is represented in the form (4.2).

Conversely, let f be regular in Ω^* and F is defined by the expression (4.2), where χ_0, g_0 are arbitrary harmonic functions in Ω^*. Therefore, by direct differentiation and by means of operator I^α properties we have the expression

$$\nabla \cdot \mathbf{F} = -f_0, \quad \nabla F_0 + \nabla \times \mathbf{F} = \mathbf{f},$$

i.e., F is the primitive of f. □

A concept of primitive function used by us is introduced as a solution of the inhomogeneous MTS and differs from those mentioned in [10], [13] because a concept of hyperdifferentiation is not used and it is not a regular (monogenic) function. In the complex analysis solutions of inhomogeneous CRS $dF/d\bar{z} = f$ are expressed by the Theodoresco operator (transform) [39]. Therefore, the formula (4.2) can be considered as a variant of the generalized Theodoresco transform for the MTS in a star shaped region without the use of a weakly singular integral. In [10] a brief but informative review of the different approaches to the introduction of the concept of primitive function is given, in it the inconvenience for numerical calculations of the integral Theodoresco operator due to the presence of weak singularity is emphasized. Particular attention is given to the derivation of explicit formulas for the primitives of monogenic polynomials. The representation (4.2) allowed us to get in the present paper the quaternion representation of the general solution of the Lamé equation different from the known ones. In addition, it is known the quaternionic integral operators (Cauchy, Theodoresco and Bergman) allow to express the solutions of many three-dimensional boundary value problems in closed forms [21]. Therefore it is possible the representation (4.2) will be useful in this area.

5. Three-dimensional quaternionic analogue of the Kolosov–Muskhelishvili formulae

Theorem 5.1. *The general solution of the Lamé equation* (2.4) *in* Ω^* *is expressed in terms of two regular in* Ω^* *functions* φ, ψ *in the form*

$$2\mu\mathbf{u}(\mathbf{r}) = \varkappa\Phi(\mathbf{r}) - \mathbf{r}\widetilde{\varphi(\mathbf{r})} - \widetilde{\psi(\mathbf{r})}, \ \varkappa = -\frac{3\lambda+7\mu}{\lambda+\mu}, \tag{5.1}$$

where as Φ *one can take any primitive of function* φ, *having subordinated* ψ *to the condition* $\varkappa\Phi_0 = \mathbf{r}\cdot\boldsymbol{\varphi} + \psi_0$.

Proof. Let $\mathbf{u}$ be a solution of the Lamé equation in Ω^*. We will show that there will be found regular functions φ, ψ and the primitive Φ of function φ, and $\varkappa\Phi_0 = \mathbf{r}\cdot\boldsymbol{\varphi} + \psi_0$ such that $\mathbf{u}$ is expressed in the form (5.1). We will introduce the regular function $f = f_0 + \mathbf{f} = (\lambda+2\mu)\nabla\cdot\mathbf{u} - \mu\nabla\times\mathbf{u}$. Then the function

$$\mathbf{F} = \mathbf{f} + \frac{\lambda+\mu}{2(\lambda+2\mu)}\mathbf{r}\times(\nabla f_0) - \nabla\psi_0, \tag{5.2}$$

where ψ_0 is a harmonic in Ω^* function, will be a vector part of regular in Ω^* function, since

$$\begin{aligned}
\nabla\cdot\mathbf{F} &= \nabla\cdot\mathbf{f} + \frac{\lambda+\mu}{2(\lambda+2\mu)}\nabla\cdot[\mathbf{r}\times(\nabla f_0)] - \Delta\psi_0 = 0,\\
\nabla\times\mathbf{F} &= \nabla\times\mathbf{f} + \frac{\lambda+\mu}{2(\lambda+2\mu)}\nabla\cdot[\mathbf{r}\times(\nabla f_0)]\\
&= -\nabla f_0 + \frac{\lambda+\mu}{2(\lambda+2\mu)}\left[\nabla f_0 + (\mathbf{r}\cdot\nabla)\nabla f_0 - 3\nabla f_0\right]\\
&= -\nabla\frac{\lambda+\mu}{2(\lambda+2\mu)}\left[\frac{3\lambda+5\mu}{\lambda+\mu}f_0 + (\mathbf{r}\cdot\nabla)f_0\right] = -\nabla F_0,
\end{aligned}$$

where

$$F_0 = \frac{\lambda+\mu}{2(\lambda+2\mu)}\left[\frac{3\lambda+5\mu}{\lambda+\mu}f_0 + (\mathbf{r}\cdot\nabla)f_0\right]. \tag{5.3}$$

Consequently, $F = F_0 + \mathbf{F}$ is regular.

We will introduce another regular function

$$\varphi(\mathbf{r}) = I^\gamma F(\mathbf{r}), \quad \gamma = \frac{2(\lambda+2\mu)}{\lambda+\mu} > 2, \tag{5.4}$$

from this we will find with the help of the operator I^α properties

$$\begin{aligned}\varphi_0 = I^\gamma F_0 &= \frac{\lambda+\mu}{2(\lambda+2\mu)}\lambda+\mu\left[\frac{3\lambda+5\mu}{\lambda+\mu}I^\gamma f_0 + (\mathbf{r}\cdot\nabla)I^\gamma f_0\right] \\ &= \frac{\lambda+\mu}{2(\lambda+2\mu)}\left[\left(\frac{3\lambda+5\mu}{\lambda+\mu}-\gamma-1\right)I^\gamma f_0 + f_0\right] = \frac{\lambda+\mu}{2(\lambda+2\mu)}f_0,\end{aligned} \tag{5.5}$$

and

$$F = (\gamma+1+\mathbf{r}\cdot\nabla)\varphi. \tag{5.6}$$

Now we will introduce the primitive Φ of the function φ such that $\varkappa\Phi_0 = \mathbf{r}\cdot\boldsymbol{\varphi}+\psi_0$ (it is easy to show that such the primitive exists, see Theorem 4.1), and we will show that the function ψ, defined by the relation

$$\widetilde{\psi(\mathbf{r})} = \varkappa\Phi(\mathbf{r}) - \mathbf{r}\widetilde{\varphi(\mathbf{r})} - 2\mu\mathbf{u}(\mathbf{r}), \quad \varkappa = -\frac{3\lambda+7\mu}{\lambda+\mu}, \quad \forall\mathbf{r}\in\Omega^*, \tag{5.7}$$

i.e.,

$$\begin{aligned}\psi_0 &= \varkappa\Phi_0 - \mathbf{r}\cdot\boldsymbol{\varphi}, \\ \boldsymbol{\psi} &= \varkappa\boldsymbol{\Phi} + \mathbf{r}\varphi_0 - \mathbf{r}\times\boldsymbol{\varphi} + 2\mu\mathbf{u},\end{aligned} \tag{5.8}$$

will be regular in Ω^*. Indeed, recalling the relations $(\lambda+2\mu)\nabla\cdot\mathbf{u} = f_0$, (5.5) and that $\varphi\in\mathcal{R}(\Omega^*)$, we have

$$\begin{aligned}\nabla\cdot\boldsymbol{\psi} = &-\varkappa\nabla\cdot\boldsymbol{\Phi} + \nabla\cdot(\mathbf{r}\varphi_0 - \mathbf{r}\times\boldsymbol{\varphi}) + 2\mu\nabla\cdot\mathbf{u} = \varkappa\varphi_0 + 3\varphi_0 + (\mathbf{r}\cdot\nabla)\varphi_0 \\ &+\mathbf{r}\cdot(\nabla\times\boldsymbol{\varphi}) + \frac{2\mu}{\lambda+2\mu}f_0 = (\varkappa+3)\varphi_0 + \mathbf{r}\cdot(\nabla\varphi_0) - \mathbf{r}\cdot(\nabla\varphi_0) \\ &+\frac{4\mu}{\lambda+\mu}\varphi_0 = \left(-\frac{3\lambda+7\mu}{\lambda+\mu}+3+\frac{4\mu}{\lambda\mu}\right)\varphi_0 = 0,\end{aligned}$$

similarly we find

$$\begin{aligned}\nabla\times\boldsymbol{\psi} &= -\varkappa\nabla\times\boldsymbol{\Phi} + \nabla\times(\mathbf{r}\varphi_0 - \mathbf{r}\times\boldsymbol{\varphi}) + 2\mu\nabla\times\mathbf{u} \\ &= -\varkappa(\boldsymbol{\varphi} - \nabla\boldsymbol{\Phi}_0) - \mathbf{r}\times(\nabla\varphi_0) - [\boldsymbol{\varphi} - (\mathbf{r}\cdot\nabla)\boldsymbol{\varphi} - 3\boldsymbol{\varphi}] - 2\mathbf{f} \\ &= (2-\varkappa)\boldsymbol{\varphi} + \varkappa\nabla\boldsymbol{\Phi}_0 - \mathbf{r}\times(\nabla\varphi_0) + (\mathbf{r}\cdot\nabla)\boldsymbol{\varphi} - 2\mathbf{f},\end{aligned}$$

but, in virtue of (5.2), (5.5) and (5.6)

$$\mathbf{f} = (\gamma+1+\mathbf{r}\cdot\nabla)\boldsymbol{\varphi} - \mathbf{r}\times(\nabla\varphi_0) + \nabla\psi_0,$$

and from (5.8) we will find

$$\begin{aligned}\varkappa\nabla\Phi_0 &= \nabla\psi_0 + \nabla(\mathbf{r}\cdot\boldsymbol{\varphi}) = \nabla\psi_0 + \boldsymbol{\varphi} + (\mathbf{r}\cdot\nabla)\boldsymbol{\varphi} + \mathbf{r}\times(\nabla\times\boldsymbol{\varphi}) \\ &= \nabla\psi_0 + \boldsymbol{\varphi} + (\mathbf{r}\cdot\nabla)\boldsymbol{\varphi} - \mathbf{r}\times(\nabla\varphi_0),\end{aligned}$$

hence

$$\begin{aligned}\nabla\times\boldsymbol{\psi} = &(2-\varkappa)\boldsymbol{\varphi} + \nabla\psi_0 + (\mathbf{r}\cdot\nabla)\boldsymbol{\varphi} - \mathbf{r}\times(\nabla\varphi_0) - \mathbf{r}\times(\nabla\varphi_0) + (\mathbf{r}\cdot\nabla)\boldsymbol{\varphi} \\ &- 2(\gamma+1)\boldsymbol{\varphi} - 2(\mathbf{r}\cdot\nabla)\boldsymbol{\varphi} - 2\mathbf{r}\times(\nabla\varphi_0) - 2\nabla\psi_0 = -\nabla\psi_0.\end{aligned}$$

Thus, indeed ψ is regular in Ω^*, and $\mathbf{u}$ is represented in the form (5.1).

Conversely, let $\mathbf{u}$ be defined by (5.1), where $\varphi, \psi \in \mathcal{R}(\Omega^*)$ are arbitrary and Φ is the primitive of φ, and $\varkappa\Phi_0 = \mathbf{r} \cdot \boldsymbol{\varphi} + \psi_0$. Then by direct differentiation we ascertain that $\mathbf{u}$ satisfies the Lamé equation. □

Now a general solution of the Lamé equation in Ω^* is given in terms of two regular functions obtained by V.V. Naumov (see [17]).

Theorem 5.2. *A general solution of the Lamé equation* (2.4) *in* Ω^* *is expressed in terms of regular in* Ω^* *two functions* $f, \nabla g_0$ *in the form*

$$\mathbf{u}(\mathbf{r}) = \frac{\mathbf{r}}{\mu} \times I^1 \mathbf{f} + \nabla \left\{ r^2 \left[\frac{3\lambda + 7\mu}{4\mu(\lambda + 2\mu)} I^{1/2} - \frac{1}{\mu} I^1 \right] f_0 \right\} + \nabla g_0, \tag{5.9}$$

and $f = (\lambda + 2\mu)\nabla \cdot \mathbf{u} - \mu \nabla \times \mathbf{u}$.

Proof. Let $\mathbf{u}$ be a solution of the Lamé equation in Ω^*. We will show that there will be found regular functions $f, \nabla g_0$, such that $\mathbf{u}$ is expressed in the form (5.9), and $f = (\lambda + 2\mu)\nabla \cdot \mathbf{u} - \mu \nabla \times \mathbf{u}$. Indeed, we will put

$$f_0 = (\lambda + 2\mu)\nabla \cdot \mathbf{u}, \quad \mathbf{f} = -\mu \nabla \times \mathbf{u},$$

in this case, as it was mentioned above f is regular. Then, according to Theorem 3.2, $\mathbf{u}$ is determined from the above system in the form

$$\mathbf{u} = \frac{\mathbf{r}}{\mu} \times I^1 \mathbf{f} + \nabla \left[\frac{r^2}{4} I^{1/2} \left(\frac{1}{\lambda + 2\mu} f_0 + \frac{\mathbf{r}}{\mu} \cdot I^2 \nabla \times \mathbf{f} \right) \right] + \nabla g_0, \tag{5.10}$$

where g_0 is a harmonic in Ω^* function. Transforming this expression, taking into account the regularity of f and using the operator I^α properties, we get

$$\begin{aligned}
\mathbf{u} &= \frac{\mathbf{r}}{\mu} \times I^1 \mathbf{f} + \nabla \left\{ \frac{r^2}{4} \left[\frac{1}{\lambda + 2\mu} I^{1/2} - \frac{1}{\mu} I^{1/2} (\mathbf{r} \cdot \nabla) I^1 \right] f_0 \right\} + \nabla g_0 \\
&= \frac{\mathbf{r}}{\mu} \times I^1 \mathbf{f} + \nabla \left\{ \frac{r^2}{4} \left[\frac{1}{\lambda + 2\mu} I^{1/2} f_0 - \frac{1}{\mu} I^{1/2} (f_0 - 2I^1 f_0 \right] \right\} + \nabla g_0 \\
&= \frac{\mathbf{r}}{\mu} \times I^1 \mathbf{f} + \nabla \left\{ \frac{r^2}{4} \left[-\frac{\lambda + \mu}{\mu(\lambda + 2\mu)} I^{1/2} f_0 + \frac{4}{\mu} \left(I^{1/2} - I^1 \right) f_0 \right] \right\} + \nabla g_0 \\
&= \frac{\mathbf{r}}{\mu} \times I^1 \mathbf{f} + \nabla \left\{ r^2 \left[\frac{3\lambda + 7\mu}{4\mu(\lambda + 2\mu)} I^{1/2} - \frac{1}{\mu} I^1 \right] f_0 \right\} + \nabla g_0.
\end{aligned}$$

Conversely, let $\mathbf{u}$ be defined by (5.9), where $f, \nabla g_0$ are arbitrary regular functions. Then by direct differentiation we ascertain that $\mathbf{u}$ satisfies the Lamé equation. □

General solutions (5.1) and (5.9) in Ω^* are equivalent, and if in (5.1) one introduces a regular in Ω^* function f according to the relation

$$f = \gamma \varphi_0 + (\gamma + 1 + \mathbf{r} \cdot \nabla)\boldsymbol{\varphi} - \mathbf{r} \times (\nabla \varphi_0) + \psi_0, \quad \gamma = \frac{2(\lambda + 2\mu)}{\lambda + \mu},$$

then (5.1) turns into (5.9) and inversely: if in (5.9) one introduces regular in Ω^* functions φ and ψ according to the relations

$$\varphi = \frac{1}{\gamma} f_0 + I^\gamma \left[\mathbf{f} + \frac{1}{\gamma} \mathbf{r} \times (\dot{\nabla f_0}) - \nabla \psi_0 \right], \quad \psi = \psi_0 + \mathbf{r} \times I^1 \nabla \psi_0,$$

where ψ_0 is an arbitrary harmonic in Ω^* function, and the primitive Φ of the function φ such that $\varkappa \Phi_0 = \mathbf{r} \cdot \varphi + \psi_0$, then (5.9) turns into (5.1).

6. Plane and axially symmetric deformations

In this section it is shown that general solutions (5.1) and (5.9) of the Lamé equation in Ω^*, expressed in terms of two regular functions, in the case of plane deformation goes into the general Kolosov–Muskhelishvili solution of the equations of the plane theory of elasticity, expressed in terms of two analytic functions of complex variable [29]. Also it is shown that in the case of axially symmetric deformation both quaternion representations go into the general Yu.I. Solov'ev solution of equations of the axially symmetric theory of elasticity, expressed in terms of two generalized analytic by Vekua functions of complex variable [1].

Theorem 6.1. *In the case of plane deformation* $u_x = u_x(x, y)$, $u_y = u_y(x, y)$, $u_z = 0$ *the quaternion representation* (5.9) *of the general solution of the Lamé equation goes into the general Kolosov–Muskhelishvili solution of the equations of the plane theory of elasticity*

$$2\mu(u_x + iu_y) = \varkappa\varphi(\zeta) - \zeta\overline{\varphi'(\zeta)} - \overline{\psi(\zeta)}, \quad \varkappa = \frac{\lambda + 3\mu}{2(\lambda + \mu)} = 3 - 4\nu, \tag{6.1}$$

where $\varphi(\zeta)$, $\psi(\zeta)$ *are the analytic functions of the complex variable* $\zeta = x + iy$.

Proof. In the considered case of plane deformation, when $u_x = u_x(x, y)$, $u_y = u_y(x, y)$, $u_z = 0$, from the relations(2.5) we have $f_x = f_y = 0$, $f_0 = f_0(x, y)$, $f_z = f_z(x, y)$, and, as it was noted in the introduction, the function $f(\zeta) = f_0(x, y) - i f_z(x, y)$ will be the analytic function of complex variable $\zeta = x + iy$, also

$$\begin{cases} u_{x,x} + u_{y,y} = \dfrac{1}{\lambda + 2\mu} f_0, \\ u_{y,x} + u_{x,y} = -\dfrac{f_z}{\mu}. \end{cases}$$

We will write the condition $u_z = 0$ for the representation (5.9):

$$u_z = 2z \left[\frac{3\lambda + 7\mu}{4\mu(\lambda + 2\mu)} I^{1/2} - \frac{1}{\mu} I^1 \right] f_0 + g_{0,z} = 0.$$

Integrating this equation over z, we will get

$$z^2 \left[\frac{3\lambda + 7\mu}{4\mu(\lambda + 2\mu)} I^{1/2} - \frac{1}{\mu} I^1 \right] f_0 + g_0 = \psi_0(x, y), \tag{6.2}$$

where $\psi_0(x,y)$ is so far an arbitrary function of integration. Since f_0, g_0 are harmonic, then from (6.2) we get

$$\begin{aligned}&\Delta\Delta\psi_0(x,y)=0,\\&\Delta\psi_0(x,y)=2\left[\frac{3\lambda+7\mu}{4\mu(\lambda+2\mu)}I^{1/2}-\frac{1}{\mu}I^1\right]f_0(x,y).\end{aligned}\tag{6.3}$$

Consequently, under Theorem 3.1, the general solution of the problem (6.3) has the form

$$\psi_0(x,y)=\chi_0(x,y)+\frac{x^2+y^2}{2}I^0\left[\frac{3\lambda+7\mu}{4\mu(\lambda+2\mu)}I^{1/2}-\frac{1}{\mu}I^1\right]f_0,\tag{6.4}$$

where $\chi_0(x,y)$ is an arbitrary harmonic function. Now from (6.2), taking into account (6.4), we find

$$g_0=\left(-z^2+\frac{x^2+y^2}{2}I^0\right)\left[\frac{3\lambda+7\mu}{4\mu(\lambda+2\mu)}I^{1/2}-\frac{1}{\mu}I^1\right]f_0(x,y)+\chi_0(x,y).\tag{6.5}$$

Substituting (6.5) into (5.9) for u_x, we will get

$$u_x=\frac{y}{\mu}I^1f_z+\frac{1}{2\mu}\left\{(x^2+y^2)\left[\frac{\lambda+3\mu}{2(\lambda+2\mu)}I^0-I^1\right]f_0\right\}_{,x}+\chi_{0,x}.$$

Similarly, we will get

$$u_y=\frac{x}{\mu}I^1f_z+\frac{1}{2\mu}\left\{(x^2+y^2)\left[\frac{\lambda+3\mu}{2(\lambda+2\mu)}I^0-I^1\right]f_0\right\}_{,y}+\chi_{0,y}.$$

We will introduce a new analytic function $\varphi(\zeta)$, such that

$$\begin{aligned}\varphi'(\zeta)&=\varphi_{x,x}(x,y)+i\varphi_{x,y}(x,y)=\\&=\frac{\lambda+\mu}{2(\lambda+2\mu)}f(\zeta)=\frac{\lambda+\mu}{2(\lambda+2\mu)}\left[f_0(x,y)-if_z(x,y)\right].\end{aligned}\tag{6.6}$$

Then, making transformations, for u_x and u_y we will get the expressions

$$\begin{aligned}2\mu u_x&=\varkappa\varphi_x-(x\varphi_{x,x}+y\varphi_{y,x})-\varkappa\varphi_x(0)+2\mu\chi_{0,x},\\2\mu u_y&=\varkappa\varphi_y-(y\varphi_{x,x}-x\varphi_{y,x})-\varkappa\varphi_y(0)+2\mu\chi_{0,y};\\\varkappa&=\frac{\lambda+3\mu}{\lambda+\mu}=-3-4\nu.\end{aligned}\tag{6.7}$$

Introducing another analytic function

$$\psi(\zeta)=2\mu\left(-\chi_{0,x}+i\chi_{0,y}\right)+\varkappa\left[\varphi_x(0)-i\varphi_y(0)\right],$$

from (6.7) we get (6.1). □

Theorem 6.2. *In the case of axially symmetric deformation the quaternion representation (5.9) of the general Lamé equation goes into the general Yu.I. Solov'ev solution of equations of the axially symmetric theory of elasticity, expressed in terms of two generalized analytic by Vekue functions of complex variable [1, p. 296].*

Proof. We write the representation (5.9) in the cylindric system of coordinates $\varrho,\ \varphi,\ z\ (x=\varrho\cos\varphi,\ y=\varrho\sin\varphi,\ z=z)$:

$$u=-\frac{z}{\mu}I^1f_\varphi+\left\{(\varrho^2+z^2)\left[\frac{3\lambda+7\mu}{4\mu(\lambda+2\mu)}I^{1/2}-\frac{1}{\mu}I^1\right]f_0\right\}_{,\varrho}+g_{0,\varrho},$$
$$v=\frac{1}{\mu}\left(zI^1f_\varrho-\varrho I^1f_z\right)+\frac{1}{\varrho}\left\{(\varrho^2+z^2)\left[\frac{3\lambda+7\mu}{4\mu(\lambda+2\mu)}I^{1/2}-\frac{1}{\mu}I^1\right]f_0\right\}_{,\varphi}+\frac{1}{\varrho}g_{0,\varphi},$$
$$w=\frac{\varrho}{\mu}I^1f_\varphi+\left\{(\varrho^2+z^2)\left[\frac{3\lambda+7\mu}{4\mu(\lambda+2\mu)}I^{1/2}-\frac{1}{\mu}I^1\right]f_0\right\}_{,z}+g_{0,z}. \tag{6.8}$$

Here (u,v,w) and $(f_\varrho,\ f_\varphi,\ f_z)$ are the physical projections of $\mathbf{u}$ $\mathbf{f}$ respectively. We will consider the axially symmetric deformation, when $u=u(z,\varrho),\ w=w(z,\varrho),\ v=0$. Then from (2.3) it follows that

$$f_\varrho=f_z=0,\ \ f_0=f_0(z,\varrho),\ \ f_\varphi=f_\varphi(z,\varrho) \tag{6.9}$$

and, as it was noted in the introduction, the function $f(\zeta)=f_0(z,\varrho)-if_\varphi(z,\varrho)$ will be the generalized by Vekua function of complex variable $\zeta=z+i\varrho$. From the condition $v=0$, from (6.8) and (6.9) it follows that $g_{0,\varphi}=0$, i.e., g_0 is the axially symmetric harmonic function:

$$\Delta g_0=\frac{1}{\varrho}(\varrho g_{0,\varrho})_{,\varrho}+g_{0,zz}=0. \tag{6.10}$$

It is easy to show that the function $F(\zeta)=F_z+iF_\varrho=g_{0,z}-ig_{0,\varrho}$ will be generalized analytic. Indeed, in virtue of (6.10) we have

$$F_{z,z}=g_{0,zz}=-\frac{1}{\varrho}(\varrho g_{0,\varrho})_{,\varrho}=\frac{1}{\varrho}(\varrho F_\varrho),$$

on the other hand

$$F_{\varrho,z}=g_{0,\varrho z}=-F_{\varrho,z},$$

i.e., the generalized Cauchy–Riemann conditions for $F(\zeta)$ are satisfied:

$$F_{z,z}=\frac{1}{\varrho}(\varrho F_\varrho)_{,\varrho},\ \ F_{z,\varrho}=-F_{\varrho,z}.$$

We will introduce a new generalized analytic function $\varphi(\zeta)$, such that

$$\varphi'(\zeta)=\varphi_{z,z}(z,\varrho)+i\varphi_{\varrho,z}(z,\varrho)=$$
$$=\frac{\lambda+\mu}{2(\lambda+2\mu)}f(\zeta)=\frac{\lambda+\mu}{2(\lambda+2\mu)}\left[f_0(z,\varrho)-if_\varphi(z,\varrho)\right],$$

i.e., the generalized Cauchy–Riemann conditions for $\varphi(\zeta)$ are satisfied:

$$\varphi_{z,z}=\frac{1}{\varrho}(\varrho\varphi_\varrho)_{,\varrho},\ \ \varphi_{z,\varrho}=-\varphi_{\varrho,z}.$$

We will note that if $\varphi(\zeta)$ is generalized analytic, then $\varphi'(\zeta) = \partial_z \varphi(\zeta)$ will also be generalized analytic. Taking into account all this, from (6.8) we have

$$2\mu w = -\frac{4(\lambda+2\mu)}{\lambda+\mu}\varrho I^1 \varphi_{\varrho,z} + \frac{2(3\lambda+7\mu)}{\lambda+\mu} z I^{1/2} \varphi_{z,z} - \frac{8(\lambda+2\mu)}{\lambda+\mu} z I^1 \varphi_{z,z} \quad (6.11)$$
$$+ \frac{3\lambda+7\mu}{\lambda+\mu}(\varrho^2+z^2) I^{3/2} \varphi_{z,zz} - \frac{4(\lambda+2\mu)}{\lambda+\mu}(\varrho^2+z^2) I^2 \varphi_{z,zz} + 2\mu g_{0,z}.$$

Then it can be shown that

$$2\mu w = \varkappa \varphi_z - (z\varphi_{z,z} + \varrho \varphi_{\varrho,z}) - \frac{\varkappa}{2} I^{-1/2} \varphi_z + 2\mu g_{0,z}, \quad (6.12)$$

where

$$\varkappa = \frac{3\lambda+7\mu}{2(\lambda+\mu)}.$$

In an analogical way we also have

$$2\mu u = \varkappa \varphi_\varrho - (\varrho \varphi_{z,z} - z\varphi_{\varrho,z}) + \frac{\varkappa}{2} I^{-1/2} \varphi_\varrho + 2\mu g_{0,\varrho}. \quad (6.13)$$

Introducing another generalized analytic function

$$\Psi(\zeta) = 2\mu\left(-g_{0,z} + i g_{0,\varrho}\right) + \frac{\varkappa}{2} I^{-1/2} \varphi(\zeta),$$

from (6.12) and (6.13) we will finally get the general Yu.I. Solov'ev solution of equations of the axially symmetric theory of elasticity

$$2\mu(w+iu) = \varkappa \varphi(\zeta) - \zeta \overline{\varphi'(\zeta)} - \overline{\Psi(\zeta)}, \quad (6.14)$$

where $\varphi(\zeta)$, $\Psi(\zeta)$ are the generalized analytic by Vekua functions of the complex variable $\zeta = z + i\varrho$. □

7. Normal loading of elastic sphere

In this section from the quaternion representation (5.9) a new representation of the general solution of the Lamé equation in Ω^* is obtained expressed in terms of three harmonic functions. With the help of that representation the solution of 2nd basic problem of the theory of elasticity for the sphere, when on the boundary of the ball a purely normal self-balanced load is applied, is obtained in the closed form. The solution of the problem is also expressed through a solution of the Dirichlet problem for the function, which is harmonic in the sphere.

7.1. General solution

Using Theorem 3.3 from the quaternion representation (5.9) it can be proved

Theorem 7.1. *The general solution of the Lamé equation* (2.4) *in* Ω^* *is expressed in terms of three arbitrary harmonic in* Ω^* *functions* χ_0, ψ_0, g_0 *in the form*

$$\mu \mathbf{u}(\mathbf{r}) = \mathbf{r} I^0 \chi_0 + \nabla\left\{ r^2 \left[\frac{3\lambda+7\mu}{4(\lambda+2\mu)} I^{1/2} - I^0 \right] \chi_0 \right\} + \mathbf{r} \times (\nabla \psi_0) + \nabla g_0, \quad (7.1)$$

and $\nabla \cdot \mathbf{u} = (1/(\lambda+2\mu))\chi_0$.

The components of stress tensor for this general solution in the spherical coordinates $r,\ \theta,\ \varphi$ have the forms:

$$\begin{aligned}
\sigma_{rr} &= \frac{5}{8(1-\nu)}f - \frac{r}{4-4\nu}f_{,r} - \frac{\tau}{2}I^{1/2}f + 2g_{,rr},\\
\sigma_{r\theta} &= \left[\left(I^0 - \tau I^{1/2} - \frac{1}{4-4\nu}\right)f + \frac{2}{r^2}\left(rg_{,r} - g\right)\right]_{,\theta} + \frac{1}{r\sin\theta}\left(\psi - r\psi_{,r}\right)_{,\phi}, \qquad (7.2)\\
\sigma_{r\varphi} &= \frac{1}{\sin\theta}\left[\left(I^0 - \tau I^{1/2} - \frac{1}{4-4\nu}\right)f + \frac{2}{r^2}\left(rg_{,r} - g\right)\right]_{,\varphi} - \frac{1}{r}\left(\psi - r\psi_{,r}\right)_{,\theta},
\end{aligned}$$

here

$$\tau = \frac{7-8\nu}{8(1-\nu)}, \qquad \nu = \frac{\lambda}{2(\lambda+\mu)},$$

ν is Poisson's ratio.

In the paper [12] it is shown that in a star shaped region in the Papkovich–Neuber representation without loss of generality one can set equal to zero one harmonic function, but only under certain restrictions on Poisson's ratio. In contrast to this our representation (7.1) by means of three harmonic functions is the general solution without any restrictions on the elastic constants.

7.2. The method of solution

By means the representation (7.1) all main problems of an equilibrium of an elastic sphere can be solved in the closed forms. We will denote through U and S the sphere of a radius R and its boundary. For example we will consider the 2nd basic problem of equilibrium of elastic sphere with a purely normal load on the boundary:

$$\begin{cases} L\mathbf{u}(\mathbf{r}) = 0, \quad \mathbf{r} \in U, \quad C^2(U) \cap C^1(\overline{U}) \\ \sigma_{rr}|_S = \sigma(\theta,\varphi) \in C^0(S); \quad \sigma_{r\theta}|_S = \sigma_{r\varphi}|_S = 0. \end{cases} \qquad (7.3)$$

Let the condition of the main vector of applied external forces being equal to zero be satisfied

$$\oint_S \boldsymbol{\sigma} dS = \frac{1}{R}\oint_S \sigma(\theta,\varphi)\mathbf{r}dS_r = 0, \qquad (7.4)$$

the principal moment for a purely normal load is always equal to zero. We will impose a stronger limitation

$$\sigma_{rr} \in C^0(\overline{U}), \quad \sigma_{r\theta} \in C^1(\overline{U}), \quad \sigma_{r\varphi} \in C^1(\overline{U}).$$

The solution of the problem (7.3), (7.4) we will find in the form (7.1), where the harmonic functions have the properties

$$f \in C^2(\overline{U}), \quad g, \psi \in C^3(\overline{U}). \qquad (7.5)$$

In virtue of (7.5) from (7.2) it is seen that from the zero boundary conditions for $\sigma_{r\vartheta}, \sigma_{r\varphi}$ it follows that $\psi - r\psi_{,r} \equiv \Psi(r)$ is harmonic and $\Psi = C_1 = \text{const}$ in U, and the functions f and g for $r = R$ on the boundary S of the sphere U are related by the relation:

$$\frac{2}{R^2}\left(g - rg_{,r}\right)\Big|_{r=R} - \left[\left(I^0 - \tau I^{1/2} - \frac{1}{4-4\nu}\right)f\right]\Big|_{r=R} = C_2 = \text{const}, \qquad (7.6)$$

but in (7.6) in virtue of (7.5) and the operator I^α properties the functions in the square brackets and parentheses will be regular harmonic in the sphere U. Hence, in virtue of the uniqueness of the solution of the Dirichlet problem from (7.6) we get the relation, which is true everywhere in the sphere $\overline{U}$:

$$g - rg_{,r} = \frac{R^2}{2}\left[\left(I^0 - \tau I^{1/2} - \frac{1}{4-4\nu}\right)f + C_2\right], \quad \forall \mathbf{r} \in \overline{U}. \qquad (7.7)$$

Differentiating (7.7) with respect to r, we will find

$$g_{,rr} = -\frac{R^2}{2r}\left[\left(I^0 - \tau I^{1/2} - \frac{1}{4-4\nu}\right)f\right]_{,r}, \qquad (7.8)$$

substituting (7.8) into the expression for σ_{rr} from (7.2), we will obtain that σ_{rr} is expressed only through f:

$$\sigma_{rr} = \frac{5}{8(1-\nu)}f - \frac{r}{4-4\nu}f_{,r} - \frac{\tau}{2}I^{1/2}f - \frac{R^2}{r}\left[\left(I^0 - \tau I^{1/2} - \frac{1}{4-4\nu}\right)f\right]_{,r}, \quad (7.9)$$

hereof and from (7.3) we will get the boundary condition

$$\sigma_{rr}|_{r=R} = \left[\frac{5}{8(1-\nu)}f - \frac{\tau}{2}I^{1/2}f\right]\Big|_{r=R} - R\left(I^0 f - \tau I^{1/2} f\right)_{,r}\Big|_{r=R} = \sigma(\vartheta,\varphi). \qquad (7.10)$$

We will clarify the structure of f. In follows from the condition of solvability (7.4) that the expansion of the regular in the sphere U harmonic function f into a series of spherical functions has the form

$$f = f_0 + f_2, \quad f_0 = f(0), \quad f_2 = \sum_{n=2}^{\infty}\sum_{m=-n}^{n} a_{nm}\left(\frac{r}{R}\right)^n Y_n^m(\vartheta,\varphi).$$

We will introduce a new harmonic function F:

$$F(\mathbf{r}) = \left[\frac{1}{2(1-\nu)} - 2\tau I^{1/2} + I^0\right] f. \qquad (7.11)$$

It is clear that the structure of the function F is analogous to the structure of the function f. With the help of the operator I^α properties one can show that the boundary condition (7.10) can be written in the form

$$F|_{r=R} = \sigma(\vartheta,\varphi).$$

Thereby, for the new harmonic function we got the Dirichlet problem

$$\begin{cases} \Delta F(\mathbf{r}) = 0, \quad \mathbf{r} \in U, \\ F|_{r=R} = \sigma(\vartheta,\varphi). \end{cases} \qquad (7.12)$$

With the help of Theorem 3.3 from (7.11) we find

$$\begin{aligned} f &= 2(1-\nu)\,(F + 2\,\mathrm{Re}\,AI^{\omega}F)\,,\\ \omega &= \frac{1}{2}(-1+2\nu+iq), \quad q=\sqrt{3-4\nu^2},\\ A &= \frac{1}{4}\left[3-4\nu+\frac{i}{q}(8\nu^2-6\nu-1)\right]. \end{aligned} \tag{7.13}$$

Now we will consider the equation (7.7). The homogeneous equation (7.7) $g - rg_{,r} = 0$ has a nontrivial solution $g_0 = \mathbf{r}\cdot\mathbf{a}$, where $\mathbf{a}$ is an arbitrary constant. This solution, according to (7.1), defines a rigid displacement $\mathbf{u} = \mathbf{a}$. A particular solution for it with the constant right-hand side is also a constant and it does not contribute to the solution of the original problem. Discarding these solutions, we will leave only the particular solution with the right-hand side, formed from f_2, which has the form

$$g = g_2 = \frac{R^2}{2}I^{-2}\left(\tau I^{1/2} - I^0 + \frac{1}{4-4\nu}\right)f_2.$$

Substituting here (7.13) and transforming the obtained expression we will get

$$\begin{aligned} g_2 &= \frac{R^2}{6}\left(I^{-2} + 2\,\mathrm{Re}\,BI^{\omega}\right)F_2,\\ B &= \frac{1}{4}\left[1+\frac{i}{q}(9-10\nu)\right]. \end{aligned} \tag{7.14}$$

We will consider the equation for ψ: $\psi - r\psi_{,r} = C_1$. Its solution has the form $\psi = C_1 + \mathbf{r}\cdot\mathbf{b}$, where $\mathbf{b}$ is an arbitrary constant vector. Therefore the function ψ in virtue of (7.1) defines the solution in the form of rigid rotation $\mathbf{u} = \mathbf{r}\times\mathbf{b}$.

Finally, substituting (7.13) and (7.14) in (7.1) and carrying out some transformations we have the result

Theorem 7.2. *The solution of the problem* (7.3)–(7.4) *exists, is defined up to a rigid displacement and has the form*

$$\mu\mathbf{u}(\mathbf{r}) = 2\mathbf{r}\,\mathrm{Re}(AI^{\omega}F) - r^2\nabla\,\mathrm{Re}(BI^{\omega}F) + \frac{R^2}{6}\nabla\left(I^{-2}+2\,\mathrm{Re}\,BI^{\omega}\right)F_2, \tag{7.15}$$

where

$$\begin{aligned} \omega &= \frac{1}{2}(-1+2\nu+iq);\ q=\sqrt{3-4\nu^2},\\ A &= \frac{1}{4}\left[3-4\nu+\frac{i}{q}(-1-6\nu+8\nu^2)\right],\\ B &= \frac{1}{4}\left[1+\frac{i}{q}(9-10\nu)\right],\\ F_2 &= F - F(0) - r\lim_{r\to 0}\frac{\partial}{\partial r}F, \end{aligned} \tag{7.16}$$

the function F is the solution of the Dirichlet problem (7.12).

7.3. Closed form of the solution

Next, we will obtain the solution of the original problem in quadratures. For this one needs to substitute into (7.15) the expression for the function F in the form of the Poisson integral, which is written in the form

$$\frac{1}{4\pi R^2}\oint \sigma\frac{1-t^2}{s^3}dS,\quad s^2=1-2tc+t^2,\quad t=r/R, \tag{7.17}$$

where

$$c=\cos\gamma=\cos\theta\cos\theta'+\sin\theta\sin\theta'\cos(\varphi-\varphi');\quad dS=R^2\sin\theta' d\theta' d\varphi',$$

$$\oint \Phi dS=\oint_S \Phi(\mathbf{r};\mathbf{r}')dS_{r'}=R^2\int_0^{2\pi}d\varphi'\int_0^{\pi}d\theta'\sin\theta'\Phi(\mathbf{r};R,\theta',\varphi').$$

We will note that in virtue of the equilibrium conditions (7.4) we will have the relation $\oint \sigma c dS=0$. For calculating $I^{-2}F_2$ the integral is necessary

$$I^{-2}\left(\frac{1-t^2}{s^3}-1-3ct\right)=1-5ct+\frac{2}{s}-3s-3ct\ln\frac{\zeta}{2},\quad \zeta=1-ct+s.$$

Substituting (7.17) into (7.15), taking into account the noted above considerations, using the formulas, which are easy to verify according to the operator I^α properties:

$$\begin{aligned} I^\omega\left(1-t^2\right)s^{-3}&=2s^{-1}-(2\omega+1)I^\omega s^{-1}, && \operatorname{Re}\omega>-1;\\ t\partial_t I^\omega\left(s^{-\alpha}-1\right)&=s^{-\alpha}-1-(\omega+1)I^\omega\left(s^{-\alpha}-1\right), && \operatorname{Re}\omega>-2,\end{aligned}$$

introducing new complex variables $A_1=-2(1+2\omega)[2A+(1+\omega)B]$, $3A_2=2(1+2\omega)(1+\omega)B$, $A_3=2(1+2\omega)(1+\omega)B$ we have the next

Theorem 7.3. *The solution of the problem (7.3)–(7.4) up to a rigid displacement has the form:*

$$\begin{aligned} u_r(\mathbf{r})=\frac{1}{8\pi\mu R}\oint\sigma\Big\{2(1-\nu)\frac{1+t^2}{ts}+\frac{(1-t^2)^2}{2ts^3}-\frac{s}{t}\\ -c\ln\zeta+\frac{1}{t}\operatorname{Re}\left[(A_1t^2+A_2)I^\omega\frac{1}{s}\right]\Big\}dS; \end{aligned} \tag{7.18}$$

$$\begin{aligned} \begin{Bmatrix} u_\theta(\mathbf{r})\\ u_\varphi(\mathbf{r})\end{Bmatrix}=\frac{1}{8\pi\mu R}\oint\sigma\begin{Bmatrix}\xi\\ \eta\end{Bmatrix}\Big[\frac{1-t^2}{s^3}+\frac{1+s+tcs}{\zeta s}\\ -\ln\zeta+\left(t^2-\frac{1}{3}\right)\operatorname{Re}\left(A_3I^{\omega+1}\frac{1}{s^3}\right)\Big]dS, \end{aligned}$$

where $\mathbf{r} \in U$;

$$\begin{aligned} &\zeta = 1 - tc + s, \quad \omega = \frac{1}{2}(-1 + 2\nu + iq), \quad q = \sqrt{3 - 4\nu^2}; \\ &A_1 = 2 - 6\nu + 4\nu^2 - \frac{i}{q}(3 + \nu - 12\nu^2 + 8\nu^3); \\ &A_2 = -1 - 2\nu - 4\nu^2 - \frac{i}{q}(2 - 5\nu - 4\nu^2 + 8\nu^3); \\ &A_3 = \frac{3}{2}\left[-3 + 4\nu + \frac{i}{q}(1 + 6\nu - 8\nu^2)\right]. \end{aligned} \tag{7.19}$$

We will note that the here appearing integrals of the form $I^\beta s^{-\alpha}$ are expressed in terms of the Appell hypergeometric function

$$I^\omega s^{-\alpha} = \frac{1}{\omega + 1} F_1\left(\omega + 1, \frac{\alpha}{2}, \frac{\alpha}{2}, \omega + 2, te^{i\gamma}, te^{-i\gamma}\right), \ \mathrm{Re}\ \omega > -2, \tag{7.20}$$

where $\alpha = \mathrm{const} \in \mathbb{C}$; $t \in [0, 1)$.

7.4. Case of axial symmetry

We will consider a particular case of axial symmetry, when in the boundary condition $\sigma = \sigma(\vartheta)$. The integration over $d\varphi'$ in the formulas (7.18) gives the following expressions for the displacement components ($u_\varphi = 0$):

$$\begin{aligned} u_r = \frac{R}{2\pi\mu} \int\limits_0^\pi d\vartheta' \sin\vartheta' \sigma(\vartheta') \Bigg[&\frac{1 - 2\nu}{1 + \nu} \frac{\pi t}{2} + 2(1 - \nu) \frac{1 + t^2}{t} U(t) \\ &+ \frac{1 - t^2}{2t} (2t\partial_t U(t) + U(t)) + \frac{1}{t} \mathrm{Re} \int\limits_0^1 \left(\frac{Pt^2 + Q}{y^{1+n_1}} + \frac{1}{y^2} \right) U(ty) dy \Bigg], \\ u_\theta = \frac{R}{2\pi\mu} \partial_\vartheta \int\limits_0^\pi d\vartheta' \sin\vartheta' \sigma(\vartheta') \Bigg[&(1 - t^2) U(t) \\ &+ \mathrm{Re} \int\limits_0^1 \left(\frac{St^2 + T}{y^{1+n_1}} + \frac{1}{y^2} \right) U(ty) dy \Bigg], \end{aligned} \tag{7.21}$$

where

$$n_1 + 1 = -\overline{\omega}, \quad P = \overline{A}_1, \quad Q = \overline{A}_2, \quad S = \overline{A}_3, \quad T = -\frac{1}{3}S,$$

i.e.,

$$\begin{aligned} &n_1 = \frac{1}{2}(-1 - 2\nu + iq), \quad \sqrt{3 - 4\nu^2}, \\ &P = 2 - 6\nu + 4\nu^2 + \frac{i}{q}(3 + \nu - 12\nu^2 + 8\nu^3), \\ &Q = -1 - 2\nu - 4\nu^2 + \frac{i}{q}(2 - 5\nu - 4\nu^2 + 8\nu^3), \end{aligned}$$

$$S = \frac{3}{2}\left[-3 + 4\nu - \frac{i}{q}(1 + 6\nu - 8\nu^2)\right],$$
$$T = \frac{1}{2}\left[3 - 4\nu + \frac{i}{q}(1 + 6\nu - 8\nu^2)\right]. \qquad (7.22)$$

We use notation

$$U(t) \equiv U(t, \vartheta, \vartheta') = \frac{1}{h}\mathbb{K}(k) - \frac{\pi}{2}(1 + t\cos\vartheta\cos\vartheta').$$

where $\mathbb{K}(k)$ is the complete normal Legendre elliptic integral of the first kind.

As it was expected, the expressions (7.21) coincide with the known result [8]. We will note that in the paper [8] there are given the wrong values for constants S and T, the right values for them are given here in (7.22).

We note that the method of solving the problem for the ball given here appeared only when using quaternion functions. An another more general approach for solving boundary problems of mathematical physics by means of the theory of quaternion functions are given in [22].

8. Conclusions

In this paper the theory of Moisil–Theodoresco system in terms of regular quaternionic functions of reduced quaternion variable is used. A radial integration technique is used systematically for solving the arising problems. Unlike [4], we used another notion of primitive of regular function. Therefore, we have proved the new version of three-dimensional quaternionic analogue of the complex Kolosov–Muskhelishvili formulae. Also another quaternion representation of Lamé equation solution is presented. An equivalence of these two representations is shown. It is shown that the quaternion representation in the case of plane deformation goes into the general Kolosov–Muskhelishvili solution and in the case of axially symmetric deformation goes into the general Yu.I. Solov'ev solution. As applications the problem of an elastic sphere equilibrium in the case of normal loading is solved. Using the special representation of the solution of the Lamé equation, which is a consequence of quaternion representation, the solution is expressed in terms of one harmonic function, which is the solution of the Dirichlet problem with the boundary condition as the original problem. This solution is also expressed in terms of quadratures of elementary functions and Appell hypergeometric function. In the particular case of axial symmetry the coincidence with the known result is obtained. In the general case it can be shown that all main problems of an elastic sphere equilibrium by means of the proposed method may be expressed in terms of three independent harmonic functions.

References

[1] A.Ya. Aleksandrov and Yu.I. Solov'ev, *Three-Dimensional Problems of the Theory of Elasticity* [in Russian]. Nauka, Moscow, 1979.

[2] S. Bergman and M. Schiffer, *Kernel functions and elliptic differential equations in mathematical physics.* Acad. Press, New York, 1953.

[3] V.I. Bloch, *Theory of Elasticity* [in Russian]. Khark.GU, Kharkiv, 1964.

[4] S. Bock and K. Gürlebeck, *On a spatial generalization of the Kolosov–Muskhelishvili formulae.* Mathematical Methods in the Applied Sciences, Vol. 32, Issue 2 (2009), 223–240. DOI: 10.1002/mma.1033.

[5] S. Bock and K. Gürlebeck, *On a Polynomial Basis Generated from the Generalized Kolosov–Muskhelishvili Formulae.* Advances in Applied Clifford Algebras, **19** (2009), 191–209. DOI. 10.1007/s00006-009-0156-5.

[6] F.A. Bogashov, *Representation of a Biharmonic Function in the Complex Space C2.* Russian Acad. Sci. Dokl. Math. [in Russian], 48(2) (1994), 259–262.

[7] F.A. Bogashov and A.G. Ugodchikov, *Development of Mushelishvili methodology to the spatial elasticity problems. Part II.* Nauchnye trudy [in Russian], N2, Nijegorod. univers., Nijniy Novgorod (1995), 161–178.

[8] V.F. Bondareva, *On the effect of an axisymmetric normal loading on an elastic sphere.* Journal of Applied Mathematics and Mechanics, Volume 33, Issue 6 (1969), 1001–1005. DOI 10.1016/0021-8928(69)90006-9.

[9] F. Brackx, R. Delanghe, and F. Sommen, *Clifford Analysis*, Research Notes in Mathematics, no. 76, Pitman, London, 1982.

[10] I. Cacao, K. Gürlebeck, *On monogenic primitives of monogenic functions.* Complex Variables and Elliptic Equations, vol. 52, no. 10-11 (2007), 1081–1100. DOI: 10.1080/17476930701399088.

[11] Fabrizio Colombo, Irene Sabadini and Daniele C. Struppa, *Slice monogenic functions.* Israel Journal of Mathematics, vol. 171, no. 1 (2009), 385–403. DOI 10.1007/s11856-009-0055-4.

[12] T. Trang Cong and G.P. Steven., *On the representation of elastic displacement fields in terms of three harmonic functions.* Journal of Elasticity, vol. 9, no. 3 (1979), 325–333. DOI 10.1007/BF00041103.

[13] R. Delanghe, *On primitives of monogenic functions.* Complex Variables and Elliptic Equations, vol. 51, no. 8-11 (2006), 959–970. DOI: 10.1080/17476930600667841.

[14] O.G. Goman, *On the Kolosov–Muskhelishvili analog for the three-dimensional state of stress.* Journal of Applied Mathematics and Mechanics, Volume 47, Issue 1 (1983), 65–69. DOI 10.1016/0021-8928(83)90035-7.

[15] Yu.M. Grigor'ev and V.V. Naumov, *Approximation theorems for the Moisil–Theodorescu system.* Siberian Mathematical Journal, September–October, Volume 25, Issue 5 (1984), 693–701. DOI 10.1007/BF00968681.

[16] Yu.M. Grigor'ev, *Solution of three dimensional problems of an elasticity theory by means of quaternionic functions methods. Disser. kand. fiz.-mat. nauk.* [in Russian]. Novosib. gos. univers, Novosibirsk, 1985.

[17] Yu.M. Grigor'ev, *Solution of a problem for an elastic sphere in a closed form.* Dynamics of Continuous Medium [in Russian], no. 71 (1985), Inst. Gidrodin. Sib. Otd. Akad. Nauk SSSR, Novosibirsk, 50–54.

[18] Yu.M. Grigor'ev and V.V. Naumov, *Solution of third and fourth main problems of an equilibrium of an elastic sphere in a closed form.* Dynamics of Continuous Medium [in Russian], no. 87 (1988), Inst. Gidrodin. Sib. Otd. Akad. Nauk SSSR, Novosibirsk, 54–66.

[19] Yu.M. Grigor'ev and V.V. Alekhin, *A quaternionic boundary element method.* Sib. jurn. industr. matem. [in Russian], vol. 2, no. 1 (1999), Inst. Matem. Sib. Otd. Akad. Nauk SSSR, Novosibirsk, 47–52.

[20] Yu.M. Grigor'ev, *A spatial analogue of the integral equation of Mushelishvili.* Dynamics of Continuous Medium [in Russian], no. 114 (1999), Inst. Gidrodin. Sib. Otd. Akad. Nauk SSSR, Novosibirsk, 161–165.

[21] K. Gürlebeck and W. Sprößig, *Quatemionic analysis and elliptic boundary value problems.* Birkhäuser Verlag, 1990.

[22] K. Gürlebeck, K. Habetha and W. Sprößig, *Quaternionic Calculus for Engineers and Physicists.* John Wiley &.Sons, Cinchester, 1997.

[23] K. Gürlebeck and W. Sprößig, *Holomorphic functions in the plane and n-dimensional space.* Birkhäuser Verlag, 2008.

[24] V.V. Kravchenko, *Applied quaternionic analysis.* Research and Exposition in Mathematics **28** (2003), Lemgo, Heldermann Verlag.

[25] M. Ku, U. Kähler and D.S. Wang, *Riemann boundary value problems on the sphere in Clifford analysis.* Advances in Applied Clifford Algebras, Volume 22, Issue 2 (June 2012), 365–390. DOI 10.1007/s00006-011-0308-2.

[26] V.G. Maz'ya and V.D. Sapozhnikova, *A note on regularization of a singular system of isotropic elasticity.* Vestnik Leningr. univers. Ser. mat., mekh. i astron [in Russian], V. 7, N. 2 (1964), Leningr. gos. univers., Leningrad, 165–167.

[27] P. Mel'nichenko and E.M. Pik, *Quaternion equations and hypercomplex potentials in the mechanics of a continuous medium.* Soviet Applied Mechanics, Volume 9, Issue 4 (April 1973), 383–387. DOI 10.1007/BF00882648.

[28] M. Misicu, *Representarea ecuatilor echilibrului elastic prin functii monogene de cuaterninoni.* Bull. Stiint. Acad. RPR. Sect. st. mat. fiz., V. 9, N. 2 (1957), 457–470.

[29] N.I. Muskhelishvili, *Some Basic Problems of the Mathematical Theory of Elasticity.* Springer, 1977.

[30] V.V. Naumov and Yu.M. Grigor'ev, *The Laurent series for the Moisil–Theodoresco system.* Dynamics of Continuous Medium [in Russian], no. 54 (1982), Inst. Gidrodin. Sib. Otd. Akad. Nauk SSSR, Novosibirsk, 115–126.

[31] V.V. Naumov, *Solution of two main problems of an equilibrium of an elastic sphere in a closed form.* Dynamics of Continuous Medium [in Russian], no. 54 (1986), Inst. Gidrodin. Sib. Otd. Akad. Nauk SSSR, Novosibirsk, 96–108.

[32] D.D. Penrod, *An analogue of the Kolosov–Muskhelishvili formulae in three dimensions.* Quart. of Appl. Math., V. 23, N. 4 (1966), 313–322.

[33] A.A. Pimenov and V.I. Pushkarev, *The use of quaternions to generalize the Kolosov–Muskhelishvili method to three-dimensional problems of the theory of elasticity.* Journal of Applied Mathematics and Mechanics, Volume 55, Issue 3 (1991), 343–347. DOI 10.1016/0021-8928(91)90036-T.

[34] G. Polozij, *The Theory and Application of p-Analytic and pq-Analytical Functions* [in Russian]. Naukova Dumka, Kiev, 1973.

[35] A. Sudbery, *Quaternionic analysis.* Mathematical Proceedings of the Cambridge Philosophical Society, no. 85 (1979), 199–225. DOI:10.1017/S0305004100055638.

[36] W. Sprößig and K. Gürlebeck, *A hypercomplex method of calculating stresses in three-dimensional bodies.* In: Frolik, Z. (ed.): Proceedings of the 12th Winter School on Abstract Analysis. Section of Topology. Rendiconti del Circolo Matematico di Palermo, Serie II, Supplemento no. 6 (1984), 271–284. http://dml.cz/dmlcz/701846.

[37] A.M. Tsalik, *Quaternion functions: Properties and applications to the continuum problems.* Dokl. Akad. Nauk Ukrainskoi SSR. Ser. A [in Russian], N. 12 (1986), Naukova Dumka, Kiev, 21–24.

[38] A.M. Tsalik, *Quaternionic Representation of the 3D Elastic and Thermoelastic Boundary Problems.* Mathematical methods in the Applied Sciences, vol. 18 (1995), 697–708. DOI 10.1002/mma.1670180904.

[39] I. Vekua, *Generalized Analytic Functions.* Addison Wesley, Reading Mass., 1962.

Yuri Grigor'ev
Theoretical Physics Department
North-Eastern Federal University
58, Belinsky Str.
Yakutsk, 677000 Russia
e-mail: `grigyum@yandex.ru`

Hypercomplex Analysis: New Perspectives and Applications
Trends in Mathematics, 167–180

On the Continuous Coupling of Finite Elements with Holomorphic Basis Functions

Klaus Gürlebeck and Dmitrii Legatiuk

Abstract. The main goal of this paper is to improve the theoretical basis of coupling of an analytical and a finite element solution to the Lamé–Navier equations in case of singularities caused by a crack. The main interest is to construct a continuous coupling between two solutions through the whole interaction interface. To realize this continuous coupling so-called coupling elements are introduced, which are based on a new interpolation operator. In the convergence study of the finite element method the invariant subspaces of the interpolation operator plays a crucial role. In previous studies it has been shown that for a given special distribution of interpolation nodes the corresponding interpolation problem is uniquely solvable. In this paper we extend the result to the case of an arbitrary number of nodes. This result is the basis for a well-defined interpolation operator with all properties which are needed for the convergence analysis of the coupled finite element method.

Mathematics Subject Classification (2010). Primary 74S70, 74S05, 30E05, 35J25; Secondary 30J15, 35Q74, 74R99.

Keywords. Lamé–Navier equations, complex function theory, analytical solution, Kolosov–Muskhelishvili formulae, FEM, coupling, interpolation problem, singularity.

1. Introduction

The idea of a coupling comes from engineering problems containing different types of singularities (like for instance cracks, gaps, corners). To handle such problems by numerical methods (like finite element method, finite difference method, etc.) one needs to perform some adaptations of the numerical scheme and usually it requires also a refined mesh in the region near to a singularity. At present, the finite element method is the most popular numerical method in computation mechanics.

The research of the second author is supported by the German Research Foundation (DFG).

An alternative to numerical methods for problems of linear elasticity are the methods of complex function theory. By using the Kolosov–Muskhelishvili formulae one can describes the near-field solution of a crack tip by only two holomorphic functions $\Phi(z)$ and $\Psi(z)$, $z \in \mathbb{C}$ [8]. The analytical solution based on the complex function theory gives us a high accuracy of the solution in the neighbourhood of the singularity. Because of using exact solutions of the partial differential equations all details of the mathematical model are preserved.

The disadvantage of the complex analytic approach is that the full linear elastic boundary value problem can be solved explicitly only for some elementary (simple) or canonical domains. Domains coming from practical engineering problems usually are more complicated. Therefore it makes sense to ask for a combination of an analytical and a finite element solution in one procedure. The problem of such combination comes from a coupling of two different solutions. On one hand, the analytical solution which is constructed by the complex function theory has a purely analytic form but on the other hand, the standard finite element solution is based on spline functions. Due to that fact we need to consider more carefully this coupling process.

In previous research [4, 3, 6] we have introduced the main idea of a new method of coupling an analytical and a numerical solution (FE-solution). This method allows to get a continuous coupling between analytical and finite element solutions through the whole interaction interface. Usually, analytic solutions and FE-solutions are coupled only through the nodes of the mesh [9, 10]. This pointwise coupling leads to a simple integration of the extra elements into the typical finite element scheme but out of the nodes the numerical solution will have jumps. Looking at the quality of the solution it is not completely satisfying that one improves the approximation of a point singularity (zero-dimensional) of the displacement field and as a result the displacement field has a one-dimensional jump.

To overcome this problem we construct a special element that contains an exact solution to the differential equation with the correct singularity and so-called coupling elements. The request for these coupling elements is to insure C^0 continuity for displacements. For that reason a special interpolation operator has been constructed that preserves the analytical solution on the coupling interface, couples it continuously with special elements which have a polynomial connection to the standard elements. In [2] following P.G. Ciarlet [1] some basic steps for convergence analysis of the proposed method have been performed. In this theory one of the most important roles is played by the unisolvence property of the interpolation operator that is used for the finite element approximation. In [4] it has been shown that for a given special distribution of interpolation nodes the corresponding interpolation problem is uniquely solvable. In this paper these results are generalized to the case of an arbitrary number of nodes. This is the necessary result to define the basis functions for the interpolation operator which permit an arbitrary refinement of the mesh and makes the method practically applicable. Based on this result the convergence and error estimates of the proposed scheme can be proved but this not the purpose of this paper.

To prepare the theorem about the new interpolation operator we will recall the basic ideas of coupling from [3, 4]. For that reason we start with the analytical solution to a crack tip problem. After that we introduce the problem of coupling and will formulate the general interpolation theorem.

2. Geometrical settings and the analytical solution

Following [3, 4], in this section a construction of the analytical solution is discussed and the general description of a domain and its decomposition for the special purposes is introduced. To construct the exact solution in the crack-tip region we are going to work in the field $\mathbb{C}$ of one complex variable, where we identify each point of the complex plane $\mathbb{C}$ with the ordered pair $z = (x_1, x_2) \in \mathbb{R}^2$, $x_1, x_2 \in \mathbb{R}$ or equivalently with the complex number $z = x_1 + ix_2 \in \mathbb{C}$, where i denotes the imaginary unit.

Let now $\Omega \subset \mathbb{C}$ be a bounded simply connected domain containing a crack. To describe the behaviour of the continuum near the crack-tip correctly we are going to model more precisely the near-field domain, called Ω_{SE} (see Figure 1). The domain Ω_{SE} can be interpreted as a special element in the triangulation $\mathcal{F}_h$ over the domain Ω. The special element is always located at the crack tip, i.e., at the origin of a Cartesian coordinate system.

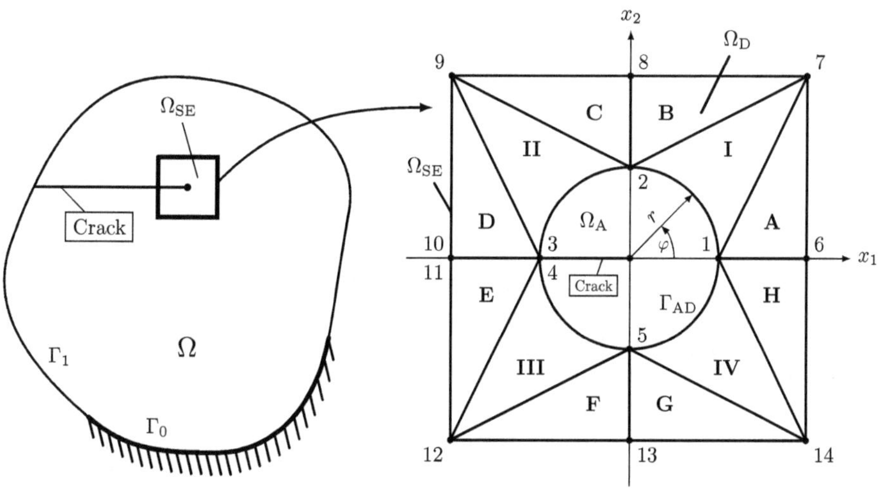

FIGURE 1. Geometrical setting of special element

The domain Ω_{SE} is decomposed in two sub-domains, $\Omega_{\text{SE}} = \Omega_{\text{A}} \cup \Omega_{\text{D}}$, separated by the fictitious joint interface $\Gamma_{\text{AD}} = \overline{\Omega}_{\text{A}} \cap \overline{\Omega}_{\text{D}}$. The discrete "numerical" domain, denoted by Ω_{D}, is modelled by two different kinds of elements: the CST-elements with $C^0(\Omega)$ continuity (elements A–H in Figure 1) and the Coupling-elements with $C^0(\Omega)$ continuity to the CST-elements, and with $C^\infty(\Omega)$ continuity on the interface Γ_{AD} (elements I–IV in Figure 1), which couple the "numerical" domain Ω_{D} with the "analytical" domain Ω_{A}. The $C^{\infty}(\Omega)$ continuity on the interface Γ_{AD} should be understood in the sense, that the interpolation functions are infinitely differentiable on the interface. But, this does not mean automatically that the connection between elements will be better than $C^{0}(\Omega)$. For that we would need to introduce some additional conditions.

The nodes $3, 10$ and $4, 11$ belong to the upper and lower crack faces, respectively. We call the sub-domain Ω_{A} analytical in that sense, that the constructed solutions are exact solutions to the differential equation in Ω_{A}. Analogously, the numerical sub-domain Ω_{D} means, that the constructed solutions are based on the finite element approximation. The idea behind this special element is to get the continuous connection through the interface Γ_{AD} by modifying the shape functions over the curved triangles I–IV.

As a result, we introduce a triangulation $\mathcal{F}_h$ over the domain Ω by three families of finite elements

$$\Omega = \cup_{\mathrm{K_A}\in\mathcal{F}_h}\mathrm{K_A} \cup_{\mathrm{K_{CE}}\in\mathcal{F}_h} \mathrm{K_{CE}} \cup_{\mathrm{K_{CST}}\in\mathcal{F}_h} \mathrm{K_{CST}},$$

where the $\mathrm{K_A}$-element is based on the analytical solution in Ω_{A}, $\mathrm{K_{CE}}$ are the Coupling-elements and $\mathrm{K_{CST}}$ are the classical CST-elements. A connection between the elements $\mathrm{K_A}$, $\mathrm{K_{CE}}$ and $\mathrm{K_{CST}}$ is defined by common sets of degrees of freedom. Additionally, the connection between $\mathrm{K_A}$ and $\mathrm{K_{CE}}$ is supplemented by continuous connection through the interface Γ_{AD}.

We will solve the following boundary value problem in the domain Ω:

$$\begin{cases} -\mu\,\Delta\mathbf{u} - (\lambda+\mu)\mathrm{grad}\;\mathrm{div}\,\mathbf{u} = \mathbf{f} \quad \text{in} \quad \Omega, \\ \mathbf{u} = 0 \quad \text{on} \quad \Gamma_0, \\ \sum_{j=1}^{2} \sigma_{ij}(\mathbf{u})v_j = g_i \quad \text{on} \quad \Gamma_1, \quad 1 \le i \le 2. \end{cases} \tag{2.1}$$

System (2.1) of equations of linear elasticity in the plane describes the state of an elastic body in the case of two dimensions and in this article we concentrate ourselves to the plane strain state, i.e., $u_3 = 0$, $\varepsilon_{3j} = 0, j = 1, \dots, 3$. In Figure 1 the domain Ω represents a volume, that is occupied by a solid body. The boundaries Γ_0 and Γ_1 are defined with Dirichlet and Neumann boundary conditions, respectively. Surface forces of density $\mathbf{g}$ and volume forces with the density $\mathbf{f}$ are given, $\mathbf{u}$ denotes the displacements, v_j are components of the unit outer normal, λ and μ are material constants (see, e.g., [1]).

The crack tip produces a singularity of the solution in the domain Ω. Due to that fact, on the one hand it must be handled in a proper way to get the right behavior of the solution near the singularity. But on the other hand in the part of the domain Ω which is free of singularities one can use the standard finite element method. For that reasons we will construct the analytical solution to the crack tip problem near the singularity by using complex function theory and couple this solution with a finite element solution for the part of domain without singularity.

To construct the analytical solution to the crack-tip problem we are going to work with the Kolosov–Muskhelishvili formulae (see [8]), which are given by

$$\begin{aligned} 2\mu(u_1 + i\,u_2) &= \kappa\,\Phi(z) - z\,\overline{\Phi'(z)} - \overline{\Psi(z)}, \\ \sigma_{11} + \sigma_{22} &= 2\left[\Phi'(z) + \overline{\Phi'(z)}\right], \\ \sigma_{22} - \sigma_{11} + 2i\,\sigma_{12} &= 2[\bar{z}\,\Phi''(z) + \Psi'(z)], \end{aligned} \tag{2.2}$$

where $\Phi(z)$ and $\Psi(z)$, $z \in \mathbb{C}$ are two holomorphic functions. The factor κ is the Kolosov constant, which is defined by

$$\kappa = \begin{cases} 3 - 4\nu & \text{for plane strain,} \\ \dfrac{3-\nu}{1+\nu} & \text{for plane stress.} \end{cases}$$

The crack faces are assumed to be traction free [7], i.e., the normal stresses $\sigma_{\varphi\varphi}$ and the shear stresses $\sigma_{r\varphi}$ on the crack faces vanish for $\varphi = \pi$ or $\varphi = -\pi$, where φ and r are polar coordinates (see Figure 1). Corresponding to [8] the Kolosov–Muskhelishvili formulae in polar coordinates read as follows:

$$\begin{aligned} 2\mu(u_r + i\,u_\varphi) &= e^{-i\varphi}\left(\kappa\,\Phi(z) - z\,\overline{\Phi'(z)} - \overline{\Psi(z)}\right) \\ \sigma_{rr} + \sigma_{\varphi\varphi} &= 2\left[\Phi'(z) + \overline{\Phi'(z)}\right], \\ \sigma_{\varphi\varphi} - \sigma_{rr} + 2i\,\sigma_{r\varphi} &= 2e^{2i\varphi}\left[\bar{z}\,\Phi''(z) + \Psi'(z)\right]. \end{aligned} \tag{2.3}$$

By adding the last two equations of (2.3) we get the following equation which connects the stresses $\sigma_{\varphi\varphi}$ and $\sigma_{r\varphi}$

$$\sigma_{\varphi\varphi} + i\,\sigma_{r\varphi} = \Phi'(z) + \overline{\Phi'(z)} + e^{2i\varphi}[\bar{z}\,\Phi''(z) + \Psi'(z)]. \tag{2.4}$$

The functions $\Phi(z)$ and $\Psi(z)$ will be written as series expansions

$$\Phi(z) = \sum_{k=0}^{\infty} a_k z^{\lambda_k}, \qquad \Psi(z) = \sum_{k=0}^{\infty} b_k z^{\lambda_k}, \tag{2.5}$$

where a_k and b_k are unknown coefficients, which should be determined through the boundary conditions for the global problem and the powers λ_k describe the behaviour of the displacements and stresses near the crack tip and should be determined through the boundary conditions on the crack faces.

After substituting (2.5) into (2.4) one can calculate the powers λ_k and obtains a relation between the unknown coefficients a_k and b_k. Finally we have the following expressions for displacements and stresses in Cartesian coordinates

$$\begin{aligned} 2\mu(u + i\,v) = &\sum_{n=1,3,\ldots}^{\infty} r^{\frac{n}{2}}\left[a_n\left(\kappa\,e^{i\varphi\frac{n}{2}} - e^{-i\varphi\frac{n}{2}}\right) + \frac{n}{2}\bar{a}_n\left(e^{-i\varphi\frac{n}{2}} - e^{-i\varphi(\frac{n}{2}-2)}\right)\right] \\ &+ \sum_{n=2,4,\ldots}^{\infty} r^{\frac{n}{2}}\left[a_n\left(\kappa\,e^{i\varphi\frac{n}{2}} + e^{-i\varphi\frac{n}{2}}\right) + \frac{n}{2}\bar{a}_n\left(e^{-i\varphi\frac{n}{2}} - e^{-i\varphi(\frac{n}{2}-2)}\right)\right]. \end{aligned} \tag{2.6}$$

The displacement field (2.6) satisfies all the conditions on the crack faces. The asymptotic behaviour at the crack tip is controlled by half-integer powers. To avoid unboundedness for ($n < 0$) and discontinuity (for $n = 0$) of the functions (2.6) at the origin the series must begin with $n = 1$. Now we are going to solve the problem of coupling in a way to get a continuous displacement field through the boundary Γ_{AD}.

3. The interpolation problem for the coupling

The purpose of this section is to define an interpolation operator on the joint interface Γ_{AD} and to prove the unique solvability of the corresponding interpolation problem. How to get the desired continuous coupling through the interface is shown already in [4] and will not be repeated here. Let us consider n nodes on the interface Γ_{AD} belonging to the interval $[-\pi, \pi]$ (see Fig. 2).

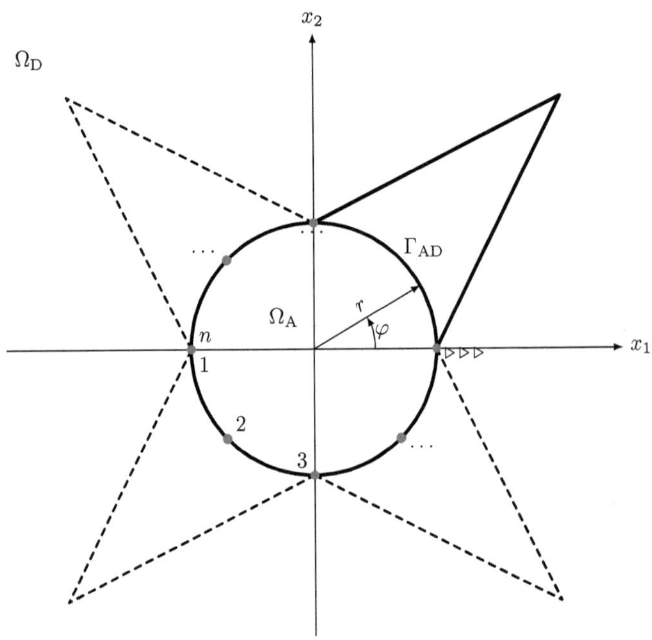

FIGURE 2. The coupling problem

As the interpolation function $f_n(\varphi)$ we use partial sums of the analytical solution (2.6) restricted to the interface Γ_{AD} (i.e., $r = r_{\mathrm{A}}$). Additionally, to be able to represent by this interpolation function all polynomials up to a certain degree, we add a constant to our ansatz and we have

$$
\begin{aligned}
f_n(\varphi) = & \sum_{k=0,2,\ldots}^{N_1} r_{\mathrm{A}}^{\frac{k}{2}} \left[a_k \left(\kappa\, e^{i\varphi\frac{k}{2}} + e^{-i\varphi\frac{k}{2}} \right) + \frac{k}{2} \bar{a}_k \left(e^{-i\varphi\frac{k}{2}} - e^{-i\varphi(\frac{k}{2}-2)} \right) \right] \\
+ & \sum_{k=1,3,\ldots}^{N_2} r_{\mathrm{A}}^{\frac{k}{2}} \left[a_k \left(\kappa\, e^{i\varphi\frac{k}{2}} - e^{-i\varphi\frac{k}{2}} \right) + \frac{k}{2} \bar{a}_k \left(e^{-i\varphi\frac{k}{2}} - e^{-i\varphi(\frac{k}{2}-2)} \right) \right],
\end{aligned} \tag{3.1}
$$

where the number of basis functions is related to n as follows:

$$
N_1 = n - 2^m, \text{ with } \begin{cases} m = 1 & \text{for even } n, \\ m = 0 & \text{for odd } n, \end{cases}
$$

$$
N_2 = n - 2^m, \text{ with } \begin{cases} m = 0 & \text{for even } n, \\ m = 1 & \text{for odd } n. \end{cases}
$$

In [4] it is shown that for $n = 5$ the corresponding interpolation problem at the nodes on the circle can be solved for arbitrary data. Numerical experiments for the case of a hinge were presented in [6, 11] and showed a very good performance compared with commercial finite element software. For an exact reasoning and as a basis for the convergence of the coupled FE-method we need the solvability for arbitrary number and location of nodes. Main problems are the occurrence of the half-integer powers in the set of ansatz functions and the fact that the coefficients a_k and $\bar{a}_k$ are not independent.

Now we formulate the following theorem:

Theorem 3.1. *For n given arbitrary nodes $\varphi_0, \varphi_1, \ldots, \varphi_{n-1}$ basis functions of the form* (3.1) *exist, satisfying the canonical interpolation problem*

$$
f_n^{\langle i \rangle}(\varphi_k) = \delta_{(i-1)k}, \quad k = 0, \ldots, n-1, \tag{3.2}
$$

where $i = 1, \ldots, n$ is the number of a canonical problem.

Proof. Without loss of generality we will consider here the first canonical problem for $f_n^{\langle 1 \rangle}$. In all upcoming calculations we take $r_{\mathrm{A}} = 1$. We start our proof by introducing the new variable

$$
t = e^{i\frac{\varphi}{2}}, \qquad |t| = 1.
$$

The function (3.1) can then be rewritten as

$$
\begin{aligned}
f_n(t) = & \sum_{k=0,2,\ldots}^{N_1} \left[a_k \kappa\, t^k + a_k \kappa\, t^{-k} + \frac{k}{2} \bar{a}_k\, t^{-k} - \frac{k}{2} \bar{a}_k\, t^{-k+4} \right] \\
& + \sum_{k=1,3,\ldots}^{N_2} \left[a_k \kappa\, t^k - a_k \kappa\, t^{-k} + \frac{k}{2} \bar{a}_k\, t^{-k} - \frac{k}{2} \bar{a}_k\, t^{-k+4} \right].
\end{aligned} \tag{3.3}
$$

Depending on the number n of nodes on the interface Γ_{AD} we have a different number of functions from the even and odd parts of the basis. For the case of an even number of nodes we have $N_2 = N_1 + 1$ and in the case of an odd number of nodes we have $N_1 = N_2 + 1$. This fact must be taken into account during the proof.

Let us consider at first the case when the number of nodes n is **even**. In this case we can write the interpolation function (3.3) as one finite sum

$$\begin{aligned} f_n(t) = \sum_{k=0}^{\frac{1}{2}n-1} \Bigg[& \frac{2k+1}{2}\bar{a}_{2k+1}\, t^{-2k-1} + a_{2k}\kappa\, t^{-2k} + k\bar{a}_{2k}\, t^{-2k} \\ & - k\bar{a}_{2k}\, t^{-2k+4} + a_{2k+1}\kappa\, t^{2k+1} - a_{2k+1}\kappa\, t^{-2k-1} \\ & + a_{2k}\kappa\, t^{2k} - \frac{2k+1}{2}\bar{a}_{2k+1}\, t^{-2k+3} \Bigg]. \end{aligned} \tag{3.4}$$

The interpolation problem (3.2) reads as follows:

$$f_n(t_j) = \delta_{0j}, \qquad j = 0, \ldots, n-1.$$

Analysing equation (3.4) we observe that the lowest degree is $-n+1$. Therefore, to obtain a polynomial we multiply both sides of the (3.4) by t^{n-1} and we get at the nodes

$$\begin{aligned} \sum_{k=0}^{\frac{1}{2}n-1} \Bigg[& \frac{2k+1}{2}\bar{a}_{2k+1}\, t_j^{n-2k-2} + a_{2k}\kappa\, t_j^{n-2k-1} + k\bar{a}_{2k}\, t_j^{n-2k-1} \\ & - k\bar{a}_{2k}\, t_j^{n-2k+3} + a_{2k+1}\kappa\, t_j^{n+2k} - a_{2k+1}\kappa\, t_j^{n-2k-2} \\ & + a_{2k}\kappa\, t_j^{n+2k-1} - \frac{2k+1}{2}\bar{a}_{2k+1}\, t_j^{n-2k+2} \Bigg] = \delta_{0j}\, t_j^{n-1}. \end{aligned}$$

Now, we introduce a new right-hand side

$$w_j := \delta_{0j}\, t_j^{n-1}, \qquad j = 0, \ldots, n-1.$$

Collecting all summands with the same degree we can write the polynomial with new coefficients α_l

$$\begin{aligned} \sum_{l=0}^{2n-2} \alpha_l t^l := \sum_{k=0}^{\frac{1}{2}n-1} \Bigg[& \frac{2k+1}{2}\bar{a}_{2k+1}\, t^{n-2k-2} + a_{2k}\kappa\, t^{n-2k-1} \\ & + k\bar{a}_{2k}\, t^{n-2k-1} - k\bar{a}_{2k}\, t^{n-2k+3} \\ & + a_{2k+1}\kappa\, t^{n+2k} - a_{2k+1}\kappa\, t^{n-2k-2} \\ & + a_{2k}\kappa\, t^{n+2k-1} - \frac{2k+1}{2}\bar{a}_{2k+1}\, t^{n-2k+2} \Bigg]. \end{aligned} \tag{3.5}$$

Thus we get the following equivalent interpolation problem

$$\sum_{l=0}^{2n-2} \alpha_l t_j^l = w_j, \qquad j = 0, \ldots, n-1. \tag{3.6}$$

Due to the fact that the polynomial (3.5) contains also shifted powers t^{n-2k+3} and t^{n-2k+2}, the equations relating the new coefficients α_l with the original coefficients

a_k will change the form with increasing number of nodes. We will consider here only the case $n > 6$, because then these relating equations take their general form and the proof applies for all $n > 6$. The remaining three cases $n = 2$, $n = 4$, and $n = 6$ can be easily obtained directly from (3.6) and will not influence the generality of the proof.

For $n > 6$ we can separate the following four groups of equations between α_l and a_k

$$
\begin{aligned}
&\text{(I)} && \begin{cases} \alpha_{2j} = \frac{n-1-2j}{2}\bar{a}_{n-1-2j} - \kappa\, a_{n-1-2j}, \\ \alpha_{2j+1} = \kappa\, a_{n-2-2j} + \left(\frac{n}{2} - 1 - j\right)\bar{a}_{n-2-2j}, \end{cases} \qquad j = 0, 1 \\
&\text{(II)} && \begin{cases} \alpha_{2j} = \frac{n-1-2j}{2}\bar{a}_{n-1-2j} - \kappa\, a_{n-1-2j} - \frac{n+3-2j}{2}\bar{a}_{n+3-2j}, \\ \qquad j = 2, \ldots, \frac{n-2}{2} \\ \alpha_{2j+1} = \kappa\, a_{n-2-2j} + \left(\frac{n}{2} - 1 - j\right)\bar{a}_{n-2-2j} - \left(\frac{n}{2} + 1 - j\right) a_{n+2-2j}, \\ \qquad j = 2, \ldots, \frac{n-4}{2} \end{cases} \\
&\text{(III)} && \begin{cases} \alpha_{2j} = \kappa\, a_{2j-n+1} + \frac{2j-n-3}{2}\bar{a}_{n+3-2j}, \\ \alpha_{2j-1} = \left(\frac{n}{2} - j + 2\right)\kappa\, a_{2j-n} - \left(\frac{n}{2} - j + 2\right)\bar{a}_{2j-n}, \end{cases} \qquad j = \frac{n}{2}, \frac{n+2}{2} \\
&\text{(IV)} && \begin{cases} \alpha_{2j} = \kappa\, a_{2j+1-n}, \\ \alpha_{2j-1} = \kappa\, a_{2j-n} \end{cases} \qquad j = \frac{n+4}{2}, \ldots, n-1.
\end{aligned}
\tag{3.7}
$$

From the equations (3.7) we can calculate explicitly all of the original coefficients a_l. The group (IV) leads to the following equations:

$$
\begin{cases} a_{2j-n} = \frac{\alpha_{2j-1}}{\kappa}, \\ a_{2j+1-n} = \frac{\alpha_{2j}}{\kappa}, \end{cases} \qquad j = \frac{n}{2} + 2, \ldots, n-1. \tag{3.8}
$$

This group of equations includes all of the coefficients a_l for $l = 4, \ldots, n-1$. Therefore we need only to add equations for the four remaining coefficients. We can calculate the coefficients a_0, a_1, a_3 from group (III) of equations (3.7). The coefficient a_2 can be obtained from the sum of the third equation in group (III) and equation $n-6$ from group (II). These coefficients are given by

$$
\begin{cases} a_0 = \frac{\alpha_{n-1}}{2\kappa} + \frac{\bar{\alpha}_{n+3}}{\kappa^2}, \\ a_1 = \left[\frac{\alpha_n}{\kappa} + \frac{3\bar{\alpha}_{n+2}}{2\kappa^2}\right]\left(1 - \frac{3}{4\kappa^2}\right)^{-1}, \\ a_2 = \frac{\alpha_{n-3} + \alpha_{n+1}}{2\kappa} + \frac{3\bar{\alpha}_{n+5}}{2\kappa^2}, \\ a_3 = \frac{\alpha_{n+2}}{\kappa} + \frac{1}{2\kappa}\left[\frac{\bar{\alpha}_n}{\kappa} + \frac{3\alpha_{n+2}}{2\kappa^2}\right]\left(1 - \frac{3}{4\kappa^2}\right)^{-1}. \end{cases} \tag{3.9}
$$

The interpolation problem (3.6) contains $2n-1$ unknown coefficients α_l, but from the interpolation nodes we can get only n equations. Therefore, we formulate $n-1$ additional equations to determine all coefficients α_l. For that reason we extend (3.5) to the whole complex plane and add $n-1$ Hermite-type interpolation conditions

$$\sum_{l=0,1}^{2n-2} \alpha_l \left(\frac{\partial^p t^l}{\partial t^p} \right)_{t=t_*} = w_j^*, \qquad j = 0, \dots, 2n-2, \tag{3.10}$$

where for simplicity we take the additional node t_* at 0 and the values w_j^* are defined as follows:

$$w_j^* = \begin{cases} 0, & j = 0, \dots, n-5, \\ \beta_j, & j = n-4, \dots, n-2, \end{cases} \tag{3.11}$$

with arbitrary complex numbers β_j.

The obtained "extended" interpolation problem (3.2), (3.10) is always solvable (for more details, see for instance [5]). The solution of this interpolation problem will give us all coefficients α_l, which are needed to define the original coefficients a_k.

To insure that the coefficients a_k satisfy the original interpolation problem (3.2) we need to satisfy compatibility conditions for the coefficients α_k. By using formulae (3.8)–(3.9) we obtain these conditions from the groups (I) and (II) of equations (3.7). The first group gives the following four equations

$$\begin{aligned} \alpha_{2j} + \alpha_{2n-2-2j} - (n-2j-1)\frac{\bar{\alpha}_{2n-2-2j}}{2\kappa} &= 0, \\ \alpha_{2j+1} - \alpha_{2n-3-2j} - \left(\frac{n-2}{2} - j\right)\frac{\bar{\alpha}_{2n-3-2j}}{\kappa} &= 0, \end{aligned} \tag{3.12}$$

for $j = 0, 1$. The second group leads to the remaining $n-5$ equations

$$\begin{aligned} \alpha_{2j} + \alpha_{2n-2-2j} - (n-2j-1)\frac{\bar{\alpha}_{2n-2-2j}}{2\kappa} + (n-2j+3)\frac{\bar{\alpha}_{2n-2(j-1)}}{2\kappa} &= 0, \\ \alpha_{2j+1} - \alpha_{2n-3-2j} - \left(\frac{n-2}{2} - j\right)\frac{\bar{\alpha}_{2n-3-2j}}{\kappa} + \left(\frac{n-2}{2} + 2 - j\right)\frac{\bar{\alpha}_{2n-2j+1}}{\kappa} &= 0, \end{aligned} \tag{3.13}$$

for $j = 2, \dots, \frac{n-6}{2}$.

$$\begin{aligned} \alpha_{n+2} + \frac{1}{2}\left[\frac{\bar{\alpha}_n}{\kappa} + \frac{3\alpha_{n+2}}{2\kappa^2}\right]\left(1 - \frac{3}{4\kappa^2}\right)^{-1} + \alpha_{n-4} + \frac{7\bar{\alpha}_{n+6}}{2\kappa} \\ - \frac{3\bar{\alpha}_{n+2}}{2\kappa} - \frac{3}{4\kappa}\left[\frac{\alpha_n}{\kappa} + \frac{3\bar{\alpha}_{n+2}}{2\kappa^2}\right]\left(1 - \frac{3}{4\kappa^2}\right)^{-1} = 0, \end{aligned} \tag{3.14}$$

$$\alpha_{n-3} - \frac{\alpha_{n-3} + \alpha_{n+1}}{2} - \frac{3\bar{\alpha}_{n+5}}{2\kappa} - \frac{\bar{\alpha}_{n-3} + \bar{\alpha}_{n+1}}{2\kappa} - \frac{3\alpha_{n+5}}{2\kappa^2} + \frac{3\bar{\alpha}_{n+5}}{\kappa} = 0,$$

$$\begin{aligned} \kappa\left[\frac{\alpha_n}{\kappa} + \frac{3\bar{\alpha}_{n+2}}{2\kappa^2}\right]\left(1 - \frac{3}{4\kappa^2}\right)^{-1} + \alpha_{n-2} + \frac{5\bar{\alpha}_{n+4}}{2\kappa} \\ - \frac{1}{2}\left[\frac{\bar{\alpha}_n}{\kappa} + \frac{3\alpha_{n+2}}{2\kappa^2}\right]\left(1 - \frac{3}{4\kappa^2}\right)^{-1} = 0. \end{aligned} \tag{3.15}$$

Our goal is to show that there exists a set of complex numbers β_j such that the compatibility conditions (3.12)–(3.15) are satisfied. Considering the first $n-4$

equations of (3.10) with the values (3.11) we get immediately that

$$\alpha_l = 0, \qquad i = 0, \ldots, n-5. \tag{3.16}$$

Applying these results to first $n-4$ compatibility conditions (3.12)–(3.13) we obtain

$$\alpha_l = 0, \qquad i = n+3, \ldots, 2n-2. \tag{3.17}$$

Therefore first $n-4$ compatibility conditions are satisfied, and we need to check the remaining three equations (3.14)–(3.13). Taking into account the remaining three values of w_j^* we get the following equations

$$\alpha_{n+2} + \frac{1}{2}\left[\frac{\bar{\alpha}_n}{\kappa} + \frac{3\alpha_{n+2}}{2\kappa^2}\right]\left(1 - \frac{3}{4\kappa^2}\right)^{-1} + \beta_{n-4} - \frac{3\bar{\alpha}_{n+2}}{2\kappa}$$
$$- \frac{3}{4\kappa}\left[\frac{\alpha_n}{\kappa} + \frac{3\bar{\alpha}_{n+2}}{2\kappa^2}\right]\left(1 - \frac{3}{4\kappa^2}\right)^{-1} = 0,$$
$$\beta_{n-3} - \frac{\beta_{n-3} + \alpha_{n+1}}{2} - \frac{\bar{\beta}_{n-3} + \bar{\alpha}_{n+1}}{2\kappa} = 0,$$
$$\kappa\left[\frac{\alpha_n}{\kappa} + \frac{3\bar{\alpha}_{n+2}}{2\kappa^2}\right]\left(1 - \frac{3}{4\kappa^2}\right)^{-1} + \beta_{n-2} - \frac{1}{2}\left[\frac{\bar{\alpha}_n}{\kappa} + \frac{3\alpha_{n+2}}{2\kappa^2}\right]\left(1 - \frac{3}{4\kappa^2}\right)^{-1} = 0.$$

The second equation can be satisfied only if $\alpha_{n+1} = \beta_{n-3} = 0$. The solution of the two other equations is given by

$$\alpha_n = -\frac{4\kappa^2\,\Re\big[\hat{\beta}_{n-2}\big]}{(4\kappa^2-3)\left(1-\frac{1}{2\kappa}\right)} + \frac{6\kappa\,\Re\big[\hat{\beta}_{n-4}\big]}{(4\kappa^2-3)\left(1-\frac{3}{2\kappa}\right)} - \frac{4i\,\kappa^2\,\Im\big[\hat{\beta}_{n-2}\big]}{(4\kappa^2-3)\left(1+\frac{1}{2\kappa}\right)} - \frac{6i\,\kappa\,\Im\big[\hat{\beta}_{n-4}\big]}{(4\kappa^2-3)\left(1+\frac{3}{2\kappa}\right)},$$
$$\alpha_{n+2} = -\frac{4\kappa^2\,\Re\big[\hat{\beta}_{n-4}\big]}{(4\kappa^2-3)\left(1-\frac{3}{2\kappa}\right)} + \frac{2\kappa\,\Re\big[\hat{\beta}_{n-2}\big]}{(4\kappa^2-3)\left(1-\frac{1}{2\kappa}\right)} - \frac{4i\,\kappa^2\,\Im\big[\hat{\beta}_{n-4}\big]}{(4\kappa^2-3)\left(1+\frac{3}{2\kappa}\right)} - \frac{2i\,\kappa\,\Im\big[\hat{\beta}_{n-2}\big]}{(4\kappa^2-3)\left(1+\frac{1}{2\kappa}\right)},$$

where

$$\hat{\beta}_{n-2} := \beta_{n-2}\left(1 - \frac{3}{4\kappa^2}\right), \qquad \hat{\beta}_{n-4} := \beta_{n-4}\left(1 - \frac{3}{4\kappa^2}\right).$$

Finally, we obtained that all compatibility conditions are satisfied. Thus we have shown, that such set of complex numbers β_j exists, and the statement of the theorem is true for the case of even number of nodes.

Now we will complete the proof by considering the case when the number of nodes n is **odd**. We will omit some details which are similar to the even case. In the case of an odd number of nodes we need to keep the structure of the interpolation function (3.3) with two separate sums

$$f_n(t) = \sum_{k=0,1}^{\frac{n-1}{2}} \left[a_{2k}\kappa\, t^{2k} + a_{2k}\kappa\, t^{-2k} + k\bar{a}_{2k}\, t^{-2k} - k\bar{a}_{2k}\, t^{-2k+4}\right]$$
$$+ \sum_{k=1,2}^{\frac{n-1}{2}} \left[a_{2k-1}\kappa\, t^{2k-1} - a_k\kappa\, t^{-2k+1} + \tfrac{2k-1}{2}\bar{a}_{2k-1}\, t^{-2k+1} - \tfrac{2k-1}{2}\bar{a}_{2k-1}\, t^{-2k+5}\right].$$

To simplify the above function we extract the term for $k = 0$ from the first sum, collect common terms and get

$$f_n(t) = 2a_0\kappa + \sum_{k=1,2}^{\frac{n-1}{2}} \Big[t^{-2k}(a_{2k}\kappa + k\,\bar{a}_{2k}) + t^{-2k+1}\left(\frac{2k-1}{2}\bar{a}_{2k-1} - a_{2k-1}\kappa\right) - k\,\bar{a}_{2k}t^{-2k+4} - \frac{2k-1}{2}\bar{a}_{2k-1}t^{-2k+5} + a_{2k-1}\kappa\, t^{2k-1} + a_{2k}\kappa\, t^{2k} \Big]. \tag{3.18}$$

The lowest degree is $-n+1$. Therefore, as in the previous case we multiply both sides of the interpolation problem by t^{n-1}, and we get the following equivalent interpolation problem

$$\sum_{l=0}^{2n-2} \alpha_l t_j^l = w_j,$$

for $j = 0, \dots, n-1$.

Since the polynomial basis contains also shifted powers t^{n-2k+3} and t^{n-2k+4} the equations relating the new coefficients α_l with the original coefficients a_k will change with increasing n and take their general form for $n > 7$. We will consider only this case. The remaining three cases $n = 3$, $n = 5$, and $n = 7$ can be easily obtained directly from the interpolation problem and will not influence the generality of the proof.

Similar to the even case, for $n > 7$ we can get four groups of equations between α_k and a_k

$$\text{(I)} \quad \begin{cases} \alpha_{2j} = \kappa\, a_{n-1-2j} + \left(\frac{n-1}{2} - j\right)\bar{a}_{n-1-2j}, \\ \alpha_{2j+1} = \frac{n-2-2j}{2}\bar{a}_{n-2-2j} - \kappa\, a_{n-2-2j}, \end{cases} \quad j = 0, 1$$

$$\text{(II)} \quad \begin{cases} \alpha_{2j} = \kappa\, a_{n-1-2j} + \left(\frac{n-1}{2} - j\right)\bar{a}_{n-1-2j} - \left(\frac{n+3}{2} - j\right) a_{n+3-2j}, \\ \alpha_{2j+1} = \frac{n-2-2j}{2}\bar{a}_{n-2-2j} - \kappa\, a_{n-2-2j} - \frac{n+2-2j}{2}\bar{a}_{n+2-2j}, \\ j = 2, \dots, \frac{n-3}{2} \end{cases}$$

$$\text{(III)} \quad \begin{cases} \alpha_{2j} = \left(\frac{n-1}{2} - j + 2\right)\kappa\, a_{2j-n+1} - \frac{n+3-2j}{2}\bar{a}_{n+3-2j}, \\ \alpha_{2j-1} = \kappa\, a_{2j-n+2} + \frac{2j-n-2}{2}\bar{a}_{n+2-2j}, \end{cases} \quad j = \frac{n-1}{2}, \frac{n+1}{2}$$

$$\text{(IV)} \quad \begin{cases} \alpha_{2j} = \kappa\, a_{2j-n+1}, \\ \alpha_{2j-1} = \kappa\, a_{2j-n+2}, \end{cases} \quad j = \frac{n+3}{2}, \dots, n-1. \tag{3.19}$$

Analogously to the even case, from equations (3.19) we get the explicit formulae for the coefficients a_k. From group (IV) we get the following equations for a_k for $k = 4, \dots, n-1$

$$\begin{cases} a_{2j-n+1} = \frac{\alpha_{2j}}{\kappa}, \\ a_{2j-n+2} = \frac{\alpha_{2j-1}}{\kappa}, \end{cases} \quad j = \frac{n+3}{2}, \dots, n-1. \tag{3.20}$$

Formulae for the remaining coefficients a_0, a_1, a_2, a_3 are completely the same as in the even case, and they are given by (3.9).

Applying the same ideas as in the case of even number of nodes we introduce some additional Hermite-type conditions (3.10). The remaining task is to prove the compatibility conditions for the case of an odd number of nodes. From the first group we get the following equations

$$\begin{cases} \alpha_{2j} - \alpha_{2n-2-2j} - \left(\frac{n-1}{2} - j\right) \frac{\bar{\alpha}_{2n-2-2j}}{\kappa} = 0, \\ \alpha_{2j+1} + \alpha_{2n-3-2j} - (n-2j-2)\frac{\bar{\alpha}_{2n-3-2j}}{2\kappa} = 0, \end{cases} \quad j = 0, 1. \tag{3.21}$$

The second group leads to the following $n-5$ equations:

$$\begin{cases} \alpha_{2j} - \alpha_{2n-2j-2} - \left(\frac{n-1}{2} - j\right) \frac{\bar{\alpha}_{2n-2-2j}}{\kappa} \\ \quad + \left(\frac{n-1}{2} - j + 2\right) \frac{\bar{\alpha}_{2n-2(j-1)}}{\kappa} = 0, \quad j = 2, \ldots, \frac{n-5}{2} \\ \alpha_{2j+1} - \alpha_{2n-2j-3} - (n-2j-2)\frac{\bar{\alpha}_{2n-3-2j}}{2\kappa} \\ \quad + (n-2j+2)\frac{\bar{\alpha}_{2n-2j+1}}{2\kappa} = 0, \quad j = 2, \ldots, \frac{n-7}{2} \\ \alpha_{n+2} + \frac{1}{2}\left[\frac{\bar{\alpha}_n}{\kappa} + \frac{3\alpha_{n+2}}{2\kappa^2}\right]\left(1 - \frac{3}{4\kappa^2}\right)^{-1} + \alpha_{n-4} + \frac{7\bar{\alpha}_{n+6}}{2\kappa} \\ \quad - \frac{3\bar{\alpha}_{n+2}}{2\kappa} - \frac{3}{4\kappa}\left[\frac{\alpha_n}{\kappa} + \frac{3\bar{\alpha}_{n+2}}{2\kappa^2}\right]\left(1 - \frac{3}{4\kappa^2}\right)^{-1} = 0, \\ \alpha_{n-3} - \frac{\alpha_{n-3}+\alpha_{n+1}}{2} - \frac{3\bar{\alpha}_{n+5}}{2\kappa} - \frac{\bar{\alpha}_{n-3}+\bar{\alpha}_{n+1}}{2\kappa} - \frac{3\alpha_{n+5}}{2\kappa^2} + \frac{3\bar{\alpha}_{n+5}}{\kappa} = 0, \\ \kappa\left[\frac{\alpha_n}{\kappa} + \frac{3\bar{\alpha}_{n+2}}{2\kappa^2}\right]\left(1 - \frac{3}{4\kappa^2}\right)^{-1} + \alpha_{n-2} + \frac{5\bar{\alpha}_{n+4}}{2\kappa} \\ \quad - \frac{1}{2}\left[\frac{\bar{\alpha}_n}{\kappa} + \frac{3\alpha_{n+2}}{2\kappa^2}\right]\left(1 - \frac{3}{4\kappa^2}\right)^{-1} = 0. \end{cases} \tag{3.22}$$

In a similar way as for the case of an even number of nodes it can be shown, that there exists a set of complex numbers β_j such that the compatibility conditions are satisfied and the statement of the theorem is true for the case of an odd number of nodes. □

4. Conclusions and outlook

It has been shown that the canonical interpolation problems (3.2) can be solved for an arbitrary number of nodes. Due to complexity of the basis function it was necessary to consider separately the cases for even and odd number of interpolation nodes. The obtained result is very important for the convergence analysis of the proposed method. Particularly, the number on nodes on the interface and the number of the coupling elements can have a significant influence on the error estimate of the proposed scheme.

Acknowledgement

The research of the second author is supported by the German Research Foundation (DFG) via Research Training Group "Evaluation of Coupled Numerical

Partial Models in Structural Engineering (GRK 1462)", which is gratefully acknowledged.

References

[1] Philippe G. Ciarlet, *The finite element method for elliptic problems*, North-Holland Publishing Company, 1978.

[2] S. Bock, K. Gürlebeck, D. Legatiuk, *Convergence of the finite element method with holomorphic functions.* AIP Conference proceedings 1558, 513 (2013).

[3] S. Bock, K. Gürlebeck, D. Legatiuk, *On a special finite element based on holomorphic functions*, AIP Conference proceedings 1479, 308 (2012).

[4] S. Bock, K. Gürlebeck, D. Legatiuk, *On the continuous coupling between analytical and finite element solutions*, Le Hung Son & Wolfgang Tutschke, eds. *Interactions between real and complex analysis*, pp. 3–19. Science and Technics Publishing House, Hanoi, 2012.

[5] Philip J. Davis *Interpolation and Approximation*, Dover Publications, Inc., 1975.

[6] D. Legatiuk, K. Gürlebeck, G. Morgenthal, *Modelling of concrete hinges through coupling of analytical and finite element solutions.* Bautechnik Sonderdruck, ISSN 0932-8351, A 1556, April 2013.

[7] H. Liebowitz, *Fracture, an advanced treatise. Volume II: Mathematical fundamentals*, Academic Press, 1968.

[8] N.I. Mußchelischwili, *Einige Grundaufgaben der mathematischen Elastizitätstheorie*, VEB Fachbuchverlag Leipzig, 1971.

[9] R. Piltner, *Some remarks on finite elements with an elliptic hole*, Finite elements in analysis and design, Volume 44, Issues 12-13, 2008.

[10] R. Piltner, *Special finite elements with holes and internal cracks*, International journal for numerical methods in engineering, Volume 21, 1985.

[11] A. Schumann, *Untersuchung und Beurteilung des Rissverhaltens eines Betongelenkes anhand unterschiedlicher Methoden*, Bachelorarbeit Nr. BB/2013/8. Supervised by K. Gürlebeck and D. Legatiuk.

Klaus Gürlebeck
Chair of Applied Mathematics
Bauhaus-University Weimar
D-99423 Weimar, Germany
e-mail: `klaus.guerlebeck@uni-weimar.de`

Dmitrii Legatiuk
Research Training Group 1462
Bauhaus-University Weimar
D-99425 Weimar, Germany
e-mail: `dmitrii.legatiuk@uni-weimar.de`

Hypercomplex Analysis: New Perspectives and Applications
Trends in Mathematics, 181–189

On ψ-hyperholomorphic Functions and a Decomposition of Harmonics

Klaus Gürlebeck and Hung Manh Nguyen

Abstract. Additive decompositions of harmonic functions play an important role in function theory and for the solution of partial differential equations. One of the best known results is the decomposition of harmonic functions as a sum of a holomorphic and an anti-holomorphic function. This decomposition can be generalized also to the analysis of quaternion-valued harmonic function, where the summands are then monogenic or anti-monogenic, respectively. For paravector-valued functions, sometimes called $\mathcal{A}$-valued functions, this decomposition is not possible. The main purpose of the paper is to show that one can find three different ψ-Cauchy–Riemann operators such that the null spaces of these operators define an additive decomposition for harmonic functions, mapping $\mathbb{R}^3$ to $\mathcal{A}\hat{=}\mathbb{R}^3$.

Mathematics Subject Classification (2010). 30G35, 42C05, 33E10.

Keywords. Quaternion analysis, ψ-hyperholomorphic functions.

1. Introduction

The theory of monogenic functions as theory of the null solutions of a Dirac operator or a generalized Cauchy–Riemann operator can be seen as a refinement of harmonic analysis or as a generalization of complex analysis. One of the most important properties of monogenic (or holomorphic) functions is that they are harmonic functions in all components of the vector functions. Already in 1989 in the thesis by Stern [17] (see also [18]) the question was asked which properties of a first-order partial differential operator ensure that all null solutions of this operator are harmonic in all components. It was shown that the coefficients (matrices in this work) must satisfy the multiplication rules of a Clifford algebra.

Independent on this research Shapiro and Vasilevski introduced in the late 1980's the theory of so-called ψ-hyperholomorphic quaternion-valued functions

The second author acknowledges the financial support of MOET-Vietnam & DAAD.

(see [14] and later [15]). In this theory the standard basis vectors from quaternionic analysis are replaced by a more general structural set. Seen as vectors from $\mathbb{R}^4$ the elements of the structural set must be an orthonormal set with respect to the standard inner product in $\mathbb{R}^4$. The authors used this approach to study some singular integral operators in spaces of quaternion-valued functions. In particular a generalized Π-operator was studied and relations of this Π-operator to the Bergmann projection could be proved. The work on the Π-operator continued some earlier work by Shevchenko [16] who studied special Π operators based on modified generalized Cauchy–Riemann operators which are covered by the theory of ψ-hyperholomorphic functions.

In 1998 it was shown in [5] that the class of ψ-hyperholomorphic functions is more than what we get by rotations from the class of monogenic functions. In this line are also the results in [7] where it could be shown that a special Π operator is invertible in L_2 and how the mapping properties of the operator change with the structural set. Recently this topic was studied again in [1] for Π operators defined on domains with fractal boundaries.

A second line of the research on ψ-hyperholomorphic functions is related to their geometric mapping properties. In [8] and [9] it was observed that ψ-hyperholomorphic functions can be connected with certain conformal mappings. Later on Malonek introduced in [11] the concept of M-conformal mappings that is also related to ψ-hyperholomorphic mappings

The third line of research is concerned with the refinement of harmonic analysis. This refinement is based on the factorization of the Laplacian by Dirac operators or by generalized Cauchy–Riemann operators. One of the basics of the theory of ψ-hyperholomorphic functions is that the structural sets must be chosen in a way that this factorization also holds for ψ-Cauchy–Riemann operators. A second question is to find additive decompositions of harmonic functions. It is well known in complex analysis that a harmonic function can be decomposed as a sum of a holomorphic and an anti-holomorphic function. An analogous result holds for $\mathbb{H}$-valued harmonic functions which can be represented as a sum of a monogenic and an anti-monogenic $\mathbb{H}$-valued function.

Recently the theory of $\mathcal{A}$-valued monogenic and harmonic functions found some interest. Motivated by applications in $\mathbb{R}^3$ and the observation that $\mathcal{A}$-valued functions share more properties with holomorphic functions [12, 13] than general $\mathbb{H}$-valued monogenic functions the question of additive decompositions was studied again for harmonic functions in $\mathbb{R}^3$. Alvarez and Porter [2] made the surprising observation that $\mathcal{A}$-valued functions cannot be written as a sum of a monogenic and an anti-monogenic $\mathcal{A}$-valued function.

They found that in the $6n+3$-dimensional subspace of homogeneous harmonic polynomials of degree n there is a $2n-1$-dimensional subspace orthogonal to the sum of monogenic and anti-monogenic polynomials of the same degree, called contragenic functions.

It will be shown in the paper that contragenic functions cannot be solutions of a first-order system of partial differential equations. So, the main question is if there are other first-order systems such that we can decompose harmonic functions as a sum of three subspaces of null solutions of first-order systems of partial differential equations with the property that all solutions of those systems are harmonic in all coordinates.

To answer the fundamental question of the existence of such additive decompositions we will study ψ-hyperholomorphic functions, ψ-anti-holomorphic functions and construct another structural set θ such that the corresponding null spaces give us the desired decomposition. We will not consider the problem to find all possible decomposition here.

2. Preliminaries

Let $\mathbb{H}$ be the algebra of real quaternions generated by the basis $\{\mathbf{1}, \mathbf{e_1}, \mathbf{e_2}, \mathbf{e_3}\}$ subject to the multiplication rules

$$\begin{aligned} \mathbf{e_i e_j} + \mathbf{e_j e_i} &= -2\delta_{ij}, \quad i,j = 1,2,3 \\ \mathbf{e_1 e_2} &= \mathbf{e_3}. \end{aligned}$$

Each quaternion can be represented in the form $q = q_0 + q_1\mathbf{e_1} + q_2\mathbf{e_2} + q_3\mathbf{e_3}$ where q_j $(j = 0, \dots, 3)$ are real numbers. The real and vector parts of q are denoted by $\mathrm{Sc}\,(q) := q_0$ and $\mathrm{Vec}\,(q) := q_1\mathbf{e_1} + q_2\mathbf{e_2} + q_3\mathbf{e_3}$. The real vector space $\mathbb{R}^3$ will be embedded in $\mathbb{H}$ by identifying the element $x = (x_0, x_1, x_2) \in \mathbb{R}^3$ with the reduced quaternion $x = x_0 + x_1\mathbf{e_1} + x_2\mathbf{e_2}$. The set of all reduced quaternions is denoted by $\mathcal{A}$ which is a $\mathbb{R}$-linear subspace of $\mathbb{H}$. Let $\mathbb{B}$ be the unit ball in $\mathbb{R}^3$. $L_2(\mathbb{B}, \mathcal{A})$ is called the right $\mathbb{R}$-linear Hilbert space of all square integrable $\mathcal{A}$-valued functions in $\mathbb{B}$, endowed with the real-valued inner product

$$\langle f, g \rangle = \int_{\mathbb{B}} \mathrm{Sc}\,(\overline{f} g)\, d\omega \tag{2.1}$$

where $d\omega$ is the Lebesgue measure in $\mathbb{R}^3$. The generalized Cauchy–Riemann operator and its adjoint operator are given by

$$\begin{aligned} \overline{\partial} &:= \frac{\partial}{\partial x_0} + \mathbf{e_1}\frac{\partial}{\partial x_1} + \mathbf{e_2}\frac{\partial}{\partial x_2}, \\ \partial &:= \frac{\partial}{\partial x_0} - \mathbf{e_1}\frac{\partial}{\partial x_1} - \mathbf{e_2}\frac{\partial}{\partial x_2}. \end{aligned}$$

Definition 2.1. A function $f \in C^1(\mathbb{B}, \mathcal{A})$ is called *(left-) monogenic* in $\mathbb{B} \subset \mathbb{R}^3$ if $\overline{\partial} f = 0$ in $\mathbb{B}$.

Definition 2.2. Let $f \in C^1(\mathbb{B}; \mathcal{A})$ be a continuous, real differentiable function and monogenic in $\mathbb{B}$. The expression $\left(\frac{1}{2}\partial f\right)$ is called *hypercomplex derivative* of f in $\mathbb{B}$.

To be short, we simply present the hypercomplex derivative as a definition. In the paper [6], it is proved that monogenicity and hypercomplex derivability are equivalent in all dimensions, and that the hypercomplex linearization of the monogenic function f is exactly given by $\frac{1}{2}\partial f$. A complete survey on this topic can be found also in [10].

3. Quaternionic ψ-hyperholomorphic functions

The definition of quaternion-valued ψ-hyperholomorphic functions was studied by M.V. Shapiro and N.L. Vasilevski in [15] as a generalization of monogenic functions with respect to the basis $\{\mathbf{1}, \mathbf{e_1}, \mathbf{e_2}, \mathbf{e_3}\}$. One can also find researches on ψ-hyperholomorphic functions by R. Delanghe, R.S. Kraußhar and H.R. Malonek [9]; R.S. Kraußhar and H.R. Malonek [8]. Following the same ideas, we consider the case in $\mathbb{R}^3$.

Let $\psi := \{\psi^0, \psi^1, \psi^2\} \subset \mathcal{A}$ and $\overline{\psi} := \{\overline{\psi^0}, \overline{\psi^1}, \overline{\psi^2}\}$. The generalized Cauchy–Riemann operator ${}^{\psi}D$ is defined by

$$ {}^{\psi}D[f] := \psi^0 \partial_0 f \,+\, \psi^1 \partial_1 f \,+\, \psi^2 \partial_2 f. $$

To fulfil the Laplacian factorization $\Delta_{\mathbb{R}^3} = {}^{\psi}D^{\overline{\psi}}D = {}^{\overline{\psi}}D^{\psi}D$, the following condition holds

$$ \psi^j \overline{\psi^k} + \psi^k \overline{\psi^j} = 2\delta_{jk} \tag{3.1} $$

for $j, k = 0, \ldots, 2$. A set ψ satisfying the relation (3.1) will be called a *structural set.* This can be interpreted geometrically. Suppose that we have a structural set $\psi = \{\psi^0, \psi^1, \psi^2\} \subset \mathcal{A}$ which is identified with a vector set $\overrightarrow{\psi}$ in $\mathbb{R}^3$. As a consequence, the vector set $\overrightarrow{\psi}$ forms an orthonormal basis in $\mathbb{R}^3$. Particularly, if we write

$$ \begin{aligned} \psi^0 &= \psi^0_0 + \psi^0_1 \mathbf{e_1} + \psi^0_2 \mathbf{e_2} \\ \psi^1 &= \psi^1_0 + \psi^1_1 \mathbf{e_1} + \psi^1_2 \mathbf{e_2} \\ \psi^2 &= \psi^2_0 + \psi^2_1 \mathbf{e_1} + \psi^2_2 \mathbf{e_2} \end{aligned} $$

then one gets a formal matrix representation

$$ \begin{pmatrix} \psi^0 & \psi^1 & \psi^2 \end{pmatrix} = \begin{pmatrix} \mathbf{1} & \mathbf{e_1} & \mathbf{e_2} \end{pmatrix} \underbrace{\begin{pmatrix} \psi^0_0 & \psi^1_0 & \psi^2_0 \\ \psi^0_1 & \psi^1_1 & \psi^2_1 \\ \psi^0_2 & \psi^1_2 & \psi^2_2 \end{pmatrix}}_{\Psi}. $$

By virtue of (3.1), the matrix Ψ is an orthogonal matrix, i.e., $\Psi\Psi' = \Psi'\Psi = I$ (where I is the 3×3- unit matrix). Correspondingly, $\{\overrightarrow{\psi}\}$ is an orthonormal basis of $\mathbb{R}^3$.

A $C^1(\mathbb{B}, \mathcal{A})$ function f is called an ($\mathcal{A}$-valued) ψ-hyperholomorphic function in $\mathbb{B}$ if it satisfies ${}^{\psi}Df(x) = 0$ for all $x \in \mathbb{B}$. The monogenic case corresponds with the standard structural set $\{\mathbf{1}, \mathbf{e_1}, \mathbf{e_2}\}$. We refer readers to [15] for a survey on ψ-hyperholomorphic functions, the ψ-hypercomplex derivative and Cauchy integrals. To this end, ${}^{\psi}\mathcal{M}(\mathbb{B}, \mathcal{A})$ stands for the L_2-space of ψ-hyperholomorphic functions in $\mathbb{B}$ and ${}^{\psi}\mathcal{M}(\mathbb{B}, \mathcal{A}, n)$ is its subspace of homogeneous ψ-hyperholomorphic polynomials of degree n. For the standard structural set, we use the notation $\mathcal{M}$ only. The notation $\overline{\mathcal{M}}$ means the space of conjugations of functions in $\mathcal{M}$. In fact, that is the space of anti-monogenic functions shown in [3].

4. Contragenic functions revisited

It is well known that a $Cl_{n,0}$-valued harmonic function can be decomposed into the sum of a ($Cl_{n,0}$-valued) monogenic and an anti-monogenic function, see [4]. However this is not the case for $\mathcal{A}$-valued harmonic functions. Particularly, due to [2], $\mathcal{H}(\mathbb{B},\mathcal{A},n)$ stands for the space of $\mathcal{A}$-valued homogeneous harmonic polynomials of degree n in $\mathbb{B}$, then with $n>0$

$$\dim \mathcal{H}(\mathbb{B},\mathcal{A},n) = 6n+3$$

while

$$\dim \left(\mathcal{M}(\mathbb{B},\mathcal{A},n) + \overline{\mathcal{M}}(\mathbb{B},\mathcal{A},n)\right) = 4n+4.$$

It shows that there are harmonic functions which can not be the sum of a monogenic and an anti-monogenic function. The orthogonal complement in the space $\mathcal{H}(\mathbb{B},\mathcal{A},n)$, denoted by

$$\mathcal{N}(\mathbb{B},\mathcal{A},n) := \left(\mathcal{M}(\mathbb{B},\mathcal{A},n) + \overline{\mathcal{M}}(\mathbb{B},\mathcal{A},n)\right)^{\perp}$$

is called the space of homogeneous *contragenic polynomials* of degree n. It yields

$$\dim \mathcal{N}(\mathbb{B},\mathcal{A},n) = 2n-1.$$

Having in mind that the following $2n+1$ spherical harmonic functions form a well-known orthogonal basis of the space of real-valued homogeneous harmonic polynomials of degree $n \geq 0$

$$\widehat{U}_0^n, \widehat{U}_1^n, \ldots, \widehat{U}_n^n, \widehat{V}_1^n, \ldots, \widehat{V}_n^n$$

where $\widehat{U}_m^n, \widehat{V}_l^n$ are defined in spherical coordinates $x_0 = r\cos\theta$, $x_1 = r\sin\theta\cos\varphi$, $x_2 = r\sin\theta\sin\varphi$, via

$$\begin{aligned} \widehat{U}_m^n &= r^n P_n^m(\cos\theta)\cos(m\varphi), \qquad & m &= 0,\ldots,n \\ \widehat{V}_l^n &= r^n P_n^l(\cos\theta)\sin(l\varphi), \qquad & l &= 1,\ldots,n. \end{aligned}$$

P_n^m is the associated Legendre function

$$P_n^m(x) = (-1)^m(1-x^2)^{m/2}(d^m/dx^m)P_n(x)$$

with P_n is the Legendre polynomial of degree n corresponding to $m=0$. According to [2, 12], we have explicit orthogonal bases of the spaces $\mathcal{M}(\mathbb{B},\mathcal{A},n)$ and $\mathcal{N}(\mathbb{B},\mathcal{A},n)$.

Proposition 4.1. *Denote by $c_m^n = (n+m)(n+m+1)/4$. For each $n \geq 1$, the following functions form an orthogonal basis of $\mathcal{M}(\mathbb{B},\mathcal{A},n)$*

$$\begin{aligned} X_0^n &= \frac{n+1}{2}\widehat{U}_0^n + \frac{1}{2}\widehat{U}_1^n\mathbf{e_1} + \frac{1}{2}\widehat{V}_1^n\mathbf{e_2} \\ X_m^n &= \frac{n+m+1}{2}\widehat{U}_m^n + \left(\frac{1}{4}\widehat{U}_{m+1}^n - c_m^n\widehat{U}_{m-1}^n\right)\mathbf{e_1} + \left(\frac{1}{4}\widehat{V}_{m+1}^n + c_m^n\widehat{V}_{m-1}^n\right)\mathbf{e_2} \\ Y_m^n &= \frac{n+m+1}{2}\widehat{V}_m^n + \left(\frac{1}{4}\widehat{V}_{m+1}^n - c_m^n\widehat{V}_{m-1}^n\right)\mathbf{e_1} - \left(\frac{1}{4}\widehat{U}_{m+1}^n + c_m^n\widehat{U}_{m-1}^n\right)\mathbf{e_2}, \end{aligned}$$

where $1 \leq m \leq n+1$.

Proposition 4.2. *Let $n \geq 1$ and $d_m^n = (n-m)(n-m+1)$. The following $2n-1$ functions*

$$Z_0^n = \widehat{V}_1^n \mathbf{e_1} - \widehat{U}_1^n \mathbf{e_2}$$
$$Z_{m,+}^n = \left(d_m^n \widehat{V}_{m-1}^n + \widehat{V}_{m+1}^n\right) \mathbf{e_1} + \left(d_m^n \widehat{U}_{m-1}^n - \widehat{U}_{m+1}^n\right) \mathbf{e_2}$$
$$Z_{m,-}^n = \left(d_m^n \widehat{U}_{m-1}^n + \widehat{U}_{m+1}^n\right) \mathbf{e_1} + \left(-d_m^n \widehat{V}_{m-1}^n + \widehat{V}_{m+1}^n\right) \mathbf{e_2}$$

for $1 \leq m \leq n-1$, form an orthogonal basis of $\mathcal{N}(\mathbb{B}, \mathcal{A}, n)$.

Let the *ambigenic function* space be the sum of monogenic and anti-monogenic spaces. Finally, an $\mathcal{A}$-valued harmonic function can be orthogonally decomposed into an ambigenic and a contragenic function. Details can be found in [2].

The contragenic function space is defined formally as the orthogonal complement of the ambigenic function space. Of course, contragenic functions are also harmonic functions. It means that they satisfy the Laplace equation as well as monogenic functions. We know that monogenic functions are null solutions of the (generalized) Cauchy–Riemann operator. The question is which first-order linear partial differential operator characterizes contragenic functions. Particularly, it would be useful if we can find a structural set ψ such that contragenic functions are solutions of the operator ${}^\psi D$. The answer will lead to another decomposition for harmonic functions.

5. An additive decomposition of $\mathcal{A}$-valued harmonic functions

In this section, we will prove that an $\mathcal{A}$-valued harmonic function can be decomposed into the sum of three different ψ-hyperholomorphic functions. Firstly, one can see that there does not exist any structural set ψ such that contragenic functions are null-solutions of the corresponding Cauchy–Riemann operator. Indeed, consider a Cauchy–Riemann operator

$${}^\psi D = \psi^0 \frac{\partial}{\partial x_0} + \psi^1 \frac{\partial}{\partial x_1} + \psi^2 \frac{\partial}{\partial x_2}.$$

Apply ${}^\psi D$ to two first contragenic basis functions in [2]

$$Z_0^1 = -x_2 \mathbf{e_1} + x_1 \mathbf{e_2}$$
$$Z_0^2 = 3x_0(-x_2 \mathbf{e_1} + x_1 \mathbf{e_2})$$

and let the results be zero, one gets $\psi^0 = 0$. This contradicts the definition of ψ. Moreover, one can prove that contragenic functions are not solutions of any first-order linear partial differential operator.

Now, we are looking for a structural set ψ such that the sum of the spaces of (corresponding) ψ-hyperholomorphic, monogenic and anti-monogenic functions is the ($\mathcal{A}$-valued) harmonic function space. It can be proved that if $f := f_0 +$

$f_1\mathbf{e_1} + f_2\mathbf{e_2}$ is a monogenic function then ${}^{\psi}f := \overline{\psi^0} f_0 - \overline{\psi^1} f_1 - \overline{\psi^2} f_2$ is a ψ-hyperholomorphic function, i.e., ${}^{\psi}D^{\psi} f = 0$. Therefore, we have the following lemma.

Lemma 5.1. *Let $\psi = \{\mathbf{1},\ \mathbf{e_2},\ -\mathbf{e_1}\}$. The following functions form an orthogonal basis for ${}^{\psi}\mathcal{M}(\mathbb{B}, \mathcal{A}, n)$*

$$
\begin{aligned}
{}^{\psi}X_0^n &= \frac{n+1}{2}\widehat{U}_0^n + \frac{1}{2}\widehat{U}_1^n\mathbf{e_2} - \frac{1}{2}\widehat{V}_1^n\mathbf{e_1} \\
{}^{\psi}X_m^n &= \frac{n+m+1}{2}\widehat{U}_m^n + \left(\frac{1}{4}\widehat{U}_{m+1}^n - c_m^n\widehat{U}_{m-1}^n\right)\mathbf{e_2} - \left(\frac{1}{4}\widehat{V}_{m+1}^n + c_m^n\widehat{V}_{m-1}^n\right)\mathbf{e_1} \\
{}^{\psi}Y_m^n &= \frac{n+m+1}{2}\widehat{V}_m^n + \left(\frac{1}{4}\widehat{V}_{m+1}^n - c_m^n\widehat{V}_{m-1}^n\right)\mathbf{e_2} + \left(\frac{1}{4}\widehat{U}_{m+1}^n + c_m^n\widehat{U}_{m-1}^n\right)\mathbf{e_1}
\end{aligned}
$$

with $1 \le m \le n+1$.

Remark that with $\psi = \{\mathbf{1},\ \mathbf{e_2},\ -\mathbf{e_1}\}$, we have

$$
\begin{aligned}
&{}^{\psi}X_{n+1}^n = Y_{n+1}^n & &{}^{\psi}Y_{n+1}^n = -X_{n+1}^n \\
&{}^{\psi}X_n^n = \frac{1}{2}\left(X_n^n + \overline{X_n^n}\right) + \frac{1}{2}\left(Y_n^n - \overline{Y_n^n}\right) & &{}^{\psi}Y_n^n = \frac{1}{2}\left(-X_n^n + \overline{X_n^n}\right) + \frac{1}{2}\left(Y_n^n + \overline{Y_n^n}\right).
\end{aligned}
$$

Next, we prove an additive decomposition of $\mathcal{A}$-valued harmonic functions.

Theorem 5.2. *Let $\psi = \{\mathbf{1},\ \mathbf{e_2},\ -\mathbf{e_1}\}$. Every $\mathcal{A}$-valued harmonic function $\mathbf{u}$ can be decomposed into the form*

$$\mathbf{u} = \mathbf{f} + \mathbf{f_1} + \mathbf{f_2}$$

where $\mathbf{f},\ \mathbf{f_1},\ \mathbf{f_2}$ are $\mathcal{A}$-valued monogenic, anti-monogenic and ψ-hyperholomorphic functions, respectively.

Proof. Remind that an $\mathcal{A}$-valued harmonic function is the sum of monogenic, anti-monogenic and contragenic functions (see [2]). Therefore, in order to prove this theorem, we will show that each contragenic function can be linearly represented by monogenic, anti-monogenic and ψ-hyperholomorphic functions. This in turn is restricted in the case of polynomials of degree $n \ge 1$. Indeed, for $m = 0$ we have

$$Z_0^n = -2\ {}^{\psi}X_0^n + X_0^n + \overline{X_0^n}.$$

The contragenic polynomial $Z_{m,+}^n$, $1 \le m \le n-1$, can be rewritten as follows:

$$Z_{m,+}^n = \widehat{V}_{m+1}^n\mathbf{e_1} - \widehat{U}_{m+1}^n\mathbf{e_2} + d_m^n\left(\widehat{V}_{m-1}^n\mathbf{e_1} + \widehat{U}_{m-1}^n\mathbf{e_2}\right).$$

This suggests that $Z_{m,+}^n$ can be described in the form

$$Z_{m,+}^n = \alpha_{m,+}^n\left\{{}^{\psi}X_m^n - \frac{1}{2}\left(X_m^n + \overline{X_m^n}\right) - \beta_{m,+}^n\left(Y_m^n - \overline{Y_m^n}\right)\right\}.$$

Straightforward calculations lead to the system

$$
\begin{cases}
\alpha_{m,+}^n\left(-\frac{1}{4} - \frac{1}{2}\beta_{m,+}^n\right) = 1 \\
\alpha_{m,+}^n\left(-c_m^n + 2c_m^n\beta_{m,+}^n\right) = d_m^n.
\end{cases}
$$

This system has the solution

$$\begin{cases} \alpha^n_{m,+} &= -\frac{4c^n_m+d^n_m}{2c^n_m} &= -\frac{4(n^2+m^2+n)}{(n+m)(n+m+1)} \\ \beta^n_{m,+} &= \frac{4c^n_m-d^n_m}{2(4c^n_m+d^n_m)} &= \frac{m(2n+1)}{2(n^2+m^2+n)}. \end{cases}$$

Analogously, we can also find the representation of $Z^n_{m,-}$ in terms of X^n_m, $\overline{X^n_m}$, Y^n_m, $\overline{Y^n_m}$ and ${}^{\psi}Y^n_m$

$$Z^n_{m,-} = \alpha^n_{m,-}\left\{{}^{\psi}Y^n_m - \beta^n_{m,-}\left(X^n_m - \overline{X^n_m}\right) - \frac{1}{2}\left(Y^n_m + \overline{Y^n_m}\right)\right\}$$

where

$$\begin{cases} \alpha^n_{m,-} &= \frac{d^n_m+4c^n_m}{2c^n_m} &= \frac{4(n^2+m^2+n)}{(n+m)(n+m+1)} \\ \beta^n_{m,-} &= \frac{d^n_m-4c^n_m}{2(d^n_m+4c^n_m)} &= -\frac{m(2n+1)}{2(n^2+m^2+n)}. \end{cases}$$

This completes our proof. □

Different from the decomposition by means of contragenic functions, this decomposition is not orthogonal. However, every component in the decomposition shares the same structure as monogenic functions. The advantage is that now in each subspace of the decomposition all tools from quaternionic analysis like integral representations and kernel functions are available.

6. Conclusion

The theory of ψ-hyperholomorphic functions shows a structural analogy with classical monogenic functions. It helps to have a better understanding about characteristics of monogenic and harmonic functions such as geometric mapping properties and harmonic decompositions. The question how to find all possible harmonic decompositions is still open.

References

[1] R. Abreu Blaya, J. Bory Reyes, A. Guzman Adan, and U. Kaehler, On some structural sets and a quaternionic (ϕ, ψ)-hyperholomorphic function theory, submitted to Mathematische Nachrichten.

[2] C. Álvarez-Peña and R. Michael Porter, *Contragenic Functions of Three Variables*, Complex Anal. Oper. Theory, **8** (2014), 409–427.

[3] S. Bock, *On a three-dimensional analogue to the holomorphic z-powers: power series and recurrence formulae*, Complex Variables and Elliptic Equations: An International Journal, **57** (2012), 1349–1370.

[4] F. Brackx, R. Delanghe and F. Sommen, *Clifford Analysis*, Pitman Publishing, Boston-London-Melbourne, 1982.

[5] K. Gürlebeck: On some classes of Pi-operators, in Dirac operators in analysis, (eds. J. Ryan and D. Struppa), Pitman Research Notes in Mathematics, No. 394, 1998.

[6] K. Gürlebeck and H. Malonek, *A hypercomplex derivative of monogenic functions in* $\mathbb{R}^{n+1}$ *and its applications*, Complex Variables, **39**, No. 3 (1999), 199–228.

[7] K. Gürlebeck, U. Kähler, M. Shapiro: On the Pi-operator in hyperholomorphic function theory, Advances in Applied Clifford Algebras, Vol. 9(1), 1999, pp. 23–40

[8] R.S. Kraußhar, H.R. Malonek, A characterization of conformal mappings in $\mathbb{R}^4$ by a formal differentiability condition. Bull. Soc. R. Sci. Liege 70, No. 1, 35–49 (2001).

[9] R. Delanghe, R.S. Kraußhar, H.R. Malonek, Differentiability of functions with values in some real associative algebras: approaches to an old problem. Bull. Soc. R. Sci. Liège 70, No. 4-6, 231–249 (2001).

[10] M.E. Luna Elizarrarás and M. Shapiro, *A survey on the (hyper)derivates in complex, quaternionic and Clifford analysis*, Millan J. of Math, **79** (2011), 521–542.

[11] H.R. Malonek, Contributions to a geometric function theory in higher dimensions by Clifford analysis method: Monogenic functions and M-conformal mappings. In Brackx, F. (ed.) et al., Clifford analysis and its applications. Proceedings of the NATO advanced research workshop, Prague, Czech Republic, October 30–November 3, 2000. Dordrecht: Kluwer Academic Publishers. NATO Sci. Ser. II, Math. Phys. Chem. 25, 213–222 (2001).

[12] J. Morais and K. Gürlebeck, *Real-Part Estimates for Solutions of the Riesz System in* $\mathbb{R}^3$, Complex Var. Elliptic Equ., **57** (2012), 505–522.

[13] J. Morais, K. Gürlebeck: Bloch's Theorem in the Context of Quaternion Analysis, Computational Methods and Function Theory, Vol. 12 (2012), No. 2, 541–558.

[14] Vasilevskij, N.L.; Shapiro, M.V., On Bergmann kernel functions in quaternion analysis. Russ. Math. 42, No. 2, 81–85 (1998); translation from Izv. Vyssh. Uchebn. Zaved., Mat. 1998, No. 2, 84–88 (1998).

[15] M.V. Shapiro and N.L. Vasilevski, *Quaternionic ψ-hyperholomorphic functions, singular integral operators and boundary value problems. I. ψ-hyperholomorphic function theory*, Complex Variables, **27** (1995), 17–46.

[16] V.I. Shevchenko, A local homeomorphism of 3-space realizable by the solution of a certain elliptic system, Dokl. Acad. Nauk 146 (1962), 1035–1038.

[17] I. Stern, Randwertaufgaben für verallgemeinerte Cauchy–Riemann-Systeme im Raum. *Dissertation A, Martin-Luther-Universität Halle-Wittenberg* 1989.

[18] I. Stern, *Boundary value problems for generalized Cauchy–Riemann systems in the space.* In: Kühnau R.; Tutschke, W. (eds.): Boundary value and initial value problems in complex analysis. Pitman Res. Notes Math. 256 (1991): 159–183.

Klaus Gürlebeck and Hung Manh Nguyen
Inst. f. Math. u. Phys.
Coudraystr. 13B
D-99423 Weimar, Deutschland
e-mail: `klaus.guerlebeck@uni-weimar.de`
`hung.manh.nguyen@uni-weimar.de`

Hypercomplex Analysis: New Perspectives and Applications
Trends in Mathematics, 191–201

Fractional Clifford Analysis

Uwe Kähler and Nelson Vieira

Abstract. In this paper we present the basic tools of a fractional function theory in higher dimensions by means of a fractional correspondence to the Weyl relations. A Fischer decomposition, Almansi decomposition, fractional Euler and Gamma operators, monogenic projection, and basic fractional homogeneous powers will be constructed.

Mathematics Subject Classification (2010). Primary: 30G35. Secondary: 26A33; 30A05; 31B05; 30G20.

Keywords. Fractional monogenic polynomials; Fischer decomposition; Almansi decomposition; fractional Dirac operator; Caputo derivatives.

1. Introduction

The use of fractional calculus in mathematical modeling has become popular in recent years. This popularity arises naturally because on the one hand different problems can be considered in the framework of fractional derivatives like, for example, in optics and quantum mechanics, and on the other hand fractional calculus gives us a new degree of freedom which can be used for more complete characterization of an object or as an additional encoding parameter. Fractional calculus, for example, is used for phase retrieval [2], signal characterization [5], space-variant filtering [1], encryption [15], watermaking [10], and creation of neural networks [4].

The connections between fractional calculus and physics are, in some sense, relatively new one but, and more important for the community, a subject of strong interest. In [17] the author proposed a fractional Dirac equation of order $2/3$ and established the relation between the corresponding γ_α^μ-matrix algebra and generalized Clifford algebras. This approach was generalized in [18], where the author found that relativistic covariant equations generated by taking the nth root of the d'Alembert operator are fractional wave equations with an inherent $\mathrm{SU}(n)$ symmetry. It is clear that the study of fractional problems is a subject of current and strong investigations, in particular, the study of the fractional Dirac operator due to its physical and geometrical interpretations. Physically, this fractional differential operator is related with some aspects of fractional quantum mechanics such as the derivation of the fractal Schrödinger type wave equation, the resolution of the gauge hierarchy problem, and the study of super-symmetries. Geometrically, the fractional classical part of this operator may be identified to the scalar curvature in Riemannian geometry. The major problem with most of the fractional approaches treated is the presence of non-local fractional differential operators. Furthermore, the adjoint of a fractional differential used to describe the dynamics is non-negative itself. Other complicated problems arise during the mathematical manipulations, as the appearance of a very complicated rule which replaces the Leibniz rule for product of functions in the case of the classic derivative. Also we have a lack of any sufficiently good analogue of the chain rule. It is important to remark that there several definitions for fractional derivatives (Riemann–Liouville, Caputo, Riesz, Feller,...), however there are not many of these allow our approach. For the purposes of this work, the definition of fractional derivatives in the sense of Caputo is the most appropriate and applicable.

Although these difficulties create problems in the establishment of a fractional Clifford analysis, there is one approach which can be relatively easy adapted. Over the last decades F. Sommen and his collaborators developed a method for establishing a higher dimension function theory based on the so-called Weyl relations [7, 8, 9]. In more restrictive settings it is nowadays called Howe dual pair technique (see [16]). Its focal point is the construction of an operator algebra (classically $\mathfrak{osp}(1|2)$) and the resulting Fischer decomposition.

The traditional Fischer decomposition in harmonic analysis yields an orthogonal decomposition of the space $\mathcal{P}_l$ of homogeneous polynomials on $\mathbb{R}^d$ of given homogeneity l in terms of spaces of harmonic homogeneous polynomials. In classical continuous Clifford analysis one obtains a refinement yielding an orthogonal decomposition with respect to the so-called Fischer inner product of homogeneous polynomials, given by

$$\langle P(x), Q(x) \rangle = \mathrm{Sc}\left[\overline{P(\partial_x)}\, Q(x)\right],$$

in terms of spaces of monogenic polynomials, i.e., null solutions of the considered Dirac operator (see [9]). Here, the notation $\mathrm{Sc}[\cdot]$ stands for taking the scalar part of a Clifford algebra-valued expression, while $\overline{P(\partial_x)}$ is a differential operator obtained by replacing in the polynomial P each variable x_j by the corresponding partial derivative ∂_{x_j} and applying Clifford conjugation. This Fischer inner product results from a duality argument, called Fischer duality, between the algebra of vector variables and the algebra of operators. Generalizations as well as refinements of the Fischer decomposition in other Clifford analysis frameworks can be found, for example, in [6, 7, 8, 11, 14, 16].

The aim of this paper is to present a Fischer decomposition, when considering the fractional Dirac operator defined via fractional Caputo derivatives, where the fractional parameter α belongs to the interval $]0, 1[$. We remark that the cases where α is outside this range can be reduced to the considered one. In fact, for $\alpha \in \mathbb{R}$ we have that $\alpha = [\alpha] + \tilde{\alpha}$, with $[\alpha]$ the integer part of α and $\tilde{\alpha} \in]0,1[$. To this end we have to establish the fractional Weyl decomposition and the notion of fractional homogeneous polynomials. As an example we can consider $\alpha = \frac{1}{2}$ in our approach which allows us to establish a proper factorization of the transport operator $\partial_t - D$, where D represents the Dirac operator and ∂_t the partial derivative with respect to the variable t (see [3]).

The outline of the paper reads as follows. In the Preliminaries we recall some basic facts about Clifford analysis, fractional Caputo derivatives and fractional Dirac operators. In Section 3 we introduce the corresponding Weyl relations for this fractional setting and we introduce the notion of a fractional homogeneous polynomial. Moreover, we present the main result of this work, namely, the fractional correspondence to the Fischer decomposition and its extension to a fractional Almansi decomposition. In the end of the paper we construct the projection of a given fractional homogeneous polynomial into the space of fractional homogeneous monogenic polynomials. We also calculate the dimension of the space of fractional homogeneous monogenic polynomials.

2. Preliminaries

We consider the d-dimensional vector space $\mathbb{R}^d$ endowed with an orthonormal basis $\{e_1, \ldots, e_d\}$. We define the universal real Clifford algebra $\mathbb{R}_{0,d}$ as the 2^d-dimensional associative algebra which obeys the multiplication rules $e_i e_j + e_j e_i = -2\delta_{i,j}$. A vector space basis for $\mathbb{R}_{0,d}$ is generated by the elements $e_0 = 1$ and $e_A = e_{h_1} \cdots e_{h_k}$, where $A = \{h_1, \ldots, h_k\} \subset M = \{1, \ldots, d\}$, for $1 \leq h_1 < \cdots < h_k \leq d$. The Clifford conjugation is defined by $\overline{1} = 1$, $\overline{e_j} = -e_j$ for all $j = 1, \ldots, d$, and we have $\overline{ab} = \overline{b}\overline{a}$. An important property of algebra $\mathbb{R}_{0,d}$ is that each non-zero vector $x \in \mathbb{R}_1^d$ has a multiplicative inverse given by $\frac{\overline{x}}{||x||^2}$. Now, we introduce the complexified Clifford algebra $\mathbb{C}_d$ as the tensor product

$$\mathbb{C} \otimes \mathbb{R}_{0,d} = \left\{ w = \sum\nolimits_A w_A e_A, \; w_A \in \mathbb{C}, A \subset M \right\},$$

where the imaginary unit i of $\mathbb{C}$ commutes with the basis elements, i.e., $ie_j = e_j i$ for all $j = 1, \ldots, d$. To avoid ambiguities with the Clifford conjugation, we denote the complex conjugation mapping a complex scalar $w_A = a_A + ib_A$, with real components a_A and b_A, onto $\overline{w}_A = a_A - ib_A$ by $\sharp$. The complex conjugation leaves the elements e_j invariant, i.e., $e_j^\sharp = e_j$ for all $j = 1, \ldots, d$. We also have a pseudonorm on $\mathbb{C}$ viz $|w| := \sum_A |w_A|$ where $w = \sum_A w_A e_A$, as usual. Notice also that for $a, b \in \mathbb{C}_d$ we only have $|ab| \leq 2^d |a||b|$. The other norm criteria are fulfilled.

A $\mathbb{C}_d$-valued function f over $\Omega \subset \mathbb{R}_1^d$ has representation $f = \sum_A e_A f_A$, with components $f_A : \Omega \to \mathbb{C}$. Properties such as continuity will be understood component-wisely. Next, we recall the Euclidean Dirac operator $D = \sum_{j=1}^d e_j \, \partial_{x_j}$, which factorizes the d-dimensional Euclidean Laplacian, i.e.,

$$D^2 = -\Delta = -\sum_{j=1}^{d} \partial_{x_j}^2 .$$

A $\mathbb{C}_d$-valued function f is called *left-monogenic* if it satisfies $Du = 0$ on Ω (resp. *right-monogenic* if it satisfies $uD = 0$ on Ω). For more details about Clifford algebras and monogenic function we refer to [9].

An important subspace of the real Clifford algebra $\mathbb{R}_{0,d}$ is the so-called space of paravectors $\mathbb{R}_1^d = \mathbb{R} \bigoplus \mathbb{R}^d$, being the sum of scalars and vectors. An an element $\underline{x} = (x_0, x_1, \ldots, x_d)$ of $\mathbb{R}^d$ will be identified by $\underline{x} = x_0 + x$, with $x = \sum_{i=1}^{d} e_i x_i$. From now until the end of the paper, we will consider paravectors of the form $\underline{x}^\alpha = x_0^\alpha + x^\alpha$, where

$$x_j^\alpha = \begin{cases} \exp(\alpha \ \ln|x_j|); & x_j > 0 \\ 0; & x_j = 0 \\ \exp(\alpha \ \ln|x_j| + i\alpha\pi); & x_j < 0 \end{cases},$$

with $0 < \alpha < 1$, and $j = 0, 1, \ldots, d$.

The fractional Dirac operator will correspond to the fractional differential operator $D^\alpha = \sum_{j=1}^{d} e_j \, {}_+^C\partial_j^\alpha$, where ${}_+^C\partial_j^\alpha$ is the fractional Caputo derivative with respect to x_j^α defined as (see [13])

$$\left({}_+^C\partial_j^\alpha f\right)(x^\alpha) = \frac{1}{\Gamma(1-\alpha)} \int_0^{x_j^\alpha} \frac{1}{(x_j^\alpha - u)^\alpha} \, f_u'(x_1^\alpha, \ldots, x_{j-1}^\alpha, u, x_{j+1}^\alpha, \ldots, x_n^\alpha) \, du. \tag{2.1}$$

A $\mathbb{C}_n$-valued function f is called *fractional left-monogenic* if it satisfies $D^\alpha u = 0$ on Ω (resp. *fractional right-monogenic* if it satisfies $uD^\alpha = 0$ on Ω). For more details about fractional calculus and applications we refer [13]. We remark that due to the definition of the fractional Caputo derivative (2.1) we have that $D^\alpha c = 0$, where c denotes a constant, i.e., a fractional monogenic function.

3. Weyl relations and fractional Fischer decomposition

The aim of this section is to provide the basic tools for a function theory for the fractional Dirac operator.

3.1. Fractional Weyl relations

Here we introduce the fractional correspondence, via Caputo derivatives, of the classical Euler and Gamma operators. Moreover, we will show that the two natural operators D^α and x^α, considered as odd elements, generate a finite-dimensional Lie super-algebra in the algebra of endomorphisms generated by the partial Caputo derivatives, the basic *vector variables* x_j^α (seen as multiplication operators), and the basis of the Clifford algebra e_j. Furthermore, we will introduce the definition of fractional homogeneous polynomials.

In order to achieve our aims, we will use some standard technique in higher dimensions, namely we study the commutator and the anti-commutator between x^α and D^α. We start proposing the following Weyl relations

$$\left[{}_{+}^{C}\partial_i^\alpha, x_i^\alpha \right] = {}_{+}^{C}\partial_i^\alpha \, x_i^\alpha - x_i^\alpha \, {}_{+}^{C}\partial_i^\alpha = \frac{\alpha\pi}{\sin(\alpha\pi)\,\Gamma(1-\alpha)} =: K_\alpha, \tag{3.1}$$

with $i = 1, \ldots, d$, $0 < \alpha < 1$, and ${}_{+}^{C}\partial_i^\alpha$ is the fractional Caputo derivative (2.1). This leads to the following relations for our fractional Dirac operator

$$\{D^\alpha, x^\alpha\} = D^\alpha x^\alpha + x^\alpha D^\alpha = -2\mathbb{E}^\alpha - K_\alpha d, \tag{3.2}$$

$$[x^\alpha, D^\alpha] = x^\alpha D^\alpha - D^\alpha x^\alpha = -2\Gamma^\alpha + K_\alpha d, \tag{3.3}$$

where $\mathbb{E}^\alpha$, Γ^α are, respectively, the fractional Euler and Gamma operators. The expressions for $\mathbb{E}^\alpha$ and Γ^α are

$$\begin{aligned} \mathbb{E}^\alpha &= \sum_{i=1}^{d} x_i^\alpha \, {}_{+}^{C}\partial_i^\alpha, \\ \Gamma^\alpha &= \sum_{i<j} e_i e_j (x_i^\alpha \, {}_{+}^{C}\partial_j^\alpha - {}_{+}^{C}\partial_j^\alpha \, x_i^\alpha). \end{aligned} \tag{3.4}$$

This can be easily checked by

$$\begin{aligned} \{D^\alpha, x^\alpha\} &= \sum_{i=1}^{d}\sum_{j=1}^{d} \left(e_i e_j \, {}_{+}^{C}\partial_i^\alpha \, x_j^\alpha + e_j e_i \, x_j^\alpha \, {}_{+}^{C}\partial_i^\alpha \right) \\ &= -\sum_{i=1}^{d} \left({}_{+}^{C}\partial_i^\alpha \, x_i^\alpha + x_i^\alpha \, {}_{+}^{C}\partial_i^\alpha \right) \\ &= -\sum_{i=1}^{d} \left(2x_i^\alpha \, {}_{+}^{C}\partial_i^\alpha + K_\alpha \right) \\ &= -2\mathbb{E}^\alpha - K_\alpha d, \end{aligned}$$

and

$$\begin{aligned} [x^\alpha, D^\alpha] &= \sum_{i=1}^{d}\sum_{j=1}^{d} \left(e_i e_j \, x_i^\alpha \, {}_{+}^{C}\partial_j^\alpha - e_j e_i \, {}_{+}^{C}\partial_j^\alpha \, x_i^\alpha \right) \\ &= \sum_{i\neq j}^{d} \left(e_i e_j \, x_i^\alpha \, {}_{+}^{C}\partial_j^\alpha - e_j e_i \, {}_{+}^{C}\partial_j^\alpha \, x_i^\alpha \right) - \sum_{i=1}^{d} \left(x_i^\alpha \, {}_{+}^{C}\partial_i^\alpha - {}_{+}^{C}\partial_i^\alpha \, x_i^\alpha \right) \\ &= -2\sum_{i<j}^{d} e_i e_j (x_i^\alpha \, {}_{+}^{C}\partial_j^\alpha - {}_{+}^{C}\partial_j^\alpha \, x_i^\alpha) - \sum_{i=1}^{d} (-K_\alpha) \\ &= -2\Gamma^\alpha + K_\alpha d. \end{aligned}$$

From relations (3.4) we derive, via straightforward calculations, the following relations

$$\begin{aligned}
&\mathbb{E}^\alpha + \Gamma^\alpha = -x^\alpha D^\alpha, && [\mathbb{E}^\alpha, \Gamma^\alpha] = 0,\\
&[x^\alpha, \mathbb{E}^\alpha] = -K_\alpha x^\alpha, && [D^\alpha, \mathbb{E}^\alpha] = K_\alpha D^\alpha,\\
&\{x^\alpha, x^\alpha\} = -2|x^\alpha|^2, && \{D^\alpha, D^\alpha\} = 2\Delta^{2\alpha},\\
&[x^\alpha, |x^\alpha|^2] = 0, && [x^\alpha, \Delta^{2\alpha}] = -2K_\alpha D^\alpha,\\
&[D^\alpha, |x^\alpha|^2] = 2K_\alpha x^\alpha, && [D^\alpha, \Delta^{2\alpha}] = 0,\\
&[\mathbb{E}^\alpha, \Delta^{2\alpha}] = -2K_\alpha \Delta^{2\alpha}, && [\mathbb{E}^\alpha, |x^\alpha|^2] = 2K_\alpha |x^\alpha|^2,\\
&[|x^\alpha|^2, \Delta^{2\alpha}] = -4K_\alpha \left(\mathbb{E}^\alpha + \frac{K_\alpha d}{2}\right),
\end{aligned} \tag{3.5}$$

where $\Delta^{2\alpha} = -D^\alpha D^\alpha$. The previous relations show that we have a finite-dimensional Lie superalgebra generated by x^α and D^α, isomorphic to $\mathfrak{osp}(1|2)$. Indeed, the normalization

$$\begin{aligned}
&H^\alpha = \frac{1}{2}\left(\mathbb{E}^\alpha + \frac{K_\alpha d}{2}\right),\\
&(E^\alpha)^+ = \frac{1}{2}\, K_\alpha |x^\alpha|^2, \qquad (E^\alpha)^- = -\frac{1}{2}\, K_\alpha \Delta^{2\alpha},\\
&(F^\alpha)^+ = \frac{1}{2\sqrt{2}}\, iK_\alpha x^\alpha, \qquad (F^\alpha)^- = \frac{1}{2\sqrt{2}}\, iK_\alpha D^\alpha,
\end{aligned}$$

leads to the standard commutation relations for $\mathfrak{osp}(1|2)$ (see [12])

$$\begin{aligned}
&[H^\alpha, (E^\alpha)^\pm] = \pm K_\alpha (E^\alpha)^\pm, && [(E^\alpha)^+, (E^\alpha)^-] = 2K_\alpha^3 H^\alpha,\\
&[H^\alpha, (F^\alpha)^\pm] = \pm\frac{1}{2}\, K_\alpha (F^\alpha)^\pm, && \{(F^\alpha)^+, (F^\alpha)^-\} = \frac{1}{2}\, K_\alpha^2 H^\alpha,\\
&[(E^\alpha)^\pm, (F^\alpha)^\mp] = -K_\alpha^2 (F^\alpha)^\pm, && \{(F^\alpha)^\pm, (F^\alpha)^\pm\} = \pm\frac{1}{2}\, K_\alpha (E^\alpha)^\pm.
\end{aligned} \tag{3.6}$$

Now we introduce the definition of fractional homogeneity of a polynomial by means of the fractional Euler operator.

Definition 3.1. A polynomial P is called fractional homogeneous of degree $l \in \mathbb{N}_0$, if and only if $\mathbb{E}^\alpha P = K_{l,\alpha}\, l\, P$, where $K_{l,\alpha} = \frac{\alpha\, \Gamma(\alpha l)}{\Gamma(1+\alpha(l-1))}$.

We remark that from the previous definition the basic fractional homogeneous powers are given by $(x^\alpha)^\beta = (x_1^\alpha)^{\beta_1} \cdots (x_d^\alpha)^{\beta_d}$, with $l = \beta_1 + \cdots + \beta_d$. In combination with the third relation in (3.5)

$$[x^\alpha, \mathbb{E}^\alpha] = -K_\alpha x^\alpha,$$

this definition also implies that the multiplication of a fractional homogeneous polynomial of degree l by x^α, will result in a fractional homogeneous polynomial of degree $l+1$, and thus may be seen as a raising operator. Moreover, we can also ensure that for a fractional homogeneous polynomial P_l of degree l, $D^\alpha P_l$ is a fractional homogeneous polynomial of degree $l-1$. Furthermore, Weyl relations (3.1) will now enable us to construct fractional homogeneous polynomials, recursively.

3.2. Fractional Fischer decomposition

In the present fractional context, a Fischer inner product of two fractional homogeneous polynomials P and Q would have the following form

$$\langle P(x^\alpha), Q(x^\alpha)\rangle = \mathrm{Sc}\left[\overline{P(\partial_{x^\alpha})}\; Q(x^\alpha)\right], \tag{3.7}$$

where ∂_{x^α} represents ${}_{+}^{C}\partial_i^\alpha$, i.e., the fractional Caputo derivative (2.1) with respect to x_j, and $P(\partial_{x^\alpha})$ is a differential operator obtained by replacing in the polynomial P each variable x_j by the corresponding fractional Caputo derivative. From (3.7) we have that for any polynomial P_{l-1} of homogeneity $l-1$ and any polynomial Q_l of homogeneity l

$$\langle x^\alpha\ P_{l-1},\ Q_l\rangle = \langle P_{l-1}, D^\alpha Q_l\rangle\,. \tag{3.8}$$

This fact allows us to prove the following result:

Theorem 3.2. *For each $l \in \mathbb{N}_0$ we have $\Pi_l = \mathcal{M}_l + x^\alpha\ \Pi_{l-1}$, where Π_l denotes the space of fractional homogeneous polynomials of degree l and $\mathcal{M}_l$ denotes the space os fractional monogenic homogeneous polynomials of degree l. Furthermore, the subspaces $\mathcal{M}_k$ and $x^\alpha\Pi_{l-1}$ are orthogonal with respect to the Fischer inner product* (3.7).

Proof. Since

$$\Pi_l = x^\alpha\ \Pi_{l-1} + (x^\alpha\ \Pi_{l-1})^\perp,$$

it suffices to prove that $(x^\alpha\Pi_{l-1})^\perp = \mathcal{M}_{l-1}$. To this end, assume that, for some $P_l \in \Pi_l$ we have

$$\langle x^\alpha\ P_{l-1},\ P_l\rangle = 0, \qquad \text{for all } P_{l-1} \in \Pi_{l-1}.$$

From (3.8) we then have that

$$\langle P_{l-1},\ D^\alpha P_l\rangle = 0, \qquad \text{for all } P_{l-1} \in \Pi_{l-1}.$$

As $D^\alpha P_l \in \Pi_{l-1}$ we obtain that $D^\alpha P_l = 0$, or that $P_l \in \mathcal{M}_l$. This means that $(x^\alpha\ \Pi_{l-1})^\perp \subset \mathcal{M}_{l-1}$. Conversely, take $P_l \in \mathcal{M}_l$. Then we have, for any $P_{l-1} \in \Pi_{l-1}$, that

$$\langle x^\alpha\ P_{l-1},\ P_l\rangle = \langle P_{l-1},\ D^\alpha P_l\rangle = \langle P_{l-1},\ 0\rangle = 0,$$

from which it follows that $\mathcal{M}_{l-1} \subset (x^\alpha\ \Pi_{l-1})^\perp$, and therefore $\mathcal{M}_{l-1} = (x^\alpha\ \Pi_{l-1})^\perp$. □

As a result we obtain the fractional Fischer decomposition with respect to the fractional Dirac operator D^α.

Theorem 3.3. *Let P_l be a fractional homogeneous polynomial of degree l. Then*

$$P_l = M_l + x^\alpha\, M_{l-1} + (x^\alpha)^2\, M_{l-2} + \cdots + (x^\alpha)^l\, M_0, \tag{3.9}$$

where each M_j denotes the fractional homogeneous monogenic polynomial of degree j.

The polynomials represented in (3.9) are orthogonal to each other with respect to the Fischer inner product (3.7). This a consequence of the construction of the fractional Euler operator $\mathbb{E}^\alpha$, and in particular of (3.2) and (3.3). Moreover, from the previous theorem we have the following direct extension to the fractional case of the Almansi decomposition:

Theorem 3.4. *For any fractional polyharmonic polynomial P_l of degree $l \in \mathbb{N}_0$ in a starlike domain D in $\mathbb{R}^d$ with respect to 0, i.e., $\Delta^{2\alpha} P_l = 0$, there exist uniquely fractional harmonic functions $P_0, P_1, \ldots, P_{l-1}$ such that*

$$P_l = P_0 + |x^\alpha|^2 P_1 + \cdots + |x^\alpha|^{2(l-1)} P_{l-1}, \qquad \forall x^\alpha \in D.$$

3.3. Explicit formulae

In this subsection we obtain an explicit formula for the projection $\pi_{\mathcal{M}}(P_l)$ of a given fractional homogeneous polynomial P_l into the space of fractional homogeneous monogenic polynomials. We start proving the following auxiliary result:

Theorem 3.5. *For any fractional homogeneous polynomial P_l and any positive integer s, we have:*

$$D^\alpha (x^\alpha)^s P_l = g_{s,l} (x^\alpha)^{s-1} P_l + (-1)^s (x^\alpha)^s D^\alpha P_l,$$

where $g_{2k,l} = -2kK_\alpha$ and $g_{2k+1,l} = -(2(kK_\alpha + K_{l,\alpha} l) + K_\alpha d)$.

Proof. The proof follows, by induction and making straightforward calculations, from the commutation between D^α and $(x^\alpha)^s$ using the relations

$$D^\alpha x^\alpha = -2\mathbb{E}^\alpha - K_\alpha d - x^\alpha D^\alpha, \qquad \mathbb{E}^\alpha x^\alpha = x^\alpha \mathbb{E}^\alpha + K_\alpha x^\alpha. \qquad \square$$

Let us now compute an explicit form of the projection $\pi_{\mathcal{M}}(P_l)$.

Theorem 3.6. *Consider the constants $c_{j,l}$ defined by*

$$c_{0,l} = 1, \qquad c_{j,l} = \frac{(-1)^j}{(2\,[j/2])!! \prod_{i=0}^{[j/2]} g_{2i+1,l-(2i+1)}},$$

where $j = 1, \ldots, l$ and $[\cdot]$ represents the integer part. Then the map $\pi_{\mathcal{M}}$ given by

$$\pi_{\mathcal{M}}(P_l) := P_l + c_{1,l} x^\alpha D^\alpha P_l + c_{2,l} (x^\alpha)^2 (D^\alpha)^2 P_l + \cdots + c_{l,l} (x^\alpha)^l (D^\alpha)^l P_l$$

is the projection of the fractional homogeneous polynomial P_l into the space of fractional homogeneous monogenic polynomials.

Proof. Let us consider the linear combination

$$r = a_0 P_l + a_1 x^\alpha D^\alpha P_l + a_2 (x^\alpha)^2 (D^\alpha)^2 P_l + \cdots + a_k (x^\alpha)^l (D^\alpha)^l P_l,$$

with $a_0 = 1$. If there are constants a_j, $j = 1, \ldots, l$, such that $r \in \mathcal{M}_l$, then r is equal to $\pi_{\mathcal{M}}(P_l)$. Indeed, we know that $P_l = \mathcal{M}_l \oplus x^\alpha P_{l-1}$ and

$$r = P_l + Q_{l-1}, \qquad \text{with} \qquad Q_{l-1} = \sum_{i=1}^{l} a_i (x^\alpha)^i (D^\alpha)^i P_l.$$

Applying Theorem 3.5, we get

$$\begin{aligned} 0 &= D^\alpha(\pi_{\mathcal{M}}(P_l)) \\ &= D^\alpha P_l + a_1 D^\alpha x^\alpha D^\alpha P_l + a_2 D^\alpha (x^\alpha)^2 (D^\alpha)^2 P_l \\ &\quad + \cdots + a_l D^\alpha (x^\alpha)^l (D^\alpha)^l P_l \\ &= (1 + a_1 g_{1,l-1}) D^\alpha P_l + ((-1)^1 a_1 + a_2 g_{2,l-2}) x^\alpha (D^\alpha)^2 P_l \\ &\quad + ((-1)^2 a_2 + a_3 g_{3,l-3})(x^\alpha)^2 (D^\alpha)^3 P_l \\ &\quad + \cdots + ((-1)^{l-1} a_{l-1} + a_l g_{l,0})(x^\alpha)^{l-1} (D^\alpha)^l P_l. \end{aligned}$$

Hence if the relation $(-1)^{j-1} a_{j-1} + a_i g_{j,l-j} = 0$ holds for each $j = 1, \ldots, l$, then the function r is fractional monogenic. By induction we get

$$a_j = \frac{(-1)^j}{(2\,[j/2])!!\, \prod_{i=0}^{[j/2]} g_{2i+1,l-(2i+1)}}. \qquad \square$$

Theorem 3.7. *Each polynomial P_l can be written in a unique way as*

$$P_l = \sum_{j=0}^{l} (x^\alpha)^j M_{l-j}(P_l),$$

where

$$M_{l-j}(P_l) = c'_j \sum_{n=0}^{j} c_{j,l-n} (x^\alpha)^n (D^\alpha)^n (D^\alpha)^{l-j} P_l, \qquad j = 0, \ldots, l,$$

and the coefficients c'_j are defined by

$$c'_j = \frac{(-1)^j}{(2\,[j/2])!!\, \prod_{i=0}^{[j/2]} g_{2i+1,l-(2i+1)}}.$$

Proof. We know that for any P_l, there is a unique decomposition

$$P_l = (x^\alpha)^l M_0 + (x^\alpha)^{l-1} M_1 + \cdots + x^\alpha M_{l-1} + M_l,$$

where $M_{l-j} \in \mathcal{M}_{l-j}$, with $j = 0, \ldots, l$. To compute a component M_{l-j} explicitly, we apply $(D^\alpha)^j$ to both sides of the previous equality

$$\begin{aligned} (D^\alpha)^j P_l &= (D^\alpha)^j (x^\alpha)^l M_0 + (D^\alpha)^j (x^\alpha)^{l-1} M_1 \\ &\quad + \cdots + (D^\alpha)^j (x^\alpha)^{j+1} M_{l-j-1} + (D^\alpha)^j (x^\alpha)^j M_{l-j}. \end{aligned}$$

The summands on the right-hand side belong, in turn, to the spaces

$$(x^\alpha)^{l-j}\mathcal{M}_0, \qquad (x^\alpha)^{l-j-1}\mathcal{M}_1, \qquad \cdots \qquad x^\alpha\mathcal{M}_{l-j-1}, \qquad \mathcal{M}_{l-j}.$$

Hence, $(D^\alpha)^j(x^\alpha)^j M_{l-j}$ is equal to $\pi_{\mathcal{M}}((D^\alpha)^j P_l)$. We can now use the expression for the harmonic projection proved above. So to get our result, it is sufficient to show that

$$(D^\alpha)^j(x^\alpha)^j M_{l-j} = \frac{1}{c_j} M_j.$$

By Theorem 3.5, we get by induction that $c_j' = a_j$, and, therefore, the proof is finished. □

As a consequence, we also obtain the dimension of the space of fractional homogeneous monogenic polynomials of degree l. Indeed, from the Fischer decomposition (3.9) we get

$$\dim(\mathcal{M}_l) = \dim(\Pi_l) - \dim(\Pi_{l-1}),$$

with the dimension of the space of fractional homogeneous polynomials of degree l given by

$$\dim(\Pi_l) = \frac{(k+d-1)!}{k!\ (d-1)!}.$$

This leads to the following theorem:

Theorem 3.8. *The space of fractional homogeneous monogenic polynomials of degree l has dimension*

$$\dim(\mathcal{M}_k) = \frac{(l+d-1)! - l(l+d-2)!}{l!\ (d-1)!} = \frac{(l+d-2)!}{l!\ (d-2)!}.$$

Acknowledgment

This work was supported by Portuguese funds through the CIDMA – Center for Research and Development in Mathematics and Applications, and the Portuguese Foundation for Science and Technology ("FCT – Fundação para a Ciência e a Tecnologia"), within project PEst-OE/MAT/UI4106/2014.

References

[1] O. Akay and G.F. Boudreaux-Bartels, *Fractional convolution and correlation via operator methods and an application to detection of linear FM signals*, IEEE Trans. Signal Processing, **49**, No. 5, (2001), 979–993.

[2] T. Alieva, M.J. Bastiaans and L. Stanković, *Signal reconstruction from two close fractional Fourier power spectra*, IEEE Trans. Signal Processing, **51**, No. 1, (2003), 112–123.

[3] V. Agoshkov, *Boundary value problems for transport equations*, Modeling and Simulation in Science, Engineering and Technology, Birkhäuser, Boston, MA, 1998.

[4] B. Barshan and B. Ayrulu, *Comparative analysis of different approaches to target differentiation and localization with sonar*, Pattern Recognition, **36**, No. 5, (2003), 1213–1231.

[5] M.J. Bastiaans and T. Alieva, *Wigner distribution moments measured as intensity moments in separable first-order optical systems*, EURASIP Journal on Applied Signal Processing, **2005**, No. 10, (2005), 1535–1540.

[6] G. Bernardes, P. Cerejeiras and U. Kähler, *Fischer decomposition and Cauchy kernel for Dunkl-Dirac operators*, Adv. Appl. Clifford Algebr., **19**, No. 2, (2009), 163–171.

[7] H. De Bie and F. Sommen, *Fischer decompositions in superspace*, in: Function spaces in complex and Clifford analysis, Le Hung Son et al. (eds.), National University Publishers, Hanoi, 2008, 170–188.

[8] H. De Ridder, H. De Schepper, U. Kähler and F. Sommen, *Discrete function theory based on skew Weyl relations*, Proc. Am. Math. Soc., **138**, No. 9, (2010), 3241–3256.

[9] R. Delanghe, F. Sommen and V. Souček, *Clifford algebras and spinor-valued functions. A function theory for the Dirac operator*, Mathematics and its Applications, Vol. 53, Kluwer Academic Publishers, Dordrecht etc., 1992.

[10] I. Djurovic, S. Stankovic and I. Pitas, *Digital watermarking in the fractional Fourier transformation domain*, Journal of Network and Computer Applications, **24**, No. 2, (2001), 167–173.

[11] D. Eelbode, *Stirling numbers and spin-Euler polynomials*, Exp. Math., **16**, No. 1, (2007), 55–66.

[12] L. Frappat, A. Sciarrino and P. Sorba, *Dictionary on Lie algebras and superalgebras*, Academic Press, San Diego, CA, 2000.

[13] A. Kilbas, H.M. Srivastava and J.J. Trujillo, *Theory and applications of fractional differential equations*, North-Holland Mathematics Studies 204, Elsevier, Amsterdam, 2006.

[14] H.R. Malonek and D. Peña-Peña, *Fischer decomposition by inframonogenic functions*, Cubo, **12**, No. 2, (2010), 189–197.

[15] N.K. Nishchal, J. Joseph and K. Singh, *Fully phase encryption using fractional Fourier transform*, Optical Engineering, **42**, No. 6, (2003), 1583–1588.

[16] B. Ørsted, P. Somberg and V. Souček, *The Howe duality for the Dunkl version of the Dirac operator*, Adv. Appl. Clifford Algebr., **19**, No. 2, (2009), 403–415.

[17] A. Raspini, *Simple solutions of the fractional Dirac equation of order* 2/3, Physica Scripta, **64**, No. 1, (2001), 20–22.

[18] P.J. Zàvada, *Relativistic wave equations with fractional derivatives and pseudodifferential operators*, J. Appl. Math., **2**, No. 4, (2002), 163–197.

Uwe Kähler and Nelson Vieira
CIDMA – Center for Research and Development
in Mathematics and Applications
Department of Mathematics, University of Aveiro
Campus Universitário de Santiago
3810-193 Aveiro, Portugal
e-mail: ukaehler@ua.pt
nloureirovieira@gmail.com

Hypercomplex Analysis: New Perspectives and Applications
Trends in Mathematics, 203–211

Spectral Properties of Differential Equations in Clifford Algebras

Yakov Krasnov

Abstract. The aim of this work is to establish a number of elementary properties about the topology and algebra of real quadratic homogeneous mappings and Riccati type ODEs occurring in Clifford Algebras. We construct an example of the quadratic vector field to show the impact of the spectral properties on qualitative theory.

Mathematics Subject Classification (2010). Primary 17A01, 34G20, 11E04, 35E20; Secondary 35F05.

Keywords. Bilinearity, differential equations in algebra.

1. Introduction

In mathematics, the term *linearization* refers to finding a *linear approximation* of a nonlinear model. This essentially allows us to exploit methods of linear operators (linear algebra). Similarly, the term *bilinearization* refers to finding a *quadratic approximation* of a nonlinear model, which in turn allows the use of the methods of nonassociative commutative binary algebra.

1.1. Linearization in dynamical systems

It is well known that in the study of dynamical systems linearization is a powerful method for assessing the local stability of an equilibrium point of a system of nonlinear differential equations or discrete dynamical systems. However, sometimes the qualitative investigations of the system require that we consider the quadratic terms (bilinearization) as well.

This paper is the extended version of a talk presented in the session of the 9th ISAAC Conference held in Krakow in 2013 titled "Clifford and Quaternionic Analysis" and organized by I. Sabadini, S. Bernstein and F. Sommen.

In the study of dynamical systems, the Grobman–Hartman theorem, called also the "linearization theorem", is a result about the local behavior of a dynamical system in a neighborhood of a *hyperbolic equilibrium point* (meaning that no eigenvalue of the linearization has its real part equal to zero).

Basically the theorem states that the behavior of a dynamical system near the hyperbolic equilibrium point is qualitatively the same as the behavior of its linearization.

Therefore, the local dynamics near a hyperbolic equilibrium is completely determined by its linearization.

In particular,

- if real parts of all eigenvalues of the Jacobian matrix are negative, then the equilibrium is stable,
- if they are all positive, then the equilibrium is unstable,
- if the values are of mixed signs, then the equilibrium is a saddle,
- any conjugate pair of complex eigenvalues indicates a spiral.

If, however, the equilibrium is non-hyperbolic, then providing an effective stability analysis, it constitutes a problem of great complexity. In such a case, it is natural to consider the bilinearization.

1.2. Bilinearization

Denote by x^* an equilibrium point of a vector field $V(x)$. Then

$$V(x) = L(x - x^*) + Q(x - x^*) + \text{higher terms with respect to } ||x - x^*||^2, \quad (1.1)$$

where $L : \mathbb{R}^n \to \mathbb{R}^n$ is a linearization of $V(x)$ near x^* and $Q : \mathbb{R}^n \to \mathbb{R}^n$ is a quadratic homogeneous map obtained as the second term in an approximation of $V(x)$.

In order to take advantage of appropriate algebraic methods, define the symmetric bilinear map $B(x, y) : \mathbb{R}^n \times \mathbb{R}^n \to \mathbb{R}^n$, associated with $Q(x)$ via the polarization identity:

$$B(x, y) = \frac{1}{2}[Q(x + y) - Q(x) - Q(y)], \quad \forall x, y \in \mathbb{R}^n, \quad (1.2)$$

or, equivalently, by using the Euler formula

$$B(x, y) = \frac{1}{2}J(x)y, \quad \forall x, y \in \mathbb{R}^n, \quad (1.3)$$

where $J(x)$ is the Jacobian matrix of the quadratic map $Q(x)$ at the point $(x - x)^* \in \mathbb{R}^n$.

Note that the use of formulas (1.1)–(1.3) is the essence of the "bilinearization" procedure for a vector field V. This formula allows us to canonically assign to V a nonassociative algebra $\mathbb{A} = (\mathbb{R}^n, \circ)$ with multiplication

$$x \circ y = B(x, y). \quad (1.4)$$

Several questions relevant to a qualitative study of the field V give rise to a natural quadratic form with which one can associate the corresponding Clifford algebra $Cl(V, Q)$.

1.3. Spectral invariants

Given a polynomial vector field $V : \mathbb{R}^n \to \mathbb{R}^n$, denote by $\mathrm{Nul}(V)$ the set of all stationary points (possibly complex) of the vector field V. Take $x^* \in \mathrm{Nul}(V)$. Then (1.1) takes place with linearization L_{x^*} and bilinearization B_{x^*}.

Definition 1.1. The collection of eigenvalues of L_{x^*} will be called the *primary spectrum* of the vector field V at $x^* \in \mathrm{Nul}(V)$.

Denote by $\mathrm{Fix}(Q_{x^*})$ the set of all non-zero fixed points of Q_{x^*}.

Definition 1.2. Let $\mathbb{A}_{x^*}$ be an algebra associated with the quadratic map Q_{x^*} via the multiplication (1.4). In this algebra are defined the *Peirce numbers* (see [1]) as the collection of eigenvalues of the Jacobian matrix of Q_{x^*} at any point from $\mathrm{Fix}(Q_{x^*})$. In what follows, these numbers will be called the *secondary spectrum* of the vector field V.

Remark 1.3. Obviously, primary and secondary spectra are invariant with respect to the linear change of variables. If necessary, one can construct spectra of higher order homogeneous part.

It turns out that the Peirce numbers respect some syzygies related to the Euler–Jacobi formula.

Example. Consider the following quadratic vector field $V : \mathbb{R}^2 \to \mathbb{R}^2$:

$$V(x, y) = \{x - y + 4x^2 - 2y^2, \quad x + y - 2xy\}. \tag{1.5}$$

Then there exist four equilibria $[0, 0]$, $[0, -\frac{1}{2}]$, $[\frac{1}{2}, \frac{1}{2}]$, $[-\frac{1}{2}, \frac{1}{2}]$. The non-zero equilibria are hyperbolic. Therefore, they can be completely studied using the primary spectrum only.

Consider the origin, which, obviously, is not hyperbolic.

The quadratic term of the expansion of V at the origin is:

$$Q_{\{0,0\}}(x, y) = \{4x^2 - 2y^2, \quad -2xy\}. \tag{1.6}$$

Clearly, $Q_{\{0,0\}}$ has also three fixed points: $x_1 = [\frac{1}{4}, 0]$, $x_{2,3} = [-\frac{1}{2}, \pm\frac{\sqrt{3}}{2}]$. After straightforward calculation we obtain the following Peirce numbers:

$$\mu_1 = -1/2; \; \mu_{2,3} = -5. \tag{1.7}$$

Also the Peirce numbers respect the following syzygy:

$$\frac{1}{\mu_1 - 1} + \frac{1}{\mu_2 - 1} + \frac{1}{\mu_3 - 1} + 1 = 0. \tag{1.8}$$

1.4. Subject of the paper

In this note, we are mainly concerned with studying several connections of primary and secondary spectra to the qualitative properties of the vector field in question. In our opinion, the following problems are interesting:

- To study a connection between topological/geometric properties of ODEs and PDEs with the properties of an underlying algebra.

- To explain the special role of Clifford algebras in differential equations based on unitality and associativity.
- To extend the known results in Clifford analysis to algebras that are isotopy equivalent to Clifford algebras.
- To clarify the role of syzygies between Peirce numbers as found in dynamical systems.

2. Algebras and DEs

Remark 2.1. (i) If V is the polynomial vector field, then by Bézout's theorem there are no more than $2^n - 1$ non-zero fixed points of the homogeneous quadratic map in $\mathbb{R}^n$ or infinitely many.

(ii) Clearly, if $z \in \mathbb{C}^n$ is an equilibrium for Q, then so is λz for any $\lambda \in \mathbb{C}$. Denote by $\mathbb{C}[z]$ the (complex) line of equilibria of Q generated by some (non-zero) $z \in \mathbb{C}^n$.

(iii) As is well known (see, for example, [10]), the set Fix(Q) consists of finitely many connected components of (complex) fixed points. Also, all equilibria Nul(Q) constitute one connected component that is a union of a collection of complex planes passing through the origin.

Also a spectral invariant is a quantity which is determined by the spectrum of the quadratic map, and two quadratic maps are called *isospectral*, if their spectra, counting multiplicities, coincide.

Given the set $P \in \mathbb{C}^n$. The following four problems were studied in [7]:

Problem A (existence). What are the algebraic conditions on the set P providing the existence of a polynomial map V with bilinearization Q at stationary point $x^* \in \mathbb{R}^n$ such that $P = Fix(Q)$?

Problem B (uniqueness). Assume a vector field $V(x)$ required in Problem A does exist. Under which conditions on P is such a V unique (up to a linear change of variables)?

Problem C (extension). Let P, N be such that Problem A is solvable for some quadratic map f. How can one extend the sets P, N to the spectrum of f without giving an *explicit* construction of f (observe that an explicit construction of f can be a problem of formidable complexity)?

Problem D (topological and dynamical properties). How are the topological and dynamical properties of a vector field V determined by its spectrum?

2.1. Linear constant coefficient ODEs as ODEs in a unary algebra

Given a unary algebra $\mathbb{A} = (\mathbb{R}^n, \star)$ with unary multiplication rule,

$$\star e_i = \sum_{j=1}^{n} a_i^j e_j, \quad i = 1, \ldots, n. \tag{2.1}$$

Any linear constant coefficient ODE $\dot{x} = Lx$ can be rewritten as

$$\dot{x}(t) = \star x(t), \tag{2.2}$$

where $x(t) = x_1(t)e_1 + \cdots + x_n(t)e_n$ and $Lx = \star x$ in $\mathbb{A}$.

Formula (2.2) indicates the parallelism between unary algebras and linear ODEs. The universal language allowing us to study both problems is the spectral theory for linear operators.

2.2. First-order PDEs and ODEs in binary algebras

Given a binary algebra $\mathbb{A} = (\mathbb{R}^n, \circ)$ with the multiplication rule:

$$e_i \circ e_j = \sum_{j=1}^{n} a_{ij}^k e_k, \quad i, j = 1, \ldots, n. \tag{2.3}$$

Then, using this rule, any quadratic ODE $\dot{x} = Q(x)$ with homogeneous quadratic map $Q : \mathbb{R}^n \to \mathbb{R}^n$ can be rewritten as a Riccati equation in the algebra:

$$\dot{x}(t) = x(t) \circ x(t). \tag{2.4}$$

Also, any first-order linear PDE with constant coefficients can be thought of as the Dirac equation in a binary algebra as follows:

$$D \circ u(x) = 0, \tag{2.5}$$

where $D = \partial_{x_1} e_1 + \cdots + \partial_{x_n} e_n$, $u(x) = u_1(x)e_1 + \cdots + u_n(x)e_n$.

Qualitative properties of the Riccati equation (2.4) can be studied in terms of invariants of algebra isomorphisms. The universal language for this investigation is primary and secondary spectral theory.

In contrast to the case of the Riccati equation in binary algebra, the structure of solutions to the Dirac equation (2.5) refers to isotopy invariants of $\mathbb{A}$ rather than to the isomorphism [6].

2.3. Complex structure

Usually, by a complex structure one understands a linear map $J : \mathbb{R}^n \to \mathbb{R}^n$ satisfying $J^2 = -I$, where I is the identity operator (cf. [Gr]). Below we extend this concept to real commutative algebras.

Definition 2.2. Let A be a real finite-dimensional (in general, nonassociative) algebra. By a *complex structure* in A we mean the existence of a non-trivial solution to the equation $x^2 \circ y = -y$, as a generalization of the real equation $x^2 = -1$ in the unital algebra.

Our study of complex structures is based on the following three observations:

- complex structures are two-dimensional in nature (this motivates our special interest in two-dimensional subalgebras);
- in each real two-dimensional algebra, complex structures perform in at most two different ways (cf. Section 2 and Theorems A–C in[2]));

- the algebra $\mathbb{C}$ of complex numbers is the only real two-dimensional algebra where equation $x^2 \circ y = -y$ is soluble for all y.

The presence of complex structures turns out to be responsible for:

- the solubility of polynomial equations with real coefficients (at least of degree three (see Theorem A in [2]);
- the existence of continuous deformations to $\mathbb{C}$ in the class of algebras respecting these structures (see Theorem A in [2]);
- the existence of bounded solutions to the Riccati equation in algebras (see [3]);
- the ellipticity of the operator of the Dirac operator in algebras (see [6]);

and many other matters that, at first glance, do not have any connection with complex structures.

A. Albert [1] considered complex structures for unital division algebras. H. Petersson [9] related Albert isotopy classes to several canonical representatives (the so-called unital hearts). An important class of the unital hearts provide the Clifford Algebras.

2.4. Euler–Jacobi formula

Let $F(x) = \{F_1(x) \cdots F_n(x)\}$ be a polynomial vector field and $\deg F_i = m_i$. Denote by Sol_F the set of roots of a system of n polynomial equations of degrees $m_1, \dots, m_n$ in n complex unknowns, $F_1 = \cdots = F_n = 0$. Assume that the set Sol_F contains exactly $\mu = m_1, \dots, m_n$ elements. In this case, the Jacobian of the system $J = \det \frac{\partial F}{\partial x}$ does not vanish on the set Sol_F. Then, for every polynomial P of degree less than $\sum_i m_i$, we have the following Euler–Jacobi formula:

$$\sum_{a \in Sol_F} \frac{P(a)}{J(a)} = 0. \tag{2.6}$$

2.5. Invariants and syzygies

The term “syzygies” was introduced in 1853 by Sylvester to denote some relations among invariants. He discovered that various invariants of the same form sometimes satisfy rational relations among them.

Using Bézout's theorem and [10], we can write the Euler–Jacobi formula for bilinearization $Q(x)$ in (1.1) as follows:

Theorem 2.3 **([7]).** *Let* $p_i = \big(x_1^{(i)}, \dots, x_n^{(i)}\big)$ *for* $i = 1, 2, \dots, M = 2^n - 1$ *be (complex) fixed points of the quadratic homogeneous polynomial vector field* $Q(x)$ *in* (1.1). *Denote by* γ_i *the numbers* $\gamma_i = 1/D\big(V(x) - Id\big)\big|_{x=p_i}$, *where* $D(\cdot)$ *stands for the Jacobian determinant. Then the following syzygies among* p_i, γ_i *hold:*

$$\sum\nolimits_{i=1}^{M} \gamma_i + (-1)^n = 0, \tag{2.7}$$

$$\sum\nolimits_{i=1}^{M} \gamma_i p_i^{\alpha_j} = 0, \quad |\alpha_j| = j, \quad j = 1, \dots, n-1, \tag{2.8}$$

where α *is a multi-index,* $x^\alpha = x_1^{\alpha_1} \cdot \; \cdots \; \cdot x_n^{\alpha_n}$.

Proof. By Bézout's theorem in the projective space $\mathbb{CP}^n$, there are no more than 2^n solutions to equation $Q(x) = \lambda x$, (counted according to their multiplicities) provided that the number of solutions is finite. Therefore, either there exist exactly $2^n - 1$ non-proportional non-zero fixed points $Q(p) = p$ or infinitely many.

Consider separately $p = 0$ and $p \neq 0$ for $Q(p) = p$:

$$J(0) = D(-Id) = (-1)^n, \quad \lim_{a \to 0} \frac{P(p)}{J(p)} = (-1)^n P(0). \tag{2.9}$$

Using these formulas, Theorem 2.3 follows from (2.6). □

3. Isotopy invariants

The notion of isotopy is a natural generalization of an algebra isomorphism and is defined as follows:

Definition 3.1. Two algebras $(G, \circ)$ and $(H, *)$ are called isotopic (or G is an isotope of H) if there are linear maps K, L, M from G to H, such that $K(x) \circ L(y) = M(x * y)$ for all x, y.

Clearly, isotopy is an equivalence relation. If $K = L = M$ one obtains an algebraic isomorphism. To determine the properties invariant under isotopy A is a problem of great importance.

4. Clifford algebras and bounded solutions

The following properties of Clifford algebras make them important for the qualitative study of DEs:

- Unitality – the DEs in unital algebras are of evolution type.
- Associativity – it makes clear the structure of the power series solution of the Dirac equations.
- The presence of a complex structure is responsible for the existence of a bounded solution to the Riccati type ODEs.

Many real life models reduce to DEs in Clifford algebras using isotopy. As is well known, any two Clifford algebras of the same dimension can be distinguished by the signature of the related quadratic form and this leads to allocation of a different isotopy class of Clifford algebras.

The following theorem relates the unboundedness of a quadratic polynomial differential system

$$\dot{x} = L(x) + Q(x) \tag{4.1}$$

to existence of a real fixed point p of the corresponding homogeneous polynomials differential system

$$\dot{y} = Q(y) \tag{4.2}$$

where x and y are real n-vectors, $L(x)$ and $Q(x)$ are real n-vector polynomial functions, $Q(x)$ is homogeneous of degree 2, and $L(x)$ is linear (cf. [4], p. 71).

Theorem 4.1. *If system* (4.2) *has a real fixed point* $H_m(p) = p$, *then system* (4.1) *is unbounded.*

Consider now the Volterra equation in $\mathbb{R}^n$:

$$\dot{x}_i = H_i(x), \quad H_i(x) = x_i\Big(x_i + \sum_{j\neq i} a_{ij}x_j\Big), \quad i = 1, 2, \ldots, n. \tag{4.3}$$

All the basis vectors $e_1, e_2, \ldots, e_n$ are fixed points of the quadratic system (4.3). Any coordinate plane $\mathbb{R}[x_i, x_j]$ is an invariant set of (4.3). Moreover, if $a_{ij}, a_{ji} \neq 1$ and $a_{ij}a_{ji} \neq 1$, then the additional fixed point

$$p_{ij} = \frac{1 - a_{ij}}{1 - a_{ij}a_{ji}} e_i + \frac{1 - a_{ji}}{1 - a_{ij}a_{ji}} e_j \tag{4.4}$$

lies in $\mathbb{R}[x_i, x_j]$.

Denote by γ_i, γ_j, γ_{ij} the same numbers, as quantities in Theorem 2.3, restricted on $\mathbb{R}[x_i, x_j]$ and computed at the fixed points e_i, e_j, p_{ij} in $\mathbb{R}[x_i, x_j]$ correspondingly. Then by straightforward computations:

$$\gamma_i = \frac{1}{a_{ji} - 1}, \quad \gamma_j = \frac{1}{a_{ij} - 1}, \quad \gamma_{ij} = \frac{1 - a_{ij}a_{ji}}{(1 - a_{ij})(1 - a_{ji})}. \tag{4.5}$$

Using results in [2] we can prove the following

Theorem 4.2. *If there exists such an* i, j *that two of the following three numbers* γ_i, γ_j *or* γ_{ij} *are positive then there exists a bounded solution to* (4.3).

In contrast to Theorem 4.2 for existence of a bounded solution to (4.3) it is necessary for an underlying algebra to have a bounded solution:

Theorem 4.3. [2] *Let* A *be a rank three algebra without* 2*-nilpotents. Then the Riccati equation* (4.3) *admits a non-trivial solution bounded on the whole axis iff* A *contains a complete complex structure.*

As a consequent result one can obtain that the Riccati equation (4.3) in Clifford algebra $Cl^{p,q}(R)$ always has a bounded solution.

5. Conclusions

Many well-known differential systems can be considered to be sets of differential equations operating in algebraic terms. As an example, we discuss different algebraic operations acting in subspaces of the solution space of a set of DEs.

Actually, the study of complex structures acts as a particular case of the general approach adopted in this paper. Namely, we are trying to study algebras from the viewpoint of the solubility of certain important polynomial equations with real constant coefficients. Solvability of such equations in algebras leads to the stability theory for ODE's (see [2], [3]), linear isotopy (see [1]), to mention a few. At the same time many algebraic properties (for example, the number of their solutions) are not homotopy invariants and the choice of a suitable language essentially depends on the properties themselves. In fact, the choice of an appropriate language means a certain classification of algebras.

Acknowledgment

The author would like to thank Prof. Irene Sabadini for her valuable comments and suggestions to improve the quality of the paper as well as for her attention and time.

References

[1] A. Albert, *Nonassociative algebras*, Ann. of Math. 43, 1942, 685–707.

[2] Z. Balanov and Y. Krasnov, *Complex Structures in Real Algebras. I. Two-Dimensional Commutative Case*, Communications in Algebra, 31(9), (2003), 4571–4609.

[3] Z. Balanov, Y. Krasnov, A. Kononovich, *Projective dynamics of homogeneous systems: local invariants, syzygies and global residue theorem*, in Proc. of the Edinburgh Math. Soc., 55(3), (2012), 577–589.

[4] C.S. Coleman, *Systems of differential equations without linear terms*, in *Nonlinear Differential Equations and Nonlinear Mechanics*, Academic Press, (1963), 445–453.

[5] W.A. Coppel, *A survey of quadratic systems*, J. Differential Equations 2 (1966), 293–304.

[6] Y. Krasnov, *Properties of ODEs and PDEs in Algebras*, Complex Analysis and Operator Theory, 7(3), 2013, 623–634.

[7] Y. Krasnov, A. Kononovich and G. Osharovich, *On a structure of the fixed point set of homogeneous maps*, Discrete & Continuous Dynamical Systems – Series S, AIMS Journal, 6(4), (2013), 1017–1027.

[8] L. Markus, *Quadratic differential equations and non-associative algebras*, Annals of Mathematics Studies, 45, 1960, 185–213.

[9] H. Petersson, *The classification of two-dimensional nonassociative algebras*, Result in Math. 37, 2000, 120–154.

[10] I.R. Shafarevich, *Basic Algebraic Geometry I*, Springer, (1994) 317.

Yakov Krasnov
Department of Mathematics
Bar-Ilan University
52900 Ramat-Gan, Israel
e-mail: krasnov@math.biu.ac.il

Hypercomplex Analysis: New Perspectives and Applications
Trends in Mathematics, 213–227

Differential Equations in Multicomplex Spaces

Daniele C. Struppa, Adrian Vajiac and Mihaela B. Vajiac

Abstract. We use Ehrenpreis' Fundamental Principle to give different representations of holomorphic functions in multicomplex spaces, and we give the first elements of a theory of constant coefficients differential equations, of finite and infinite order, in the multicomplex setting.

Mathematics Subject Classification (2010). 30A97, 34A20, 35C10.

Keywords. Multicomplex spaces, constant coefficients differential equations.

1. Introduction

Let $\mathbb{BC}_n$ be the ring of multicomplex numbers (we refer to Section 3 for the specific definitions, but we recall that $\mathbb{BC}_1$ is the field of complex numbers and $\mathbb{BC}_2$ is the ring of bicomplex numbers). The theory of holomorphic functions from $\mathbb{BC}_n$ to itself is by now well developed [6, 15, 27, 28], and it shows in particular that a function F is multicomplex holomorphic if and only if it is a complex holomorphic map from $\mathbb{C}^{2^n}$ to itself, and its 2^n complex components satisfy a system of first-order constant coefficients differential equations known as the multicomplex Cauchy–Riemann system.

In this paper we extend to the multicomplex setting some results from [24]. In particular we will show how to use the Fundamental Principle of Ehrenpreis to give a suitable representation for holomorphic functions in $\mathbb{BC}_n$, and in this way we will recover the idempotent representation for holomorphic multicomplex functions.

From this point of view the space of holomorphic multicomplex functions, considered as the kernel of a system of differential equations on the space of holomorphic maps in suitable dimensions, ends up being an Analytically Uniform Space in the sense of Ehrenpreis.

As such, we were also able to study ordinary differential equations of infinite order in such a space and provide an exponential representation for such solutions.

The plan of the paper is as follows. In Section 2 we recall the fundamental ideas on Analytically Uniform Spaces and we recall the Fundamental Theorem of

Ehrenpreis. In Section 3 we introduce bicomplex and more generally multicomplex spaces, and we give the basic properties of holomorphic functions in those settings. The next section is devoted to the use of the Fundamental Principle to obtain exponential representations for such holomorphic functions. The final section shows how to use these same ideas to begin a study of linear constant coefficients ordinary differential equations of both finite and infinite order.

2. Analytically uniform spaces

This section is preliminary to the rest of the paper and can be skipped by the reader who is familiar with the work of Ehrenpreis. Following [10], we consider a space W of (generalized) functions on $\mathbb{R}^n$ or $\mathbb{C}^n$. We will say that W is analytically uniform (AU) if such a space satisfies the following conditions:

a) The space W is the dual of a locally convex topological space W' and the topology of W is given by uniform convergence on bounded sets of W'.
b) For all $z \in \mathbb{C}$, the exponential function $x \rightarrowtail \exp(ix \cdot z)$ belongs to W and the resulting map from $\mathbb{C}$ to W is complex analytic. Moreover linear combinations of these exponentials are dense in W. For any $S \in W'$, we can therefore define its Fourier transform $\widehat{S} := S(\exp(ix \cdot z))$; such function is defined for all $z \in \mathbb{C}$, is entire, and it determines S.
c) There exists a family K of continuous positive functions on $\mathbb{C}$, which can take the value ∞, such that for all $F \in \widehat{W'}$ and all $k \in K$, $\dfrac{|F(z)|}{k(z)} \longrightarrow 0$ as $|z| \to \infty$. Moreover, the sets N_k of functions in $\widehat{W'}$ that are bounded by k form a fundamental system of neighborhoods of the origin in $\widehat{W'}$ (i.e., every N_k is a neighborhood of 0 and any neighborhood of 0 contains an N_k).

Any space satisfying conditions a), b), and c) is called an AU-space and K is said to be an AU-structure on it. The main examples of AU-spaces are the space $\mathcal{O}$ of entire functions, the space $\mathcal{E}$ of infinitely differentiable functions, the space $\mathcal{D}'$ of distributions, and the space $\mathcal{D}'_F$ of distributions of finite order. A classical example of a space which is surprisingly not AU is the space of real analytic functions. It is also known that the space of solutions of a system of linear constant coefficients partial differential equations in any AU-space is still an AU-space.

The reason for the importance of these spaces lies in the fact that solutions of systems of linear constant coefficients differential equations in these spaces can be given a very important exponential representation known as the Fundamental Principle of Ehrenpreis–Palamodov. Strictly speaking, this is true only for a special subset of such spaces (the so-called Product Localizeable Analytically Uniform Spaces) but we will not give the additional technical details that are not necessary for the rest of this paper. We refer the reader to, e.g., [2, 9, 10, 23].

This Fundamental Principle can be stated, at least in a special case, as follows:

Theorem 2.1. *Let $P_1, \ldots, P_r$ be polynomials in n variables, and let $D = \left(-i\frac{\partial}{\partial x_1}, \ldots, -i\frac{\partial}{\partial x_n}\right)$. Then there exist algebraic varieties $V_1, \ldots, V_t$ in $\mathbb{C}^n$ and differential operators $\partial_1, \ldots, \partial_t$ with polynomials coefficients, such that every $f \in \mathcal{C}^\infty(\mathbb{R}^n)$ satisfying the system*

$$P_1(D)f = \cdots = P_r(D)f = 0, \tag{1}$$

can be represented as

$$f(x) = \sum_{\ell=1}^{t} \int_{V_\ell} \partial_\ell(e^{\mathbf{i}x\cdot z})d\nu_\ell(z)\,, \tag{2}$$

where $d\nu_\ell(z)$ are suitable Radon measures supported on V_ℓ. The collection

$$V = \{(V_1, \partial_1); \ldots; (V_t, \partial_t)\}$$

is called a multiplicity variety.

Remark 2.2. If the system (1) above consists of only one equation, then $\partial_1, \ldots, \partial_t$ have constant coefficients. The operators $\partial_1, \ldots, \partial_t$ are called, in Palamodov's terminology [14], Noetherian operators because their construction relies essentially on a theorem of M. Noether on a membership criterion for polynomial submodules. The nature of the original proof of the Fundamental Principle is essentially existential and therefore the question of the explicit construction of such operators is of great interest (see, e.g., [8] and the references therein).

Remark 2.3. As we indicated, the statement above is a special case of the Fundamental Principle. To begin with, the space of infinitely differentiable functions can be replaced with any other AU-space, but in that case the representation (2) need to be interpreted in the sense of generalized functions. Most important, for our specific concerns, the system (1) can actually be generalized to be a matrix system, so that the unknown is not a single function f but actually a vector $\vec{f}$ of functions satisfying a matrix system of differential equations. This is important to remember because the multicomplex holomorphic functions fall exactly in this category.

3. Multicomplex spaces

Without giving many details (for which we refer the reader to the fairly comprehensive recent references [6, 12, 27, 28]), we will simply say that the space $\mathbb{BC}_n$ of multicomplex numbers is the space generated over the reals by n commuting imaginary units. The algebraic properties of this space and analytic properties of multi-complex-valued functions defined on $\mathbb{BC}_n$ has been studied in [28]. It is worth noting also that the algebraic properties of the space $\mathbb{BC}_2$ and the properties of its holomorphic functions have been discussed before in [6, 7, 27], using computational algebra techniques. Other related references include [4, 16, 17, 18, 19, 20, 21, 22].

In the case of only one imaginary unit, denoted by $\mathbf{i}_1$, the space $\mathbb{BC}_1$ is the usual complex plane $\mathbb{C}$. Since, in what follows, we will have to work with different complex planes, generated by different imaginary units, we will denote such a space also by $\mathbb{C}(\mathbf{i}_1)$, in order to clarify which is the imaginary unit used in the space itself.

The next case occurs when we have two commuting imaginary units $\mathbf{i}_1$ and $\mathbf{i}_2$. This yields the bicomplex space $\mathbb{BC}_2$. This space has extensively been studied in [4, 15, 16], and more recently by the authors in [6, 7, 27], where it is referred to as as $\mathbb{BC}$.

Because of the various units in $\mathbb{BC}_2$, we have several different conjugations that can be defined naturally. Let therefore consider a bicomplex number Z, and let us write it both in terms of its complex coordinates, $Z = z_1 + \mathbf{i}_2 z_2$ with $z_1, z_2 \in \mathbb{C}(\mathbf{i}_1)$, and in terms of its real coordinates $Z = x_1 + \mathbf{i}_1 y_1 + \mathbf{i}_2 x_2 + \mathbf{i}_1\mathbf{i}_2 y_2$ with $x_\ell, y_\ell \in \mathbb{R}$. Note that $\mathbf{k}_{12} := \mathbf{i}_1\mathbf{i}_2$ is a hyperbolic unit, i.e., it squares to 1.

For a bicomplex number defined as $Z = z_1 + \mathbf{i}_2 z_2$ we mention the existence of three conjugations in $\mathbb{C}(\mathbf{i}_1)$: we will denote by $\overline{Z}^{\mathbf{i}_1}$, the one corresponding to the involution $\mathbf{i}_1 \mapsto -\mathbf{i}_1$, by $\overline{Z}^{\mathbf{i}_2}$, the one corresponding to the involution $\mathbf{i}_2 \mapsto -\mathbf{i}_2$, and finally by $\overline{Z}^{\mathbf{i}_1\mathbf{i}_2}$, that corresponding to both involutions. All these conjugations have been carefully described in [6, 7].

A function $F : \mathbb{BC}_2 \to \mathbb{BC}_2$ can be written as $F = f_1 + \mathbf{i}_2 f_2$, where $f_1, f_2 : \mathbb{BC}_2 \to \mathbb{BC}_1 = \mathbb{C}(\mathbf{i}_1)$. Following the notations from [6], we introduce the following bicomplex differential operators, written in the standard basis of $\mathbb{BC}_2$, seen as $\mathbb{C}^2$ in the variables z_1 and z_2:

$$\begin{aligned}
\frac{\partial}{\partial Z} &:= \frac{1}{2}\left(\frac{\partial}{\partial z_1} - \mathbf{i}_2\frac{\partial}{\partial z_2}\right), & \frac{\partial}{\partial \overline{Z}^{\mathbf{i}_2}} &:= \frac{1}{2}\left(\frac{\partial}{\partial z_1} + \mathbf{i}_2\frac{\partial}{\partial z_2}\right), \\
\frac{\partial}{\partial \overline{Z}^{\mathbf{i}_1}} &:= \frac{1}{2}\left(\frac{\partial}{\partial \bar{z}_1} - \mathbf{i}_2\frac{\partial}{\partial \bar{z}_2}\right), & \frac{\partial}{\partial \overline{Z}^{\mathbf{i}_1\mathbf{i}_2}} &:= \frac{1}{2}\left(\frac{\partial}{\partial \bar{z}_1} + \mathbf{i}_2\frac{\partial}{\partial \bar{z}_2}\right),
\end{aligned} \tag{3}$$

where $\frac{\partial}{\partial z_1}, \frac{\partial}{\partial z_2}$ are the usual complex derivatives. These differential operators also have multiple descriptions, for which we refer the reader to [6, 7].

$\mathbb{BC}_2$ is not a division algebra, and it has two distinguished zero divisors, $\mathbf{e}_{12}$ and $\bar{\mathbf{e}}_{12}$, which are idempotent, linearly independent over the reals, and mutually annihilating with respect to the bicomplex multiplication:

$$\mathbf{e}_{12} := \frac{1 + \mathbf{i}_1\mathbf{i}_2}{2}, \qquad \bar{\mathbf{e}}_{12} := \frac{1 - \mathbf{i}_1\mathbf{i}_2}{2}.$$

Just like $\{1, \mathbf{i}_2\}$, they form a basis of the complex algebra $\mathbb{BC}_2$, which is called the *idempotent basis*. If we define the following complex variables in $\mathbb{C}(\mathbf{i}_1)$:

$$\beta_1 := z_1 - \mathbf{i}_1 z_2, \qquad \beta_2 := z_1 + \mathbf{i}_1 z_2,$$

the *idempotent representations* for $Z = z_1 + \mathbf{i}_2 z_2$ and its conjugates are given by

$$Z = \beta_1 \mathbf{e}_{12} + \beta_2 \overline{\mathbf{e}}_{12}, \qquad \overline{Z}^{\mathbf{i}_1} = \overline{\beta_2}^{\mathbf{i}_1} \mathbf{e}_{12} + \overline{\beta_1}^{\mathbf{i}_1} \overline{\mathbf{e}}_{12},$$
$$\overline{Z}^{\mathbf{i}_2} = \beta_2 \mathbf{e}_{12} + \beta_1 \overline{\mathbf{e}}_{12}, \qquad \overline{Z}^{\mathbf{i}_1\mathbf{i}_2} = \overline{\beta_1}^{\mathbf{i}_1} \mathbf{e}_{12} + \overline{\beta_2}^{\mathbf{i}_1} \overline{\mathbf{e}}_{12}.$$

In the idempotent representation, the bicomplex differential operators defined above become:

$$\frac{\partial}{\partial Z} = \frac{\partial}{\partial \beta_1}\mathbf{e}_{12} + \frac{\partial}{\partial \beta_2}\overline{\mathbf{e}}_{12}, \qquad \frac{\partial}{\partial \overline{Z}^{\mathbf{i}_2}} = \frac{\partial}{\partial \beta_2}\mathbf{e}_{12} + \frac{\partial}{\partial \beta_1}\overline{\mathbf{e}}_{12},$$
$$\frac{\partial}{\partial \overline{Z}^{\mathbf{i}_1}} = \frac{\partial}{\partial \overline{\beta_2}^{\mathbf{i}_1}}\mathbf{e}_{12} + \frac{\partial}{\partial \overline{\beta_1}^{\mathbf{i}_1}}\overline{\mathbf{e}}_{12}, \qquad \frac{\partial}{\partial \overline{Z}^{\mathbf{i}_1\mathbf{i}_2}} = \frac{\partial}{\partial \overline{\beta_1}^{\mathbf{i}_1}}\mathbf{e}_{12} + \frac{\partial}{\partial \overline{\beta_2}^{\mathbf{i}_1}}\overline{\mathbf{e}}_{12}.$$

The following notion of a bicomplex derivative is introduced:

Definition 3.1. Let Ω be an open set in $\mathbb{BC}_2$ and let $Z_0 \in \Omega$. A function $F : \Omega \to \mathbb{BC}_2$ is called bicomplex derivable at Z_0 if the limit

$$\lim_{Z \to Z_0} (Z - Z_0)^{-1} (F(Z) - F(Z_0))$$

exists, for all Z in Ω such that $Z - Z_0$ is invertible, i.e., it is not a divisor of zero. When the limit exists, we will call it $F'(Z_0)$, and we will say that the function F has derivative equal to $F'(Z_0) \in \mathbb{BC}_2$ at Z_0.

Note that the limit in the definition above avoids the divisors of zero in $\mathbb{BC}_2$, which are the union of the two ideals generated by $\mathbf{e}_{12}$ and $\overline{\mathbf{e}}_{12}$, the so-called *cone of singularities*. The set of zero divisors together with 0 is usually denoted by $\mathfrak{S}_0$.

Functions which admit bicomplex derivative at each point in their domain are called bicomplex holomorphic, and it can be shown that this is equivalent to require that they admit a power series expansion in Z [15, Definition 15.2]. However, there are more equivalent statements of bicomplex holomorphy [27]. For example:

Theorem 3.2. *Let Ω be an open set in $\mathbb{BC}_2$ and let $F = f_1 + \mathbf{i}_2 f_2 : \Omega \to \mathbb{BC}_2$ be of class $C^1(\Omega)$. Then F is bicomplex holomorphic on Ω if and only if:*

1. *f_1 and f_2 are complex holomorphic in both complex $\mathbb{C}(\mathbf{i}_1)$ variables z_1 and z_2.*
2. $\dfrac{\partial f_1}{\partial z_1} = \dfrac{\partial f_2}{\partial z_2}$ *and* $\dfrac{\partial f_2}{\partial z_1} = -\dfrac{\partial f_1}{\partial z_2}$ *on Ω; these equations are called the* complex *Cauchy–Riemann conditions.*

Moreover,

$$F' = \frac{1}{2}\frac{\partial F}{\partial Z} = \frac{\partial f_1}{\partial z_1} + \mathbf{i}_2 \frac{\partial f_2}{\partial z_1} = \frac{\partial f_2}{\partial z_2} - \mathbf{i}_2 \frac{\partial f_1}{\partial z_2},$$

and $F'(Z)$ is invertible if and only if the corresponding Jacobian is non-zero.

In the idempotent basis, we have the following characterization of bicomplex holomorphy:

Theorem 3.3. *A bicomplex function $F : \Omega \to \mathbb{BC}_2$ is bicomplex holomorphic if and only if*

$$F = G_1(\beta_1)\mathbf{e}_{12} + G_2(\beta_2)\overline{\mathbf{e}}_{12}\,,$$

where G_1 is a complex holomorphic function on the complex domain Ω_1 defined by $\Omega_1 \cdot \mathbf{e}_{12} := \Omega \cdot \mathbf{e}_{12}$, and G_2 is a complex holomorphic function on the complex domain Ω_2 defined by $\Omega_2 \cdot \overline{\mathbf{e}}_{12} := \Omega \cdot \overline{\mathbf{e}}_{12}$.

An immediate consequence of this theorem is that a bicomplex holomorphic function F defined on Ω extends to a bicomplex holomorphic function defined on $\widetilde{\Omega} := \Omega_1 \cdot \mathbf{e}_{12} + \Omega_2 \cdot \overline{\mathbf{e}}_{12}$, which strictly includes Ω. Domains of this form are called bicomplex domains.

A further interesting characterization of holomorphicity in $\mathbb{BC}_2$ is the following result from [6]:

Theorem 3.4. *Let $\Omega \subseteq \mathbb{BC}_2$ be an open set and let $F : \Omega \to \mathbb{BC}_2$ be of class C^1 on Ω. Then F is bicomplex holomorphic if and only if F is $\overline{Z}^{\mathbf{i}_1}, \overline{Z}^{\mathbf{i}_2}, \overline{Z}^{\mathbf{i}_1\mathbf{i}_2}$-regular, i.e.,*

$$\frac{\partial F}{\partial \overline{Z}^{\mathbf{i}_1}} = \frac{\partial F}{\partial \overline{Z}^{\mathbf{i}_2}} = \frac{\partial F}{\partial \overline{Z}^{\mathbf{i}_1\mathbf{i}_2}} = 0.$$

Note that both Theorems 3.2 and 3.4 indicate that holomorphic functions on bicomplex variables can be seen as solutions of overdetermined systems of differential equations with constant coefficients. We have exploited this particularity in [6, 7, 27].

We now turn to the definition of the multicomplex spaces, $\mathbb{BC}_n$, for values of $n \geq 2$. These spaces are defined by taking n commuting imaginary units $\mathbf{i}_1, \mathbf{i}_2, \ldots, \mathbf{i}_n$, i.e., $\mathbf{i}_a^2 = -1$, and $\mathbf{i}_a\mathbf{i}_b = \mathbf{i}_b\mathbf{i}_a$ for all $a, b = 1, \ldots, n$. Since the product of two commuting imaginary units is a hyperbolic unit, and since the product of an imaginary unit and a hyperbolic unit is an imaginary unit, we see that these units will generate a set $\mathfrak{A}_n$ of 2^n units, 2^{n-1} of which are imaginary and 2^{n-1} of which are hyperbolic units. Then the algebra generated over the real numbers by $\mathfrak{A}_n$ is the multicomplex space $\mathbb{BC}_n$ which forms a ring under the usual addition and multiplication operations. As in the case $n = 2$, the ring $\mathbb{BC}_n$ can be represented as a real algebra, so that each of its elements can be written as $Z = \sum_{I \in \mathfrak{A}_n} Z_I I$, where Z_I are real numbers.

In particular, following [15], it is natural to define the n-dimensional multicomplex space as follows:

$$\mathbb{BC}_n := \left\{ Z_n = Z_{n-1,1} + \mathbf{i}_n Z_{n-1,2} \,\middle|\, Z_{n-1,1}, Z_{n-1,2} \in \mathbb{BC}_{n-1} \right\}$$

with the natural operations of addition and multiplication. Since $\mathbb{BC}_{n-1}$ can be defined in a similar way using the $\mathbf{i}_{n-1}$ unit and multicomplex elements of $\mathbb{BC}_{n-2}$, we recursively obtain, at the kth level:

$$Z_n = \sum_{|I|=n-k} \prod_{t=k+1}^{n} (\mathbf{i}_t)^{\alpha_t - 1} Z_{k,I}$$

where $Z_{k,I} \in \mathbb{BC}_k$, $I = (\alpha_{k+1}, \ldots, \alpha_n)$, and $\alpha_j \in \{1, 2\}$.

Because of the existence of n imaginary units, we can define multiple types of conjugations. Following [30] we define:

$$\overline{Z_n}^{\mathbf{i}_l} = \sum_{|I|=n-k} \sum_{i=k+1}^{n} \delta_{l,i} (-1)^{\alpha_l - 1} c_k Z_{k,I} + \sum_{|I|=n-k} \sum_{i=1}^{k} \delta_{l,i} c_k \overline{Z_{k,I}}^{\mathbf{i}_l}.$$

Just as in the case of $\mathbb{BC}_2$ we have idempotent bases in $\mathbb{BC}_n$, that will be organized at each "nested" level $\mathbb{BC}_k$ inside $\mathbb{BC}_n$ as follows. Denote by

$$\mathbf{e}_{kl} := \frac{1 + \mathbf{i}_k \mathbf{i}_l}{2}, \qquad \bar{\mathbf{e}}_{kl} := \frac{1 - \mathbf{i}_k \mathbf{i}_l}{2}.$$

Consider the following sets:

$$\begin{aligned} S_1 &:= \{\mathbf{e}_{n-1,n}, \bar{\mathbf{e}}_{n-1,n}\}, \\ S_2 &:= \{\mathbf{e}_{n-2,n-1} \cdot S_1, \bar{\mathbf{e}}_{n-2,n-1} \cdot S_1\}, \\ &\vdots \\ S_{n-1} &:= \{\mathbf{e}_{12} \cdot S_{n-2}, \bar{\mathbf{e}}_{12} \cdot S_{n-2}\}. \end{aligned}$$

At each stage k, the set S_k has 2^k idempotents. It is possible to immediately verify the following

Proposition 3.5. *In each set S_k, the product of any two idempotents is zero.*

We have several idempotent representations of $Z_n \in \mathbb{BC}_n$, as follows.

Theorem 3.6. *Any $Z_n \in \mathbb{BC}_n$ can be written as:*

$$Z_n = \sum_{j=1}^{2^k} Z_{n-k,j} \mathbf{e}_j,$$

where $Z_{n-k,j} \in \mathbb{BC}_{n-k}$ and $\mathbf{e}_j \in S_k$.

Due to the fact that the product of two idempotents is 0 at each level S_k, we will have many zero divisors in $\mathbb{BC}_n$ organized in "singular cones". The topology of the space is difficult, but just like in the case of $\mathbb{BC}_2$ we can circumvent this by avoiding the zero divisors to define the derivative of a multicomplex function as follows.

Definition 3.7. Let Ω be an open set of $\mathbb{BC}_n$ and let $Z_{n,0} \in \Omega$. A function $F : \Omega \to \mathbb{BC}_n$ has a multicomplex derivative at $Z_{n,0}$ if

$$\lim_{Z_n \to Z_{n,0}} (Z_n - Z_{n,0})^{-1} (F(Z_n) - F(Z_{n,0})) =: F'(Z_{n,0}),$$

exists whenever $Z_n - Z_{n,0}$ is invertible in $\mathbb{BC}_n$.

Just as in the case of $\mathbb{BC}_2$, functions which admit a multicomplex derivative at each point in their domain are called multicomplex holomorphic, and it can be shown that this is equivalent to require that they admit a power series expansion in Z_n [15, Section 47].

We will denote by $\mathcal{O}(\mathbb{BC}_n)$ the space of multicomplex holomorphic functions. A multicomplex holomorphic function $F \in \mathcal{O}(\Omega)$, where $\Omega \subset \mathbb{BC}_n$, can be split as $F = U + \mathbf{i}_n V$, where U, V are holomorphic functions of two $\mathbb{BC}_{n-1}$ variables. As in the case $n = 2$, there is an equivalent notion of multicomplex holomorphicity, which is more suitable for our computational algebraic purposes, and the following theorem can be proved in a similar fashion as its correspondent for the case $n = 2$ (the differential operators that appear in the statement are defined in detail later on in the paper, but it is not difficult to imagine their actual definition).

Theorem 3.8. *Let Ω be an open set in $\mathbb{BC}_n$ and let $F : \Omega \to \mathbb{BC}_n$ be such that $F = U + \mathbf{i}_n V \in C^1(\Omega)$. Then F is multicomplex holomorphic if and only if:*

1. *U and V are multicomplex holomorphic in both multicomplex $\mathbb{BC}_{n-1}$ variables $Z_{n-1,1}$ and $Z_{n-1,2}$.*
2. $\dfrac{\partial U}{\partial Z_{n-1,1}} = \dfrac{\partial V}{\partial Z_{n-1,2}}$ *and* $\dfrac{\partial V}{\partial Z_{n-1,1}} = -\dfrac{\partial U}{\partial Z_{n-1,2}}$ *on Ω; these equations are called the* multicomplex *Cauchy–Riemann conditions.*

There are several other ways to identify multicomplex holomorphic functions in $\mathbb{BC}_n$, as there are many idempotent representations with respect to each level S_k. For example, for $k = n-1$, we obtain the following characterization of $\mathbb{BC}_n$-holomorphy:

Theorem 3.9. *A multicomplex function $F : \Omega \to \mathbb{BC}_n$ is $\mathbb{BC}_n$-holomorphic if and only if*

$$F = \sum_{\ell=1}^{2^{n-1}} G_\ell(\beta_\ell)\mathbf{e}_\ell\,,$$

where $\mathbf{e}_\ell \in S_{n-1}$ and G_ℓ are complex holomorphic functions on the complex domains Ω_ℓ defined by $\Omega_\ell \cdot \mathbf{e}_\ell := \Omega \cdot \mathbf{e}_\ell$, for all $\ell = 1, \ldots, 2^{n-1}$.

As before, a multicomplex holomorphic function F defined on Ω extends to a multicomplex holomorphic function defined on

$$\widetilde{\Omega} := \sum_{\ell=1}^{2^{n-1}} \Omega_\ell \cdot \mathbf{e}_\ell$$

which strictly includes Ω. Domains of this form are called multicomplex domains.

4. Representations of bicomplex and multicomplex holomorphic functions

In the theory of bicomplex functions, the complex light cone in two dimensions, i.e., $\Gamma = \{(z_1, z_2) \,|\, z_1^2 + z_2^2 = 0\}$, plays a very important role as it coincides with the set $\mathfrak{S}_0$ of zero-divisors in $\mathbb{BC}_2$ (together with 0). The complex Laplacian plays a prominent role because it can be factored as the product of two linear operators, one of which is the bicomplex differentiation.

Indeed

$$\Delta_{\mathbb{C}^2(\mathbf{i}_1)} = 4\frac{\partial}{\partial Z}\frac{\partial}{\partial Z^{\dagger}}, \tag{4}$$

where the operators act on $\mathbb{BC}_2$-valued functions holomorphic in the sense of the complex variables z_1, z_2. Hence, the theory of $\mathbb{BC}_2$-holomorphic functions can be seen as the function theory for $\mathbb{C}(\mathbf{i}_1)$-complex Laplacian. This factorization allows us to establish direct relations between $\mathbb{BC}_2$-holomorphic functions and complex harmonic functions, that is, null solutions to the operator $\Delta_{\mathbb{C}^2(\mathbf{i})}$.

Denote by $\Omega \to \mathcal{B}_2(\Omega)$ the sheaf of bicomplex holomorphic functions (see [6], [7], [27], [28]). Then, if we use the notation $\mathcal{S}^P$ to denote the sheaf of $\mathcal{S}$ solutions to the differential operator P we have that:

$$\mathcal{B}_2(\mathbb{BC}) \simeq \left[\mathcal{O}(\mathbb{C}^2) \times \mathcal{O}(\mathbb{C}^2)\right]^{\partial_{Z^{\mathbf{i}_2}}}$$

where we recall that

$$\partial_{Z^{\mathbf{i}_2}} := \frac{\partial}{\partial Z^{\mathbf{i}_2}} = \frac{\partial}{\partial z_1} + \mathbf{i}_2\frac{\partial}{\partial z_2}$$

is the bicomplex differential operator acting on bicomplex functions $F = f_1 + \mathbf{i}_2 f_2$ by

$$f_1 + \mathbf{i}_2 f_2 \mapsto (\partial_{z_1} f_1 - \partial_{z_2} f_2, \partial_{z_2} f_1 + \partial_{z_1} f_2).$$

We now apply the formalism described thoroughly in [5] and applied to the bicomplex case in [6]. This formalism defines an associated matrix representation to a linear partial differential operator and its associated variety in the sense of Ehrenpreis. We obtain the matrix representation of $\partial_{Z^{\mathbf{i}_2}}$ as:

$$\begin{pmatrix} \theta_1 & -\theta_2 \\ \theta_2 & \theta_1 \end{pmatrix}$$

and its associated variety as

$$V = \{\Theta = (\theta_1, \theta_2) \in \mathbb{C}^2 \mid \theta_1^2 + \theta_2^2 = 0\}.$$

The variety V splits into a union $V_1 \cup V_2$, where $V_1 = \{\theta_1 - \mathbf{i}_1\theta_2 = 0\}$ and $V_2 = \{\theta_1 + \mathbf{i}_1\theta_2 = 0\}$. Using the Fundamental Principle for $f_1(z_1, z_2)$, we write:

$$\begin{aligned} f_1(z_1, z_2) &= \int_{\theta_1^2+\theta_2^2=0} e^{\mathbf{i}_1(z_1,z_2)\cdot(\theta_1,\theta_2)}\, d\theta_1\, d\theta_2 \\ &= \int_{\theta_1-\mathbf{i}_1\theta_2=0} e^{\mathbf{i}_1(z_1,z_2)\cdot(t,\mathbf{i}_1 t)}\, dt + \int_{\theta_1+\mathbf{i}_1\theta_2=0} e^{\mathbf{i}_1(z_1,z_2)\cdot(t,-\mathbf{i}_1 t)}\, dt \\ &= \int_{\mathbb{C}(\mathbf{i}_1)} e^{\mathbf{i}_1 t(z_1-\mathbf{i}_1 z_2)}\, dt + \int_{\mathbb{C}(\mathbf{i}_1)} e^{\mathbf{i}_1 t(z_1+\mathbf{i}_1 z_2)}\, dt \\ &= g_1(z_1 - \mathbf{i}_1 z_2) + g_2(z_1 + \mathbf{i}_1 z_2). \end{aligned}$$

Similarly,

$$f_2(z_1, z_2) = h_1(z_1 - \mathbf{i}_1 z_2) + h_2(z_1 + \mathbf{i}_1 z_2).$$

If we write $\beta_1 = z_1 - \mathbf{i}_1 z_2$ and $\beta_2 = z_1 + \mathbf{i}_1 z_2$, then the first equation in the Cauchy–Riemann system is enough to imply

$$h_1(\beta_1) = \mathbf{i}_1 g_1(\beta_1),$$
$$h_2(\beta_2) = -\mathbf{i}_1 g_2(\beta_2),$$

up to a complex constant. Using the two representations for f_1, f_2, we obtain

$$\begin{aligned} F = f_1 + \mathbf{i}_2 f_2 &= g_1(\beta_1)(1 + \mathbf{i}_1\mathbf{i}_2) + g_2(\beta_2)(1 - \mathbf{i}_1\mathbf{i}_2) \\ &= 2g_1(\beta_1)\mathbf{e}_{12} + 2g_2(\beta_2)\overline{\mathbf{e}}_{12}. \end{aligned}$$

The Fundamental Principle therefore captures the fact that in the idempotent representation, bicomplex holomorphic functions have the form above, where $g_1 = g_1(\beta_1)$ is a holomorphic function of one complex variable β_1, and $g_2 = g_2(\beta_2)$ is a holomorphic function of one complex variable β_2 and we recover the familiar result found in [6, 7, 15]:

Theorem 4.1. *If F is a bicomplex holomorphic function then it can be written as: $F = 2g_1(\beta_1)\mathbf{e}_{12} + 2g_2(\beta_2)\overline{\mathbf{e}}_{12}$. where g_1, g_2 are holomorphic functions of a single complex variable.*

The same result is obtained if we use directly the idempotent representation of bicomplex numbers and functions: write $Z = \beta_1\mathbf{e}_{12} + \beta_2\overline{\mathbf{e}}_{12}$ and $F = u_1\mathbf{e}_{12} + u_2\overline{\mathbf{e}}_{12}$, where

$$\beta_1 = z_1 - \mathbf{i}_1 z_2 \quad \text{and} \quad \beta_2 = z_1 + \mathbf{i}_1 z_2$$

are $\mathbb{C}(\mathbf{i}_1)$-complex numbers, and

$$u_1 := f_1 - \mathbf{i}_1 f_2 \quad \text{and} \quad u_2 := f_1 + \mathbf{i}_1 f_2$$

are complex functions in β_1 and β_2. Then the operator $\partial_{\overline{Z}^{\mathbf{i}_2}}$ is given by

$$\partial_{\overline{Z}^{\mathbf{i}_2}} = \partial_{\beta_2}\mathbf{e}_{12} + \partial_{\beta_1}\overline{\mathbf{e}}_{12}$$

acting on $F = (u_1, u_2)$.

The varieties are

$$V_1 = \{\beta_1 = 0\} \quad \text{and} \quad V_2 = \{\beta_2 = 0\},$$

which gives directly the same result as in the standard basis:

$$F = u_1(\beta_1)\mathbf{e}_{12} + u_2(\beta_2)\overline{\mathbf{e}}_{12},$$

where $u_1 = 2g_1$ and $u_2 = 2g_2$.

Note that in the idempotent coordinates, we have:

$$\mathcal{B}_2(\mathbb{BC}) \simeq \left[\mathcal{O}(\mathbb{C}^2) \times \mathcal{O}(\mathbb{C}^2)\right]^{\partial_{\beta_2}\mathbf{e}_{12} + \partial_{\beta_1}\overline{\mathbf{e}}_{12}}.$$

In an analogous way we can analyze the multicomplex case.

5. Differential equations of finite and infinite order in $\mathbb{BC}_2$ and $\mathbb{BC}_n$

In this section we show that, quite simply, linear constant coefficients differential equations on multicomplex holomorphic functions behave exactly as one would expect. We therefore begin with a simple result that gives a representation for the solutions of such homogeneous equations.

Theorem 5.1. *Let $P(Z)$ be a polynomial in $\mathbb{BC}_2$, where $P(Z) = \sum_{k=0}^{n} a_k Z^k$, with all zeroes of multiplicity 1. The there exist points $\nu_1, \dots, \nu_\ell, \dots, \nu_t$ such that any solution of the equation $P(D)F = 0$ where F is a bicomplex holomorphic function and $D = \partial_Z$ can be written as:*

$$F(Z) = \sum_{\ell=1}^{t} c_\ell (e^{\nu_\ell Z}),$$

where ν_ℓ are the zeroes of P, and the c_ℓ are suitable bicomplex numbers.

If P has zeroes $\nu_1, \dots, \nu_\ell, \dots, \nu_t$ of multiplicities m_ℓ respectively, then F becomes:

$$F(Z) = \sum_{\ell=1}^{t} p_\ell(Z)(e^{\nu_\ell Z}),$$

where $p_\ell(Z)$ are suitable bicomplex polynomials with degree $m_\ell - 1$.

Proof. We start by re-writing the polynomial $P(Z)$ and bicomplex holomorphic function F in the idempotent representation, $P(Z) = P_1(\beta_1)e + P_2(\beta_2)\mathbf{e}^*$ and $F(Z) = \Phi(\beta_1)e + \Psi(\beta_2)e^*$. Note that only a bicomplex holomorphic function can be written this way, otherwise Φ and Ψ would be functions of both β_1, β_2. Then the equation $P(D)F = 0$ is equivalent to the system:

$$P_1(\partial_{\beta_1})\Phi(\beta_1) = 0,$$
$$P_2(\partial_{\beta_2})\Psi(\beta_2) = 0.$$

The result follows from Euler's Principle for complex numbers applied to the polynomials P_1 and P_2. The varieties V_i for P are obtained from the corresponding zeroes of P_1 and P_2. □

This representation is consistent with the general Ehrenpreis' Fundamental Principle. Indeed, a bicomplex holomorphic function can be interpreted as a map whose domain is $\mathbb{R}^4$. In order for the map to be bicomplex holomorphic, its components need to be holomorphic and satisfy the bicomplex Cauchy–Riemann system. With the request that such a function is in the kernel of an additional differential equation, we now see that the variety that supports the function is discrete, and that is reflected in the theorem we just proved.

The same result holds for a polynomial of a multi-complex variable $P(Z)$ with $Z \in \mathbb{BC}_n$.

Theorem 5.2. *Let $P(Z)$ be a polynomial in $\mathbb{BC}_n$, where $P(Z) = \sum_{k=0}^{n} a_k Z^k$. Then any solution of the equation $P(D)F = 0$ where F is a bicomplex holomorphic function and $D = \partial_Z$ can be written as:*

$$F(Z) = \sum_{\ell=1}^{t} p_\ell(Z)(e^{\nu_\ell Z}),$$

where the ν_ℓ are the zeroes of P, and $p_\ell(Z)$ are polynomials of degree $m_\ell - 1$, where m_ℓ is the multiplicity of ν_ℓ.

The proof is similar to the proof for a simple bicomplex variable, it involves reducing the bicomplex holomorphic function F to its 2^n components in the idempotent representation and we shall leave it to the reader.

We now extend the results above to the case of infinite-order differential equations, by considering a series of a bicomplex variables such as $f(Z) = \sum_{k=0}^{\infty} a_k Z^k$ (analogous notations can be used in the case of $\mathbb{BC}_n$). As in the complex case we introduce the notion of infraexponential type functions:

Definition 5.3. Let f be a function from $\mathbb{BC}_2$ to $\mathbb{BC}_2$, which is $\mathbb{BC}_2$-holomorphic. We say that f is of infraexponential type if for all $\epsilon > 0$ there exists an A_ϵ such that $|f(Z)| \leq A_\epsilon e^{\epsilon |Z|}$. We will denote the space of infraexponential type functions by $\mathrm{Exp}_0(\mathbb{BC}_2)$.

The space of infraexponential type functions is a proper subspace of the space of functions of exponential type defined by

$$\mathrm{Exp}(\mathbb{BC}_2) := \{f \in \mathcal{B}(\mathbb{BC}_2) \,|\, |f(Z)| \leq A \exp(B|Z|), \text{ for some } A, B > 0\}.$$

One can easily prove the following lemma:

Lemma 5.4. *If f is a bicomplex holomorphic function in $\mathbb{BC}_2$ then $f \in \mathrm{Exp}_0(\mathbb{BC}_2)$ if and only if its components ϕ, ψ in the idempotent representation are in $\mathrm{Exp}_0(\mathbb{C})$. Similarly, $f \in \mathrm{Exp}(\mathbb{BC}_2)$ if and only if its components in the idempotent representation ϕ, ψ are in $\mathrm{Exp}(\mathbb{C})$.*

Definition 5.5. Let $P(Z) = \sum_{n=0}^{\infty} A_n Z^n$ be a function from $\mathbb{BC}_2$ to $\mathbb{BC}_2$. We say that $P\left(\frac{\partial}{\partial Z}\right)$ is a differential operator of infinite order if and only if $\lim_{n \longrightarrow \infty} n \sqrt[n]{|A_n|} = 0$.

We have the following two equivalence lemmas:

Lemma 5.6. *The operator $P(Z) = \sum_{n=0}^{\infty} A_n Z^n$ is of infinite order if and only if its components P_1, P_2 in the idempotent representation are of infinite order in $\mathbb{C}$.*

Proof. If $P(Z) = P_1(\beta_1)e_1 + P_2(\beta_2)e_2$ then $A_n = a_n e_1 + b_n e_2$ where a_n, b_n are the coefficients of the series of P_1, P_2 respectively. We want to prove that $\lim_{n\longrightarrow\infty} n\sqrt[n]{|A_n|} = 0$ if and only if $\lim_{n\longrightarrow\infty} n\sqrt[n]{|a_n|} = 0$ and $\lim_{n\longrightarrow\infty} n\sqrt[n]{|b_n|} = 0$. The direct implication is proven by a series of inequalities:

$$n\sqrt[n]{|A_n|} = n\sqrt[2n]{|a_n|^2 + |b_n|^2} \leq n\sqrt[n]{|a_n| + |b_n|} \leq n\sqrt[n]{|a_n|} + n\sqrt[n]{|b_n|}.$$

The reciprocal is trivial and left to the reader. □

Lemma 5.7. *P is of infinite order if and only if $P \in \mathrm{Exp}_0(\mathbb{BC}_2)$.*

Proof. This proof is a direct consequence of the previous two lemmas. □

Theorem 5.8. *If $P \in \mathrm{Exp}_0(\mathbb{BC}_2)$ then $P\left(\frac{\partial}{\partial Z}\right)$ is a continuous surjective endomorphism of the space of bicomplex holomorphic functions.*

Proof. The proof follows the known results in complex analysis. We have that P is in $\mathrm{Exp}_0(\mathbb{BC}_2)$ if and only if its components P_1, P_2 are of infinite order in $\mathbb{C}$. From [11] we have that $P_1(\partial_{\beta_1}), P_2(\partial_{\beta_2})$ are continuous surjective endomorphisms in the spaces of holomorphic functions in the variables β_1 and β_2, respectively, and the conclusion follows. □

Following [25], we obtain a series representation for solutions to infinite-order differential equations in the bicomplex setting as follows:

Theorem 5.9. *Let $P \in \mathrm{Exp}_0(\mathbb{BC}_2)$ and let*

$$V = \{Z \mid P(Z) = 0\} = \{(\alpha_k, m_k) \mid |\alpha_1| \leq |\alpha_2| \leq \cdots\},$$

where α_k are bicomplex numbers with the respective multiplicities m_k and $|\alpha_k|$ denotes the Euclidean norm. Then there exists a sequence of indices $k_1 < \cdots < k_n$ such that every entire bicomplex holomorphic solution of $P\left(\frac{\partial}{\partial Z}\right)F = 0$ can be written as:

$$F(Z) = \sum_{n\geq 1}\left(\sum_{k_n \leq k < k_{n+1}} P_k(Z)e^{\alpha_k Z}\right),$$

where P_k is a polynomial of degree less than m_k and the sequence of polynomials P_k satisfies the growth condition that allows the right-hand side of the sum above to converge.

Proof. The proof is a consequence of the equivalence lemmas above and the rather intricate proof of the one variable complex case found for example in [25]. □

References

[1] C.A. Berenstein, R. Gay, *Complex Analysis and Special Topics in Harmonic Analysis*, Springer-Verlag, 1995.

[2] C.A. Berenstein, D.C. Struppa, *Complex Analysis and Convolution Equations*, in Encyclopaedia of Mathematical Sciences, vol. 54, Several Complex Variables V, Springer-Verlag, Berlin-Heidelberg, 1993, pp. 1–109.

[3] C.A. Berenstein, B.A. Taylor, *A new look at interpolation theory for entire functions of one variable*, Adv. Math. **33**, 1979, pp. 109–143.

[4] K.S. Charak, D. Rochon, N. Sharma, *Normal families of bicomplex holomorphic functions*, ArXiv:0806.4403v1, 2008.

[5] F. Colombo, I. Sabadini, Sommen F., D.C. Struppa, *Analysis of Dirac Systems and Computational Algebra*, Birkhäuser, Boston, 2004

[6] F. Colombo, I. Sabadini, D.C. Struppa, A. Vajiac, M. Vajiac, *Singularities of functions of one and several bicomplex functions*, Ark. Mat. **49**, 2011, pp. 277–294.

[7] F. Colombo, I. Sabadini, D.C. Struppa, A. Vajiac, M. Vajiac, *Bicomplex hyperfunction theory*, Ann. Mat. Pura Appl. **190**, 2011, pp. 247–261.

[8] A. Damiano, I. Sabadini, D.C. Struppa, *Computational methods for the construction of a class of Noetherian operators.* Experiment. Math. 16 (2007), no. 1, 41–53.

[9] L. Ehrenpreis, *The Fundamental Principle for Linear Constant Coefficients Partial Differential Equations*, in Proc. Intern. Symp. Linear Spaces, Jerusalem, 1960, pp. 161–174.

[10] L. Ehrenpreis, *Fourier Analysis in Several Complex Variables*, Wiley Interscience, New York, 1970.

[11] A. Kaneko, *Introduction to Hyperfunctions*, Mathematics and Its Applications (Japanese Series), KTK Scientific Publishers/Tokyo, Kluwer Academic Publishers, 1988.

[12] M.E. Luna-Elizarraras, M. Shapiro, D.C. Struppa, A. Vajiac, *Bicomplex Numbers and their Elementary Functions*, CUBO: A Mathematical Journal, **14**, no. 2, 2012, pp. 61–80.

[13] M. Morimoto, K. Fujita, *Analytic functionals and entire functionals on the complex light cone*, Hiroshima Math. J. **25**, 1995, pp. 493–512.

[14] V.P. Palamodov, *Linear Differential Operators with Constant Coefficients*, Springer-Verlag, Berlin-Heidelberg-New York, 1970.

[15] G. Baley Price, *An Introduction to Multicomplex Spaces and Functions*, Marcel Dekker, New York, 1991.

[16] D. Rochon, M. Shapiro, *On algebraic properties of bicomplex and hyperbolic numbers*, An. Univ. Oradea Fasc. Mat., **11**, 2004, pp. 71–110.

[17] J. Ryan, *Complexified Clifford analysis*, Complex Variables and Elliptic Equations **1**, 1982, pp. 119–149.

[18] J. Ryan, $\mathbb{C}^2$ *extensions of analytic functions defined in the complex plane*, Adv. in Applied Clifford Algebras **11**, 2001, pp. 137–145.

[19] G. Scorza Dragoni, *Sulle funzioni olomorfe di una variabile bicomplessa*, Reale Accad. d'Italia, Mem. Classe Sci. Nat. Fis. Mat., **5**, 1934, pp. 597–665.

[20] C. Segre, *Le rappresentazioni reali delle forme complesse e gli enti iperalgebrici*, Math. Ann. **40**, 1892, pp. 413–467.

[21] N. Spampinato, *Estensione nel campo bicomplesso di due teoremi, del Levi-Civita e del Severi, per le funzioni olomorfe di due variabili bicomplesse I, II*, Reale Accad. Naz. Lincei, **22**, 1935, pp. 38–43, pp. 96–102.

[22] N. Spampinato, *Sulla rappresentazione di funzioni di variabile bicomplessa totalmente derivabili*, Ann. Mat. Pura Appli., **14**, 1936, pp. 305–325.

[23] D.C. Struppa, *The fundamental principle for systems of convolution equations*, Mem. Amer. Math. Soc. **273**, 1983.

[24] D.C. Struppa, *Analytic functionals on the complex light cone*, Advances in Hypercomplex Analysis Springer INdAM Series, Volume **1**, 2013, pp. 119–124.

[25] D.C. Struppa, *On the "grouping" phenomenon for holomorphic solutions of infinite order differential equations*, Kyoto Surikaisekikenkyusho Kokyuroku, No. 1001, 1997, pp. 22–38.

[26] D.C. Struppa, G. Kato, *Fundamentals of Algebraic Microlocal Analysis*, Chapman & Hall, CRC Pure and Applied Mathematics, first edition, January 1999.

[27] D.C. Struppa, A. Vajiac, M. Vajiac, *Remarks on Holomorphicity in three settings: Complex, Quaternionic, and Bicomplex*, in Hypercomplex Analysis and Applications, Trends in Mathematics, Birkhäuser Verlag, Basel, 2010, pp. 261–274.

[28] D.C. Struppa, A. Vajiac, M. Vajiac, *Holomorphy in multicomplex spaces*, Spectral Theory, Mathematical System Theory, Evolution Equations, Differential and Difference Equations, Volume 221, Springer ISBN 9783034802970, 2012, pp. 617–634.

[29] F. Treves, *Topological Vector Spaces, Distributions and Kernels*, Academic Press, New York, 1975.

[30] A. Vajiac, M. Vajiac, *Multicomplex Hyperfunctions*, Complex Variables and Elliptic Equations, Complex Variables and Elliptic Equations, Volume **57**, Issue 7-8, 2012, pp. 751–762.

Daniele C. Struppa, Adrian Vajiac and Mihaela B. Vajiac
Schmid College of Science and Technology
Chapman University
Orange 92866, CA, USA
e-mail: `struppa@chapman.edu`
`avajiac@chapman.edu`
`mbvajiac@chapman.edu`

MIX
Papier aus verantwortungsvollen Quellen
Paper from responsible sources
FSC® C105338

If you have any concerns about our products,
you can contact us on
ProductSafety@springernature.com

In case Publisher is established outside the EU,
the EU authorized representative is:
Springer Nature Customer Service Center GmbH
Europaplatz 3, 69115 Heidelberg, Germany

Printed by Libri Plureos GmbH
in Hamburg, Germany